# NONLINEAR SOLID MECHANICS FOR FINITE ELEMENT ANALYSIS: STATICS

Designing engineering components that make optimal use of materials requires consideration of the nonlinear static and dynamic characteristics associated with both manufacturing and working environments. The modeling of these characteristics can only be done through numerical formulation and simulation, which requires an understanding of both the theoretical background and associated computer solution techniques. By presenting both the nonlinear solid mechanics and the associated finite element techniques together, the authors provide, in the first of two books in this series, a complete, clear, and unified treatment of the static aspects of nonlinear solid mechanics.

Alongside a range of worked examples and exercises are user instructions, program descriptions, and examples for the FLagSHyP MATLAB computer implementation, for which the source code is available online.

While this book is designed to complement postgraduate courses, it is also relevant to those in industry requiring an appreciation of the way their computer simulation programs work.

JAVIER BONET is a Professor of Engineering and Head of the College of Engineering at Swansea University, Director of the Welsh Sêr Cymru National Research Network in Advanced Engineering and Materials, and a visiting professor at the Universitat Politecnica de Catalunya in Spain. He has extensive experience of teaching topics in structural mechanics, including large strain nonlinear solid mechanics, to undergraduate and graduate engineering students. He has been active in research in the area of computational mechanics for over 25 years, with contributions in modeling superplastic forming, large strain dynamic analysis, membrane modeling, and finite element technology including error estimation and meshless methods (smooth particle hydrodynamics). Since the book was completed, he has been appointed as Deputy Vice-Chancellor, Research and Enterprise, at the University of Greenwich.

ANTONIO J. GIL is an Associate Professor in the Zienkiewicz Centre for Computational Engineering at Swansea University. He has numerous publications in various areas of computational mechanics, with specific experience in the field of large strain nonlinear mechanics. His work covers the areas of computational simulation of nanomembranes, biomembranes (heart valves) and superplastic forming of medical prostheses, fluid-structure interaction, modeling of smart electro-magneto-mechanical devices, and numerical analysis of fast transient dynamical phenomena. He has received a number of prizes for his contributions to the field of computational mechanics.

RICHARD D. WOOD is an Honorary Research Fellow in the Zienkiewicz Centre for Computational Engineering at Swansea University. He has over 20 years' experience of teaching the course "Nonlinear Continuum Mechanics for Finite Element Analysis" at Swansea University, which he originally developed at the University of Arizona. Wood's academic career has focused on finite element analysis. He has written numerous papers in international journals, and many chapter contributions, on topics related to nonlinear finite element analysis.

# NONLINEAR SOLID MECHANICS FOR FINITE ELEMENT ANALYSIS: STATICS

**Javier Bonet**
*Swansea University*

**Antonio J. Gil**
*Swansea University*

**Richard D. Wood**
*Swansea University*

CAMBRIDGE
UNIVERSITY PRESS

**CAMBRIDGE**
**UNIVERSITY PRESS**

Shaftesbury Road, Cambridge CB2 8EA, United Kingdom

One Liberty Plaza, 20th Floor, New York, NY 10006, USA

477 Williamstown Road, Port Melbourne, VIC 3207, Australia

314–321, 3rd Floor, Plot 3, Splendor Forum, Jasola District Centre, New Delhi – 110025, India

103 Penang Road, #05–06/07, Visioncrest Commercial, Singapore 238467

Cambridge University Press is part of Cambridge University Press & Assessment, a department of the University of Cambridge.

We share the University's mission to contribute to society through the pursuit of education, learning and research at the highest international levels of excellence.

www.cambridge.org
Information on this title: www.cambridge.org/9781107115798



First published 2016

*A catalogue record for this publication is available from the British Library*

*Library of Congress Cataloging-in-Publication data*
Names: Bonet, Javier, 1961– author. | Gil, Antonio J. author. | Wood, Richard D., 1943– author.
Title: Nonlinear solid mechanics for finite element analysis : statics / Javier Bonet, University of Wales, Swansea, Antonio J. Gil, University of Wales, Swansea, Richard D. Wood, University of Wales, Swansea.
Description: Cambridge :Cambridge University Press is part of the University of Cambridge, [2016]
Identifiers: LCCN 2016012972 | ISBN 9781107115798
Subjects: LCSH: Continuum mechanics. | Statics. | Finite element method.
Classification: LCC QA808.2 .B658 2016 | DDC 531/.2–dc23
LC record available at https://lccn.loc.gov/2016012972

ISBN 978-1-107-11579-8 Hardback

To Catherine, Clare, Doreen and our children

**A fragment from the poem**
**"An Essay on Criticism"**
**by Alexander Pope** (1688–1744)

A little Learning is a dang'rous Thing;
Drink deep, or taste not the Pierian Spring:
There shallow Draughts intoxicate the Brain,
And drinking largely sobers us again.
Fir'd at first Sight with what the Muse imparts,
In fearless Youth we tempt the Heights of Arts,
While from the bounded Level of our Mind,
Short Views we take, nor see the lengths behind,
But more advanc'd, behold with strange Surprize
New, distant Scenes of endless Science rise!

# CONTENTS

# PREFACE

A fundamental aspect of engineering is the desire to design artifacts that exploit materials to a maximum in terms of performance under working conditions and efficiency of manufacture. Such an activity demands an increasing understanding of the behavior of the artifact in its working environment together with an understanding of the mechanical processes occurring during manufacture.

To be able to achieve these goals it is likely that the engineer will need to consider the nonlinear characteristics associated possibly with the manufacturing process but certainly with the response to working load. Currently, analysis is most likely to involve a computer simulation of the behavior. Because of the availability of commercial finite element computer software, the opportunity for such nonlinear analysis is becoming increasingly realized.

Such a situation has an immediate educational implication because, for computer programs to be used sensibly and for the results to be interpreted wisely, it is essential that the users have some familiarity with the fundamentals of nonlinear continuum mechanics, nonlinear finite element formulations, and the solution techniques employed by the software. This book seeks to address this problem by providing a unified introduction to these three topics.

The style and content of the book obviously reflect the attributes and abilities of the authors. The authors have lectured on this material for a number of years to postgraduate classes, and the book has emerged from these courses. We hope that our complementary approaches to the topic will be in tune with the variety of backgrounds expected of our readers and, ultimately, that the book will provide a measure of enjoyment brought about by a greater understanding of what we regard as a fascinating subject.

The original edition of this book, titled *Nonlinear Continuum Mechanics for Finite Element Analysis*, published in 1997, contained a chapter on a FORTRAN program implementation of the material in the text, this being freely available at `www.flagshyp.com`. In 2008 a second edition included new chapters on elasto-plastic behavior of trusses and solids and retained the FORTRAN implementation. It was envisioned that an expanded third edition could include dynamics, although this would involve substantial additional material not suitable to the needs of all readers. Consequently the subject has been divided into two complementary volumes, these being the present text, *Nonlinear Solid Mechanics for Finite Element Analysis: Statics*, and a companion volume, *Nonlinear Solid Mechanics for Finite Element Analysis: Dynamics*. These texts are both aimed at the same readership. Recognising its widespread adoption, particularly as a graduate training platform, this present statics text employs MATLAB®* for the implementation of the finite element analysis, the software being freely available at `www.flagshyp.com`.

This present text contains additional examples, and solutions to all exercises are given in the companion book, *Worked Examples in Nonlinear Continuum Mechanics for Finite Element Analysis*, published by Cambridge University Press (ISBN 9781107603615).

## READERSHIP

This book is most suited to a postgraduate level of study by those in either higher education or industry who have graduated with an engineering or applied mathematics degree. However, the material is equally applicable to first-degree students in the final year of an applied mathematics course or an engineering course containing some additional emphasis on maths and numerical analysis. A familiarity with statics and elementary stress analysis is assumed, as is some exposure to the principles of the finite element method. However, a primary objective of the book is that it be reasonably self-contained, particularly with respect to the nonlinear continuum mechanics chapters, which comprise a large portion of the content.

When dealing with such a complex set of topics it is unreasonable to expect all readers to become familiar with all aspects of the book. If the reader is prepared not to get too hung up on details, it is possible to use the book to obtain a reasonable overview of the subject. Such an approach may be suitable for someone starting to use a nonlinear computer program. Alternatively, the requirements of a research project may necessitate a deeper understanding of the concepts discussed. To assist in this latter endeavor the book provides access to a computer program for the nonlinear finite deformation finite element analysis of two- and three-dimensional solids. Such a program provides the basis for a contemporary approach to finite deformation elasto-plastic analysis.

* Mathworks, Inc.

## LAYOUT

### Chapter 1: Introduction

Here, the nature of nonlinear computational mechanics is discussed, and followed by a series of very simple examples that demonstrate various aspects of nonlinear structural behavior. These examples are intended, to an extent, to upset the reader's preconceived ideas inherited from an overexposure to linear analysis and, we hope, provide a motivation for reading the rest of the book! Nonlinear strain measures are introduced and illustrated using a simple one-degree-of-freedom truss analysis. The concepts of linearization and the directional derivative are of sufficient importance to merit a gentle introduction in this chapter. Linearization naturally leads on to the Newton–Raphson iterative solution, which is the fundamental way of solving the nonlinear equilibrium equations occurring in finite element analysis. Consequently, by way of an example, the simple truss is solved and a short MATLAB program is presented that, in essence, is the prototype for the main finite element program discussed later in the book.

### Chapter 2: Mathematical Preliminaries

Vector and tensor manipulations occur throughout the book, and these are introduced in this chapter. Although vector algebra is a well-known topic, tensor algebra is less familiar, certainly, to many approaching the subject with an engineering educational background. Consequently, tensor algebra is considered in enough detail to cover the needs of the subsequent chapters; in particular, it is hoped that readers will understand the physical interpretation of a second-order tensor. Crucial to the development of the finite element solution scheme are the concepts of linearization and the directional derivative. The introduction provided in Chapter 1 is now thoroughly developed. Finally, for completeness, some standard analysis topics are briefly presented.

### Chapter 3: Analysis of Three-dimensional Truss Structures

This chapter is largely independent of the remainder of the book and deals with the large strain elasto-plastic behavior of trusses. The chapter begins with a discussion of the nonlinear kinematics of a simple two-noded truss member, which leads to a definition of logarithmic strain. A hyperelastic stress–strain relationship is then derived and used to obtain the equilibrium equations at a node. In preparation for the variational formulation in Chapter 8 the equilibrium equations are re-derived using an energy approach. These equations are then linearized with respect to small incremental displacements to provide a Newton–Raphson solution process. The chapter then moves on to discuss a simple hyperelastic plastic model for the truss member based on the multiplicative decomposition of the total stretch into elastic and plastic components. The constitutive model is also linearized to provide a tangent modulus. The chapter concludes with some examples of the use of the formulation obtained using the FLagSHyP program.

## Chapter 4: Kinematics

This chapter deals with the kinematics of finite deformation, that is, the study of motion without reference to the cause. Central to this concept is the deformation gradient tensor, which describes the relationship between elemental vectors defining neighboring particles in the undeformed and deformed configurations of the body whose motion is under consideration. The deformation gradient permeates most of the development of finite deformation kinematics because, among other things, it enables a variety of definitions of strain to be established. Material (initial) and spatial (current) descriptions of various items are discussed, as is the linearization of kinematic quantities. Although dynamics is not the subject of this book, it is nevertheless necessary to consider velocity and the rate of deformation. The chapter concludes with a brief discussion of rigid body motion and objectivity.

## Chapter 5: Stress and Equilibrium

The definition of the true or Cauchy stress is followed by the development of standard differential equilibrium equations. As a prelude to the finite element development the equilibrium equations are recast in the weak integral virtual work form. Although initially in the spatial or current deformed configuration, these equations are reformulated in terms of the material or undeformed configuration, and as a consequence alternative stress measures emerge. Finally, stress rates are discussed in preparation for the following chapter on hyperelasticity.

## Chapter 6: Hyperelasticity

Hyperelasticity, whereby the stress is found as a derivative of some potential energy function, encompasses many types of nonlinear material behavior and provides the basis for the finite element treatment of elasto-plastic behavior. Isotropic hyperelasticity is considered both in a material and in a spatial description for compressible and incompressible behavior. The topic is extended to a general description in principal directions that is specialized for the cases of plane strain, plane stress, and uniaxial behavior.

## Chapter 7: Large Elasto-Plastic Deformations

This chapter provides an introduction to the formulation of inelastic deformation processes based on the multiplicative decomposition of the deformation gradient into recoverable and permanent components. Although only a basic Von Mises model with a radial return-mapping procedure is presented, the use of principal directions and logarithmic stretches provides a simple mechanism whereby small strain concepts can be extended to large strains. From the outset, the approach followed, to derive the kinematic rate equations necessary for the flow rule, anticipates the standard trial stress and return-map procedure required to satisfy the plasticity constraints. Such a development clarifies the kinematic rate equations in the context of the eventual incremental algorithmic procedure.

## Chapter 8: Linearized Equilibrium Equations

To establish the Newton–Raphson solution procedure the virtual work expression of equilibrium may be linearized either before or after discretization. Here the former approach is adopted. Linearization of the equilibrium equations includes consideration of deformation-dependent surface pressure loading. A large proportion of this chapter is devoted to incompressibility and to the development, via the Hu–Washizu principle, of the mean dilatation technique.

## Chapter 9: Discretization and Solution

All previous chapters have provided the foundation for the development of the discretized equilibrium and linearized equilibrium equations considered in this chapter. Linearization of the virtual work equation leads to the familiar finite element expression of equilibrium involving $\int \boldsymbol{B}^T \boldsymbol{\sigma} dv$, whereas discretization of the linearized equilibrium equations leads to the tangent matrix, which comprises constitutive and initial stress components. Discretization of the mean dilatation technique is presented in detail. The tangent matrix forms the basis of the Newton–Raphson solution procedure, which is presented as the fundamental solution technique enshrined in the computer program discussed in the following chapter. The chapter concludes with a discussion of line search and arc length enhancements to the Newton–Raphson procedure.

## Chapter 10: Computer Implementation

Here, information is presented on a nonlinear finite element computer program called FLagSHyP,[†] for the solution of finite deformation elasto-plastic finite element problems employing the neo-Hookean hyperelastic compressible and incompressible constitutive equations developed in Chapters 6 and 7. The usage and layout of the MATLAB program is discussed together with the description of the various key functions. The program is available free on the Internet from the website `www.flagshyp.com`. The website also contains the original FORTRAN version of the program which is detailed in previous editions of this text. Note that the user instructions and output layout are the same for both the FORTRAN and MATLAB versions of the program. The software can also be obtained by `email` request to any of the authors: `a.j.gil@swansea.ac.uk`, `r.d.wood@swansea.ac.uk` or `j.bonet@swansea.ac.uk`. The authors would like to acknowledge the assistance given by Dr. Rogelio Ortigosa in the development of this computer program.

A bibliography is provided that enables the reader to access the background to the more standard aspects of finite element analysis. Also listed are books and papers that have been of use in the preparation of this book or that cover similar material in greater depth.

Note on equation numbering: Typically, Equation (x.yz a, b, c, d)$_b$ refers to Equation (x.yz, b).

[†] Finite element **L**arge **S**train **Hy**perelasto-plastic **P**rogram.

# CHAPTER ONE

# INTRODUCTION

## 1.1 NONLINEAR COMPUTATIONAL MECHANICS

Two sources of nonlinearity exist in the analysis of solid continua, namely, material and geometric nonlinearity. The former occurs when, for whatever reason, the stress–strain behavior given by the constitutive relation is nonlinear, whereas the latter is important when changes in geometry, however large or small, have a significant effect on the load deformation behavior. Material nonlinearity can be considered to encompass contact friction, whereas geometric nonlinearity includes deformation-dependent boundary conditions and loading.

Despite the obvious success of the assumption of linearity in engineering analysis, it is equally obvious that many situations demand consideration of nonlinear behavior. For example, ultimate load analysis of structures involves material nonlinearity and perhaps geometric nonlinearity, and any metal-forming analysis such as forging or crash-worthiness must include both aspects of nonlinearity. Structural instability is inherently a geometric nonlinear phenomenon, as is the behavior of tension structures. Indeed the mechanical behavior of the human body itself, say in impact analysis, involves both types of nonlinearity. Nonlinear and linear continuum mechanics deal with the same subjects, including kinematics, stress and equilibrium, and constitutive behavior. But in the linear case an assumption is made that the deformation is sufficiently small to enable the effect of changes in the geometrical configuration of the solid to be ignored, whereas in the nonlinear case the magnitude of the deformation is unrestricted.

Practical stress analysis of solids and structures is unlikely to be served by classical methods, and currently numerical analysis, predominately in the form of the finite element method, is the only route by which the behavior of a complex component subject to complex loading can be successfully simulated. The study of the numerical analysis of nonlinear continua using a computer is called *nonlinear computational mechanics*, which, when applied specifically to the investigation of solid continua, comprises nonlinear continuum mechanics together with the numerical schemes for solving the resulting governing equations.

The finite element method may be summarized as follows. It is a procedure whereby the continuum behavior described at an infinity of points is approximated in terms of a finite number of points, called *nodes*, located at specific points in the continuum. These nodes are used to define regions, called *finite elements*, over which both the geometry and the primary variables in the governing equations are approximated. For example, in the stress analysis of a solid the finite element could be a tetrahedron defined by four nodes and the primary variables the three displacements in the Cartesian directions. The governing equations describing the nonlinear behavior of the solid are usually recast in a so-called *weak integral form* using, for example, the principle of virtual work or the principle of stationary total potential energy. The finite element approximations are then introduced into these integral equations, and a standard textbook manipulation yields a finite set of nonlinear algebraic equations in the primary variable. These equations are then usually solved using the Newton–Raphson iterative technique.

The topic of this book can succinctly be stated as the exposition of the nonlinear continuum mechanics necessary to develop the governing equations in continuous and discrete form and the formulation of the Jacobian or tangent matrix used in the Newton–Raphson solution of the resulting finite set of nonlinear algebraic equations.

## 1.2 SIMPLE EXAMPLES OF NONLINEAR STRUCTURAL BEHAVIOR

It is often the case that nonlinear behavior concurs with one's intuitive expectation of the behavior and that it is linear analysis that can yield the nonsensical result. The following simple examples illustrate this point and provide a gentle introduction to some aspects of nonlinear behavior. These two examples consider rigid materials, but the structures undergo finite displacements; consequently, they are classified as geometrically nonlinear problems.

### 1.2.1 Cantilever

Consider the weightless rigid bar–linear elastic torsion spring model of a cantilever shown in Figure 1.1. Taking moments about the hinge gives the equilibrium equation as

$$FL\cos\theta = M. \tag{1.1}$$

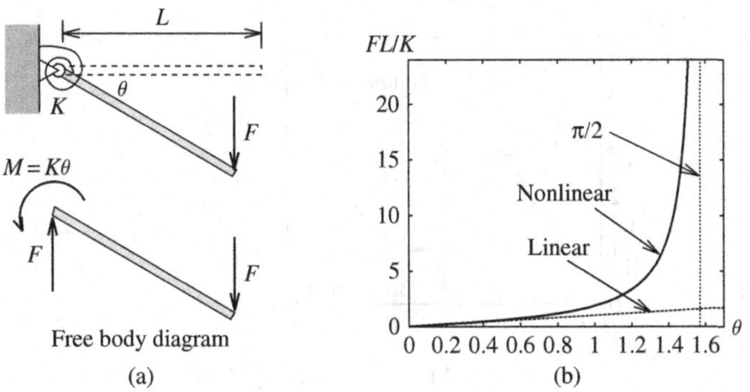


**FIGURE 1.1** Simple cantilever.

If $K$ is the torsional stiffness of the spring, then $M = K\theta$ and we obtain the following nonlinear relationship between $F$ and $\theta$:

$$\frac{FL}{K} = \frac{\theta}{\cos\theta}. \tag{1.2}$$

If the angle $\theta \to 0$, then $\cos\theta \to 1$, and the linear equilibrium equation is recovered as

$$F = \frac{K}{L}\theta. \tag{1.3}$$

The exact nonlinear equilibrium path is shown in Figure 1.1(b), where clearly the nonlinear solution makes physical sense because $\theta < \pi/2$.

### 1.2.2 Column

The same bar–spring system is now positioned vertically (Figure 1.2(a)), and again moment equilibrium about the hinge gives

$$PL\sin\theta = M \quad \text{or} \quad \frac{PL}{K} = \frac{\theta}{\sin\theta}. \tag{1.4}$$

The above equilibrium equation can have two solutions: first, if $\theta = 0$ then $\sin\theta = 0$, $M = 0$, and equilibrium is satisfied; and second, if $\theta \neq 0$ then $PL/K = \theta/\sin\theta$. These two solutions are shown in Figure 1.2(b), where the vertical axis is the equilibrium path for $\theta = 0$ and the horseshoe-shaped equilibrium path is the second solution. The intersection of the two solutions is called a *bifurcation point*. Observe that for $PL/K < 1$ there is only one solution, namely $\theta = 0$, but for $PL/K > 1$ there are three solutions. For instance, when $PL/K \approx 1.57$, either $\theta = 0$ or $\pm\pi/2$.

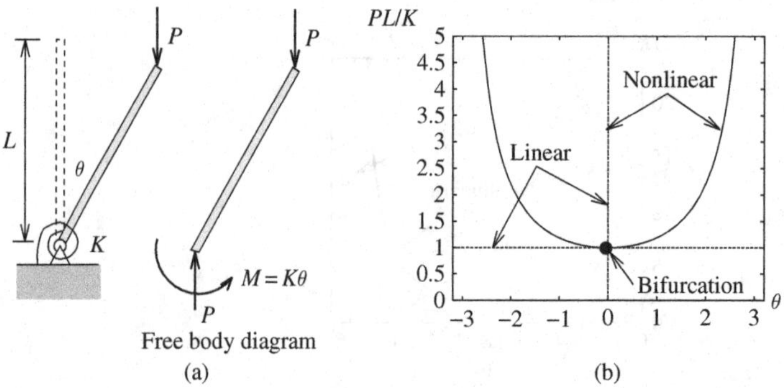


**FIGURE 1.2** Simple column.

For very small values of $\theta$, $\sin\theta \to \theta$ and Equation (1.4) reduces to the linear (in $\theta$) equation

$$(K - PL)\theta = 0. \tag{1.5}$$

Again there are two solutions: $\theta = 0$ or $PL/K = 1$ for any value of $\theta$, the latter solution being the horizontal path shown in Figure 1.2(b). Equation (1.5) is a typical *linear stability analysis* where $P = K/L$ is the elastic critical (buckling) load. Applied to a beam column, such a geometrically nonlinear analysis would yield the Euler buckling load. In a finite element context for, say, plates and shells, this would result in an eigenvalue analysis, the eigenvalues being the buckling loads and the eigenvectors being the corresponding buckling modes.

Observe in these two cases that it is only by considering the finite displacement of the structures that a complete nonlinear solution has been achieved.

## 1.3 NONLINEAR STRAIN MEASURES

In the examples presented in the previous section, the beam or column remained rigid during the deformation. In general, structural components or continuum bodies will exhibit large strains when undergoing a geometrically nonlinear deformation process. As an introduction to the different ways in which these large strains can be measured we consider first a one-dimensional truss element and a simple example involving this type of structural component undergoing large displacements and large strains. We will then give a brief introduction to the difficulties involved in the definition of correct large strain measures in continuum situations.

### 1.3.1 One-Dimensional Strain Measures

Imagine that we have a truss member of initial length $L$ and area $A$ that is stretched to a final length $l$ and area $a$ as shown in Figure 1.3. The simplest possible quantity

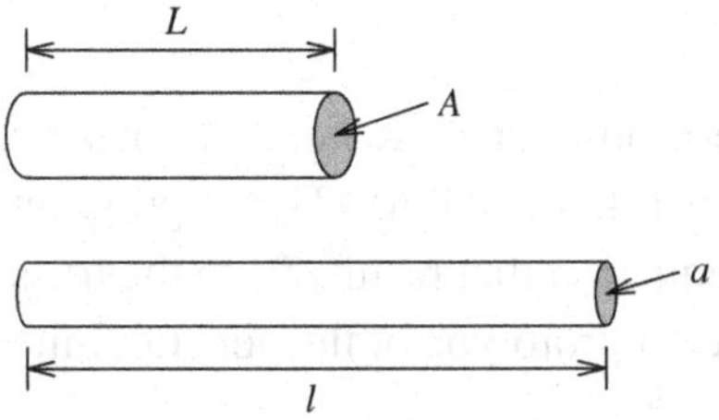


**FIGURE 1.3** One-dimensional strain.

that we can use to measure the strain in the bar is the so-called *engineering strain* $\varepsilon_E$, defined as

$$\varepsilon_E = \frac{l - L}{L}. \tag{1.6}$$

Clearly different measures of strain could be used. For instance, the change in length $\Delta l = l - L$ could be divided by the final length rather than the initial length. Whichever definition is used, if $l \approx L$ the small strain quantity $\varepsilon = \Delta l/l$ is recovered.

An alternative large strain measure can be obtained by adding up all the small strain increments that take place when the rod is continuously stretched from its original length $L$ to its final length $l$. This integration process leads to the definition of the *natural* or *logarithmic* strain $\varepsilon_L$ as

$$\varepsilon_L = \int_L^l \frac{dl}{l} = \ln \frac{l}{L}. \tag{1.7}$$

Although the above strain definitions can in fact be extrapolated to the deformation of a three-dimensional continuum body, this generalization process is complex and computationally costly. Strain measures that are much more readily generalized to continuum cases are the so-called *Green* or *Green's* strain $\varepsilon_G$ and the *Almansi* strain $\varepsilon_A$, defined as

$$\varepsilon_G = \frac{l^2 - L^2}{2L^2}; \tag{1.8a}$$

$$\varepsilon_A = \frac{l^2 - L^2}{2l^2}. \tag{1.8b}$$

Irrespective of which strain definition is used, a simple Taylor's series analysis shows that, for the case where $l \approx L$, all the above quantities converge to the small strain definition $\Delta l/l$. For instance, in the Green strain case, we have

$$\begin{aligned}\varepsilon_G(l \approx L) &\approx \frac{(l + \Delta l)^2 - l^2}{2l^2}\\ &= \frac{1}{2}\frac{l^2 + \Delta l^2 + 2l\Delta l - l^2}{l^2}\\ &\approx \frac{\Delta l}{l}.\end{aligned} \tag{1.9}$$

## 1.3.2 Nonlinear Truss Example

This example is included in order to introduce a number of features associated with finite deformation analysis. Later, in Section 1.4, a small MATLAB program will be given to solve the nonlinear equilibrium equation that results from the truss analysis. The structure of this program is, in effect, a prototype of the general finite element program presented later in this book.

We consider the truss member shown in Figure 1.4 with initial and loaded lengths, cross-sectional areas, and volumes: $L, A, V$ and $l, a, v$ respectively. For simplicity we assume that the material is incompressible and hence $V = v$ or $AL = al$. Two constitutive equations are chosen, based, without explanation at the moment, on Green's and a logarithmic definition of strain, hence the Cauchy, or true, stress $\sigma$ is either

$$\sigma = E\,\frac{l^2 - L^2}{2L^2} \quad \text{or} \quad \sigma = E \ln \frac{l}{L}, \tag{1.10a,b}$$

where $E$ is a (Young's modulus-like) constitutive constant that, in ignorance, has been chosen to be the same irrespective of the strain measure being used. Physically this is obviously wrong, but it will be shown below that for small strains it is acceptable. Indeed, it will be seen in Chapter 5 that the Cauchy stress cannot be simply associated with Green's strain, but for now such complications will be ignored.

The equation for vertical equilibrium at the sliding joint $B$, in nomenclature that will be used later, is simply

$$R(x) = T(x) - F = 0; \quad T = \sigma a \sin\theta; \quad \sin\theta = \frac{x}{l}; \tag{1.11a,b,c}$$

where $T(x)$ is the vertical component, at $B$, of the internal force in the truss member and $x$ gives the truss position. $R(x)$ is the *residual* or *out-of-balance* force,

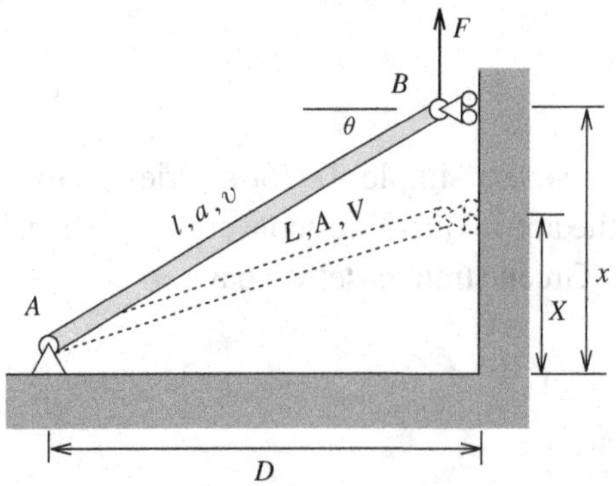


**FIGURE 1.4** Single incompressible truss member.

and a solution for $x$ is achieved when $R(x) = 0$. In terms of the alternative strain measures, $T$ is

$$T = \frac{Evx}{l^2}\left(\frac{l^2 - L^2}{2L^2}\right) \quad \text{or} \quad T = \frac{Evx}{l^2}\ln\frac{l}{L}. \tag{1.12a,b}$$

Note that in this equation $l$ is function of $x$ as $l^2 = D^2 + x^2$ and therefore $T$ is highly nonlinear in $x$.

Given a value of the external load $F$, the procedure that will eventually be used to solve for the unknown position $x$ is the Newton–Raphson method, but in this one-degree-of-freedom case it is easier to choose a value for $x$ and find the corresponding load $F$. Typical results are shown in Figure 1.5, where an initial angle of 45° has been assumed. It is clear from this figure that the behavior is highly nonlinear. Evidently, where finite deformations are involved it appears as though care has to be exercised in defining the constitutive relations because different strain choices will lead to different solutions. But, at least in the region where $x$ is in the neighborhood of its initial value $X$ and strains are likely to be small, the equilibrium paths are close.

In Figure 1.5 the local maximum and minimum forces $F$ occur at the so-called *limit points* $p$ and $q$, although in reality if the truss were compressed to point $p$ it would experience a violent movement or *snap-through behavior* from $p$ to point $p'$ as an attempt is made to increase the compressive load in the truss beyond the limit point.

By making the truss member initially vertical we can examine the large strain behavior of a rod. The typical load deflection behavior is shown in Figure 1.6, where clearly the same constant $E$ should not have been used to represent the same material characterized using different strain measures. Alternatively, by making the

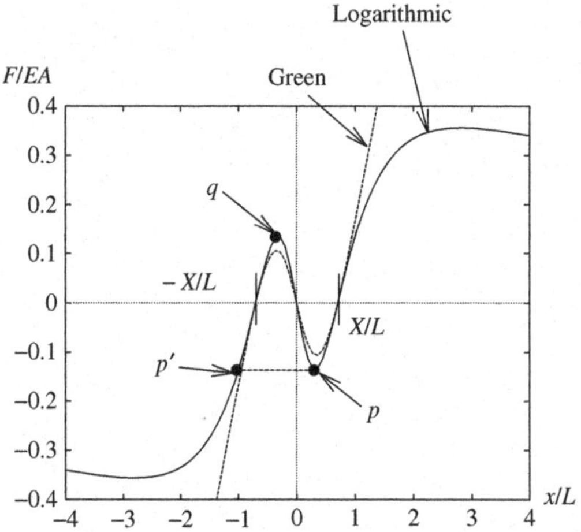


**FIGURE 1.5** Truss example – load deflection behavior.

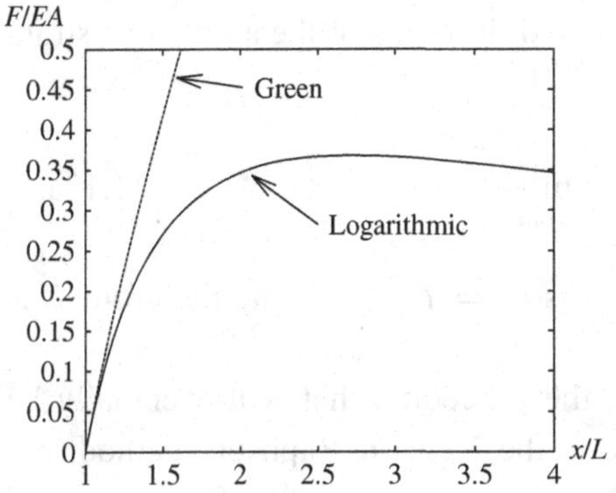


**FIGURE 1.6** Large strain rod – load deflection behavior.

truss member initially horizontal, the stiffening effect due to the development of tension in the member can be observed in Figure 1.7.

Further insight into the nature of nonlinearity in the presence of large deformation can be revealed by this simple example if we consider the vertical stiffness of the truss member at joint $B$. This stiffness is the change in the equilibrium equation, $R(x) = 0$, due to a change in position $x$, and is generally represented by $K = dR/dx$. If the load $F$ is constant, the stiffness is the change in the vertical component, $T$, of the internal force, which can be obtained with the help of Equations (1.11b,c) together with the incompressibility condition $a = V/l$ as

$$\begin{aligned} K &= \frac{dT}{dx} \\ &= \frac{d}{dx}\left(\frac{\sigma V x}{l^2}\right) \\ &= \left(\frac{ax}{l}\frac{d\sigma}{dl} - \frac{2\sigma a x}{l^2}\right)\frac{dl}{dx} + \frac{\sigma a}{l} \end{aligned}$$

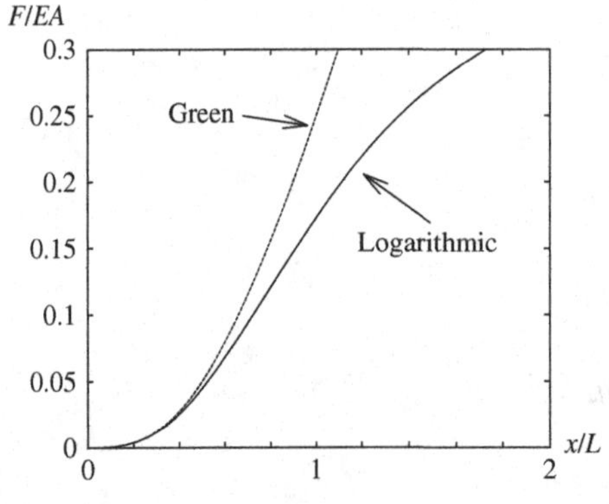


**FIGURE 1.7** Horizontal truss – tension stiffening.

$$= a\left(\frac{d\sigma}{dl} - \frac{2\sigma}{l}\right)\frac{x^2}{l^2} + \frac{\sigma a}{l}. \tag{1.13}$$

All that remains is to find $d\sigma/dl$ for each strain definition, labeled $G$ and $L$ for Green's and the logarithmic strain respectively, to give

$$\left(\frac{d\sigma}{dl}\right)_G = \frac{El}{L^2} \quad \text{and} \quad \left(\frac{d\sigma}{dl}\right)_L = \frac{E}{l}. \tag{1.14a,b}$$

Hence the stiffnesses are

$$K_G = \frac{A}{L}\left(E - 2\sigma\frac{L^2}{l^2}\right)\frac{x^2}{l^2} + \frac{\sigma a}{l}; \tag{1.15a}$$

$$K_L = \frac{a}{l}(E - 2\sigma)\frac{x^2}{l^2} + \frac{\sigma a}{l}. \tag{1.15b}$$

Despite the similarities in the expressions for $K_G$ and $K_L$, the gradient of the curves in Figure 1.5 shows that the stiffnesses are generally not the same. This is to be expected, again, because of the casual application of the constitutive relations.

Finally, it is instructive to attempt to rewrite the final term in (1.15a) in an alternative form to give $K_G$ as

$$K_G = \frac{A}{L}(E - 2S)\frac{x^2}{l^2} + \frac{SA}{L}; \quad S = \sigma\frac{L^2}{l^2}. \tag{1.15c}$$

The above expression introduces the second Piola–Kirchhoff stress $S$, which gives the force per unit undeformed area but transformed by what will become known as the *deformation gradient inverse*, that is, $(l/L)^{-1}$. It will be shown in Chapter 5 that the second Piola–Kirchhoff stress is associated with Green's strain and not the Cauchy stress, as was erroneously assumed in Equation (1.10a,b)$_a$. Allowing for the local-to-global force transformation implied by $(x/l)^2$, Equations (1.15c) illustrate that the stiffness can be expressed in terms of the initial undeformed configuration or the current deformed configuration.

The above stiffness terms show that, in both cases, the constitutive constant $E$ has been modified by the current state of stress $\sigma$ or $S$. We can see that this is a consequence of allowing for geometry changes in the formulation by observing that the $2\sigma$ term emerges from the derivative of the term $1/l^2$ in Equation (1.13). If $x$ is close to the initial configuration $X$ then $a \approx A$, $l \approx L$, and therefore $K_L \approx K_G$.

Equations (1.15) contain a stiffness term $\sigma a/l$ $(=SA/L)$ which is generally known as the *initial stress stiffness*. The same term can be derived by considering the change in the equilibrating global end forces occurring when an initially stressed rod rotates by a small amount, hence $\sigma a/l$ is also called the *geometric stiffness*. This is the term that, in general, occurs in an instability analysis because a sufficiently large negative value can render the overall stiffness singular. The geometric stiffness is unrelated to the change in cross-sectional area and is purely associated with force changes caused by rigid body rotation.

The second Piola–Kirchhoff stress will reappear in Chapter 5, and the modification of the constitutive parameters by the current state of stress will reappear in Chapter 6, which deals with constitutive behavior in the presence of finite deformation.

### 1.3.3 Continuum Strain Measures

In linear stress–strain analysis the deformation of a continuum body is measured in terms of the small strain tensor $\boldsymbol{\varepsilon}$. For instance, in a simple two-dimensional case $\boldsymbol{\varepsilon}$ has components $\varepsilon_{xx}$, $\varepsilon_{yy}$, and $\varepsilon_{xy} = \varepsilon_{yx}$, which are obtained in terms of the $x$ and $y$ components of the displacement of the body as

$$\varepsilon_{xx} = \frac{\partial u_x}{\partial x}; \tag{1.16a}$$

$$\varepsilon_{yy} = \frac{\partial u_y}{\partial y}; \tag{1.16b}$$

$$\varepsilon_{xy} = \frac{1}{2}\left(\frac{\partial u_x}{\partial y} + \frac{\partial u_y}{\partial x}\right). \tag{1.16c}$$

These equations rely on the assumption that the displacements $u_x$ and $u_y$ are very small, so that the initial and final positions of a given particle are practically the same. When the displacements are large, however, this is no longer the case and one must distinguish between initial and final coordinates of particles. This is typically done by using capital letters $X, Y$ for the initial positions and lower-case $x, y$ for the current coordinates. It would then be tempting to extend the use of the above equations to the nonlinear case by simply replacing derivatives with respect to $x$ and $y$ by their corresponding initial coordinates $X, Y$. It is easy to show that, for large displacement situations this would result in strains that contradict the physical reality. Consider for instance a two-dimensional solid undergoing a 90° rotation about the origin as shown in Figure 1.8. The corresponding displacements of any given particle are seen from the figure to be

$$u_x = -X - Y; \tag{1.17a}$$

$$u_y = X - Y; \tag{1.17b}$$

and therefore the application of the above formulas gives

$$\varepsilon_{xx} = \varepsilon_{yy} = -1; \quad \varepsilon_{xy} = 0. \tag{1.18a,b}$$

These values are clearly incorrect, as the solid experiences no strain during the rotation.

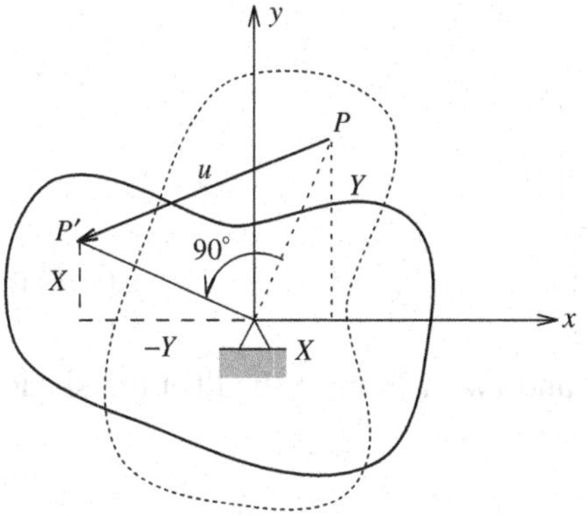


**FIGURE 1.8** 90° rotation of a two-dimensional body.

It is clearly necessary to re-establish the definition of strain for a continuum so that physically correct results are obtained when the body is subject to a finite motion or deformation process. Although general nonlinear strain measures will be discussed at length in Chapter 4, we can introduce some of the basic ideas by trying to extend the definition of Green's strain given in Equation (1.8a) to the two-dimensional case. Consider for this purpose a small elemental segment $dX$ initially parallel to the $x$ axis that is deformed to a length $ds$ as shown in Figure 1.9. The final length can be evaluated from the displacements as

$$ds^2 = \left(dX + \frac{\partial u_x}{\partial X}dX\right)^2 + \left(\frac{\partial u_y}{\partial X}dX\right)^2. \tag{1.19}$$

Based on the one-dimensional Green strain Equation (1.8a), the $x$ component of the two-dimensional Green strain can now be defined as

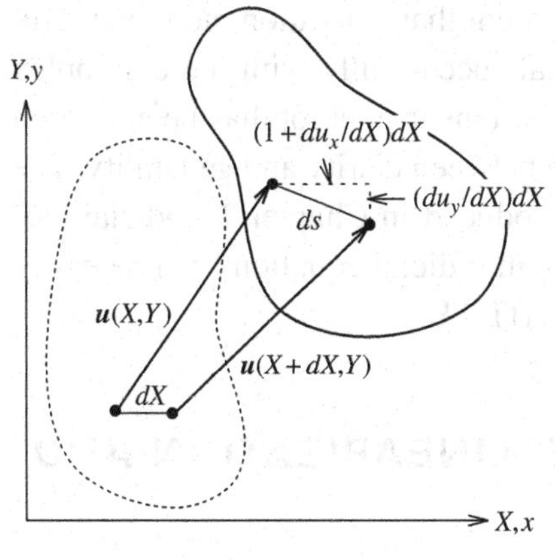


**FIGURE 1.9** General deformation of a two-dimensional body.

$$\begin{aligned}
E_{xx} &= \frac{ds^2 - dX^2}{2dX^2} \\
&= \frac{1}{2}\left[\left(1 + \frac{\partial u_x}{\partial X}\right)^2 + \left(\frac{\partial u_y}{\partial X}\right)^2 - 1\right] \\
&= \frac{\partial u_x}{\partial X} + \frac{1}{2}\left[\left(\frac{\partial u_x}{\partial X}\right)^2 + \left(\frac{\partial u_y}{\partial X}\right)^2\right].
\end{aligned} \qquad (1.20a)$$

Using similar arguments, equations for $E_{yy}$ and (with more difficulty) the shear strains $E_{xy} = E_{yx}$ are obtained as

$$E_{yy} = \frac{\partial u_y}{\partial Y} + \frac{1}{2}\left[\left(\frac{\partial u_x}{\partial Y}\right)^2 + \left(\frac{\partial u_y}{\partial Y}\right)^2\right]; \qquad (1.20b)$$

$$E_{xy} = \frac{1}{2}\left(\frac{\partial u_x}{\partial Y} + \frac{\partial u_y}{\partial X}\right) + \frac{1}{2}\left(\frac{\partial u_x}{\partial X}\frac{\partial u_x}{\partial Y} + \frac{\partial u_y}{\partial X}\frac{\partial u_y}{\partial Y}\right). \qquad (1.20c)$$

Clearly, if the displacements are small, the quadratic terms in the above expressions can be ignored and we recover Equations (1.16a,b,c). It is a simple exercise to show that, for the rigid rotation case discussed above, the Green strain components are $E_{xx} = E_{yy} = E_{xy} = 0$, which coincides with one's intuitive perception of the lack of strain in this particular type of motion.

It is clear from Equations (1.20a–c) that nonlinear measures of strain in terms of displacements can become much more intricate than in the linear case. In general, it is preferable to restrict the use of displacements as problem variables to linear situations where they can be assumed to be infinitesimal and deal with fully nonlinear cases using current or final positions $x(X, Y)$ and $y(X, Y)$ as problem variables. In a fully nonlinear context, however, linear displacements will arise again during the Newton–Raphson solution process as iterative increments from the current position of the body until final equilibrium is reached. This linearization process is one of the most crucial aspects of nonlinear analysis and will be introduced in the next section. Finally, it is apparent that a notation more powerful than the one used above will be needed to deal successfully with more complex three-dimensional cases. In particular, Cartesian tensor notation has been chosen in this book as it provides a reasonable balance between clarity and generality. The basic elements of this type of notation are introduced in Chapter 2. Indicial tensor notation is used only very sparingly, although indicial equations can be easily translated into a computer language such as MATLAB.

## 1.4 DIRECTIONAL DERIVATIVE, LINEARIZATION AND EQUATION SOLUTION

The solution to the nonlinear equilibrium equation, typified by Equation (1.11a,b,c), amounts to finding the position $x$ for a given load $F$. This is achieved in finite deformation finite element analysis by using a Newton–Raphson iteration. Generally, this involves the *linearization* of the equilibrium equations, which requires an understanding of the *directional derivative*. A directional derivative is a generalization of a derivative in that it provides the change in an item due to a small change in something upon which the item depends. For example, the item could be the determinant of a matrix, in which case the small change would be in the matrix itself.

### 1.4.1 Directional Derivative

This topic is discussed in detail in Chapter 2 but will be introduced here via a tangible example using the simple linear spring structure shown in Figure 1.10.

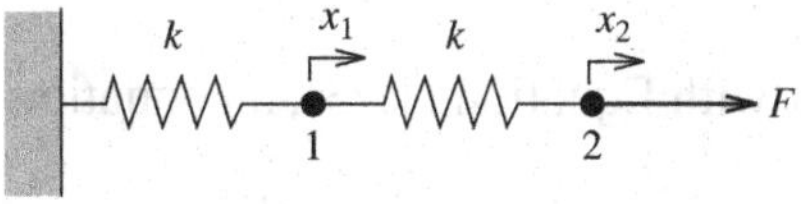


**FIGURE 1.10** Two-degrees-of-freedom linear spring structure.

The Total Potential Energy (TPE), $\Pi$, of the structure is

$$\Pi(\mathbf{x}) = \tfrac{1}{2}kx_1^2 + \tfrac{1}{2}k(x_2 - x_1)^2 - Fx_2, \tag{1.21}$$

where $\mathbf{x} = (x_1, x_2)^T$ and $x_1$ and $x_2$ are the displacements of the joints 1 and 2. Now consider the TPE due to a change in displacements given by the increment vector $\mathbf{u} = (u_1, u_2)^T$ as

$$\Pi(\mathbf{x}+\mathbf{u}) = \tfrac{1}{2}k(x_1 + u_1)^2 + \tfrac{1}{2}k(x_2 + u_2 - x_1 - u_1)^2 - F(x_2 + u_2). \tag{1.22}$$

The directional derivative represents the gradient of $\Pi$ in the direction $\mathbf{u}$ and gives a linear (or first-order) approximation to the increment in TPE due to the increment in position $\mathbf{u}$ as

$$D\Pi(\mathbf{x})[\mathbf{u}] \approx \Pi(\mathbf{x}+\mathbf{u}) - \Pi(\mathbf{x}), \tag{1.23}$$

where the general notation $D\Pi(\mathbf{x})[\mathbf{u}]$ indicates directional derivative of $\Pi$ at $\mathbf{x}$ in the direction of an increment $\mathbf{u}$. The evaluation of this derivative is illustrated in Figure 1.11 and relies on the introduction of a parameter $\epsilon$ that is used to scale the increment $\mathbf{u}$ to give new displacements $x_1 + \epsilon u_1$ and $x_2 + \epsilon u_2$ for which the TPE is

$$\Pi(\mathbf{x}+\epsilon\mathbf{u}) = \tfrac{1}{2}k(x_1 + \epsilon u_1)^2 + \tfrac{1}{2}k(x_2 + \epsilon u_2 - x_1 - \epsilon u_1)^2 - F(x_2 + \epsilon u_2). \tag{1.24}$$

Observe that for a given $\mathbf{x}$ and $\mathbf{u}$ the TPE is now a function of the parameter $\epsilon$, and a first-order Taylor's series expansion about $\epsilon = 0$ gives

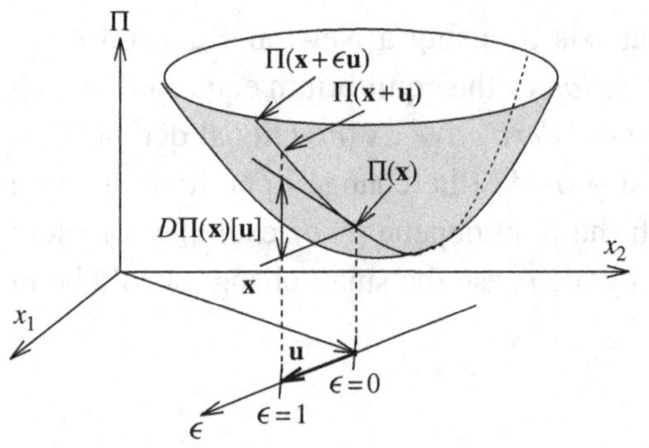


**FIGURE 1.11** Directional derivative.

$$\Pi(\mathbf{x} + \epsilon\mathbf{u}) \approx \Pi(\mathbf{x}) + \left[\frac{d}{d\epsilon}\bigg|_{\epsilon=0} \Pi(\mathbf{x} + \epsilon\mathbf{u})\right] \epsilon. \tag{1.25}$$

If we take $\epsilon = 1$ in this equation and compare it with Equation (1.23), an equation for the directional derivative emerges as

$$\begin{aligned} D\Pi(\mathbf{x})[\mathbf{u}] &= \frac{d}{d\epsilon}\bigg|_{\epsilon=0} \Pi(\mathbf{x} + \epsilon\mathbf{u}) \\ &= kx_1u_1 + k(x_2 - x_1)(u_2 - u_1) - Fu_2 \\ &= \mathbf{u}^T(\mathbf{Kx} - \mathbf{F}), \end{aligned} \tag{1.26}$$

where

$$\mathbf{K} = \begin{bmatrix} 2k & -k \\ -k & k \end{bmatrix}; \quad \mathbf{F} = \begin{bmatrix} 0 \\ F \end{bmatrix}. \tag{1.27}$$

It is important to note that, although the TPE function $\Pi(\mathbf{x})$ was nonlinear in $\mathbf{x}$, the directional derivative $D\Pi(\mathbf{x})[\mathbf{u}]$ is always linear in $\mathbf{u}$. In this sense we say that the function has been *linearized* with respect to the increment $\mathbf{u}$.

The equilibrium of the structure is enforced by requiring the TPE to be stationary, which implies that the gradient of $\Pi$ must vanish for any direction $\mathbf{u}$. This is expressed in terms of the directional derivative as

$$D\Pi(\mathbf{x})[\mathbf{u}] = 0 \quad \text{for any } \mathbf{u}, \tag{1.28}$$

and consequently the equilibrium position $\mathbf{x}$ satisfies

$$\mathbf{Kx} - \mathbf{F} = \mathbf{0}. \tag{1.29}$$

If the direction $\mathbf{u}$ in Equation (1.26) or (1.28) is interpreted as a virtual displacement $\delta\mathbf{u}$ then, clearly, the virtual work expression of equilibrium is obtained.

The concept of the directional derivative is far more general than this example implies. For example, we can find the directional derivative of the determinant of a $2 \times 2$ matrix $\mathbf{A} = [A_{ij}]$ in the direction of the change $\mathbf{U} = [U_{ij}]$, for $i, j = 1, 2$, as

$$
\begin{aligned}
D\det(\mathbf{A})[\mathbf{U}] &= \left.\frac{d}{d\epsilon}\right|_{\epsilon=0} \det(\mathbf{A} + \epsilon\mathbf{U}) \\
&= \left.\frac{d}{d\epsilon}\right|_{\epsilon=0} \big[(A_{11} + \epsilon U_{11})(A_{22} + \epsilon U_{22}) \\
&\qquad - (A_{21} + \epsilon U_{21})(A_{12} + \epsilon U_{12})\big] \\
&= A_{22}U_{11} + A_{11}U_{22} - A_{21}U_{12} - A_{12}U_{21}.
\end{aligned}
\tag{1.30}
$$

We will see in Chapter 2 that for general $n \times n$ matrices this directional derivative can be rewritten as

$$
D\det(\mathbf{A})[\mathbf{U}] = \det \mathbf{A}\,(\mathbf{A}^{-T} : \mathbf{U}), \tag{1.31}
$$

where, generally, the double contraction of two matrices is $\mathbf{A} : \mathbf{B} = \sum_{i,j=1}^{n} A_{ij} B_{ij}$.

### 1.4.2 Linearization and Solution of Nonlinear Algebraic Equations

As a prelude to finite element work, let us consider the solution of a set of *nonlinear* algebraic equations:

$$
\mathbf{R}(\mathbf{x}) = \mathbf{0}, \tag{1.32}
$$

where, for example, for a simple case with two equations and two unknowns,

$$
\mathbf{R}(\mathbf{x}) = \begin{bmatrix} R_1(x_1, x_2) \\ R_2(x_1, x_2) \end{bmatrix}; \quad \mathbf{x} = \begin{bmatrix} x_1 \\ x_2 \end{bmatrix}. \tag{1.33a,b}
$$

Typically, nonlinear equations of this type are solved using a Newton–Raphson iterative process whereby, given a solution estimate $\mathbf{x}_k$ at iteration $k$, a new value $\mathbf{x}_{k+1} = \mathbf{x}_k + \mathbf{u}$ is obtained in terms of an increment $\mathbf{u}$ by establishing the linear approximation

$$
\mathbf{R}(\mathbf{x}_{k+1}) \approx \mathbf{R}(\mathbf{x}_k) + D\mathbf{R}(\mathbf{x}_k)[\mathbf{u}] = \mathbf{0}. \tag{1.34}
$$

This directional derivative is evaluated with the help of the chain rule as

$$
\begin{aligned}
D\mathbf{R}(\mathbf{x}_k)[\mathbf{u}] &= \frac{d}{d\epsilon}\bigg|_{\epsilon=0} \mathbf{R}(\mathbf{x}_k + \epsilon\mathbf{u}) \\
&= \frac{d}{d\epsilon}\bigg|_{\epsilon=0} \begin{bmatrix} R_1(x_1 + \epsilon u_1, x_2 + \epsilon u_2) \\ R_2(x_1 + \epsilon u_1, x_2 + \epsilon u_2) \end{bmatrix} \\
&= \mathbf{K}\mathbf{u},
\end{aligned} \tag{1.35}
$$

where the *tangent matrix* $\mathbf{K}$ is

$$
\mathbf{K}(\mathbf{x}_k) = [K_{ij}(\mathbf{x}_k)]; \quad K_{ij}(\mathbf{x}_k) = \frac{\partial R_i}{\partial x_j}\bigg|_{\mathbf{x}_k}. \tag{1.36}
$$

If we substitute Equation (1.35) for the directional derivative into Equation (1.34), we obtain a linear set of equations for $\mathbf{u}$ to be solved at each Newton–Raphson iteration as

$$
\mathbf{K}(\mathbf{x}_k)\mathbf{u} = -\mathbf{R}(\mathbf{x}_k); \quad \mathbf{x}_{k+1} = \mathbf{x}_k + \mathbf{u}. \tag{1.37a,b}
$$

For equations with a single unknown $x$, such as Equation (1.11a,b,c) for the truss example seen in Section 1.3.2 where $R(x) = T(x) - F$, the above Newton–Raphson process becomes

$$
u = -\frac{R(x_k)}{K(x_k)}; \quad x_{k+1} = x_k + u. \tag{1.38a,b}
$$

This is illustrated in Figure 1.12.

In practice, the external load $F$ is applied in a series of increments as

$$
F = \sum_{i=1}^{l} \Delta F_i, \tag{1.39}
$$

and the resulting Newton–Raphson algorithm is given in Box 1.1, where boldface items generalize the above procedure in terms of column and square matrices. Note

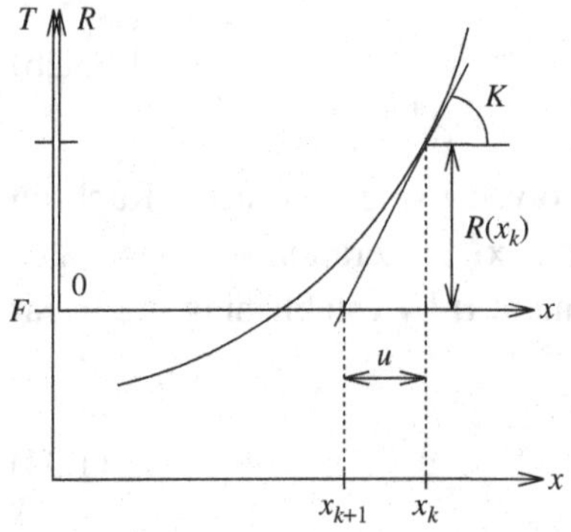


**FIGURE 1.12** Newton–Raphson iteration.

**BOX 1.1: Newton–Raphson Algorithm**

- INPUT geometry, material properties, and solution parameters
- INITIALIZE $\mathbf{F} = \mathbf{0}$, $\mathbf{x} = \mathbf{X}$ (initial geometry), $\mathbf{R} = \mathbf{0}$
- FIND initial $\mathbf{K}$ (typically (1.13))
- LOOP over load increments
  - FIND $\Delta\mathbf{F}$ (establish the load increment)
  - SET $\mathbf{F} = \mathbf{F} + \Delta\mathbf{F}$
  - SET $\mathbf{R} = \mathbf{R} - \Delta\mathbf{F}$
  - DO WHILE ($\|\mathbf{R}\|/\|\mathbf{F}\| >$ tolerance)
    - SOLVE $\mathbf{K}\mathbf{u} = -\mathbf{R}$ (typically (1.38a))
    - UPDATE $\mathbf{x} = \mathbf{x} + \mathbf{u}$ (typically (1.38b))
    - FIND $\mathbf{T}$ (typically (1.12)) and $\mathbf{K}$ (typically (1.13))
    - FIND $\mathbf{R} = \mathbf{T} - \mathbf{F}$ (typically (1.11))
  - ENDDO
- ENDLOOP

that this algorithm reflects the fact that in a general finite element program, internal forces and the tangent matrix are more conveniently evaluated at the same time. A simple MATLAB program for solving the one-degree-of-freedom truss example is given in Box 1.2. This program stops once the stiffness becomes singular, that is, at the limit point $p$. A technique to overcome this deficiency is dealt with in Section 9.6.3. The way in which the Newton–Raphson process converges toward the final solution is depicted in Figure 1.13 for the particular choice of input variables shown in Box 1.2. Note that only six iterations are needed to converge to values within machine precision. We can contrast this type of quadratic rate of convergence with a linear convergence rate, which, for instance, would result from a modified Newton–Raphson scheme based, per load increment, on using the same initial stiffness throughout the iteration process.

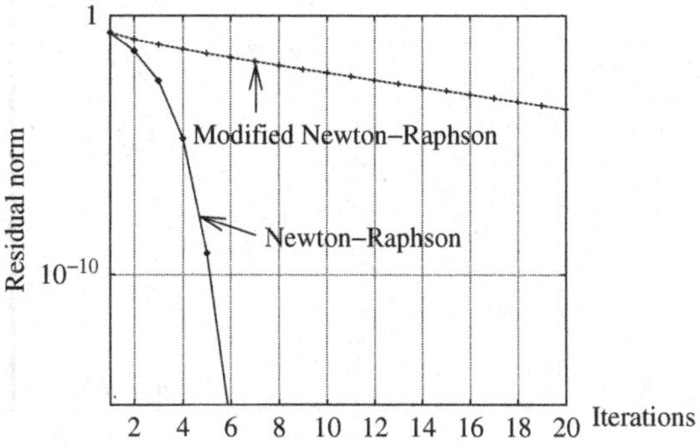


**FIGURE 1.13** Convergence rate.

**BOX 1.2: Simple Truss Program**

```
%--------------------------------------------------------
      program truss
%--------------------------------------------------------
% Newton-Raphson solverfor 1 d.o.f. truss
%-----------------------------------------------------
% input
%
% d     --> horizontal span
% x     --> initial height
% area  --> initial cross-sectional area
% E     --> Young's Modulus
% nincr --> number of load increments
% fincr --> force increment
% cnorm --> resdual force convergence norm
% miter --> maximum number of Newton-Raphson
%           iterations
%-----------------------------------------------------
% data
%
d=2500;x=2500;area=100;E=5.e5;f=0;resid=0;
nincr=30;fincr=-8e5;cnorm=1.0e-10;miter=20;
%
% initialize geometry data and stiffness
%
lzero=sqrt(x^2+d^2);vol=area*lzero;
stiff=(area/lzero)*E*(x/lzero)*(x/lzero);
%
% start load incrementation
%
for incrm=1:nincr;
    f=f+fincr;
    resid=resid-fincr;
    rnorm=cnorm*2;
    niter=0;
% Newton-Raphson iteration
%
    while ((rnorm>cnorm)&&(niter<miter))
          niter=niter+1;



%
% find geometry increment
%
          u=-resid/stiff;
          x=x+u;
```

*(continued)*

**BOX 1.2:** *(cont.)*

```
            l=sqrt(x*x+d*d);
            area=vol/l;
%
% find stress and residual force
%
            stress=E*log(l/lzero);
            t=stress*area*x/l;
            resid=t-f;
            rnorm=abs(resid/f);
            output=sprintf('% 4.2g % 4.2g % .5e % .5e % 0.5e',...
                   incrm,niter,rnorm,x,f);output
%
% find stiffness and check stability
%
            stiff=(area/l)*(E-2*stress)*(x/l)*(x/l)+(stress*area/l);
            if abs(stiff)<(1e-20)
               display('near zero stiffness - stop');
               return
            end
        end
        if(rnorm<cnorm); display('converged');end
end
```

## Exercises

1. The structure shown in Figure 1.14 comprises a rigid weightless rod $a$–$b$ supported by a spring of stiffness $k = 5$. The force $F$ is positive downwards as is the vertical displacement $v$.
   (a) Find the equilibrium equation relating $F$ to the slope angle $\theta$ and then plot $F$ against the vertical displacement $v$.

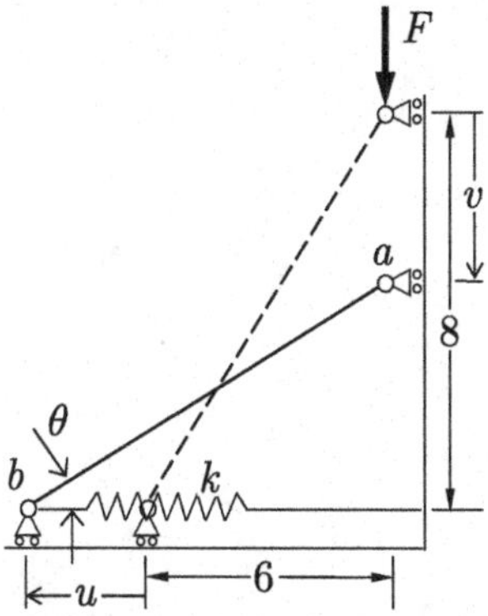


**FIGURE 1.14** Rod-spring structure.

(b) Also determine the directional derivative of $F$ with respect to a change $\beta$ in $\theta$.

This simple example illustrates the phenomenon known as *snap-through behavior.*

2. Figure 1.15 shows a weightless rigid column supported by a torsion spring at the base. In the unloaded position the column has an initial imperfection of $\theta_0$. The length of the column is 10, the torsional stiffness is 10, and the initial imperfection is $\theta_0 = 0.01$ rad. This example is a simple model of the nonlinear behavior of a vertical column under the action of an axial load.

   (a) Find the rotational equilibrium equation and plot the force $P$ against the lateral displacement $u$.

   (b) Linearize the equilibrium equation and set out in outline a Newton–Raphson procedure to solve the equilibrium equation.

   (c) Write a computer program to implement the Newton–Raphson solution.

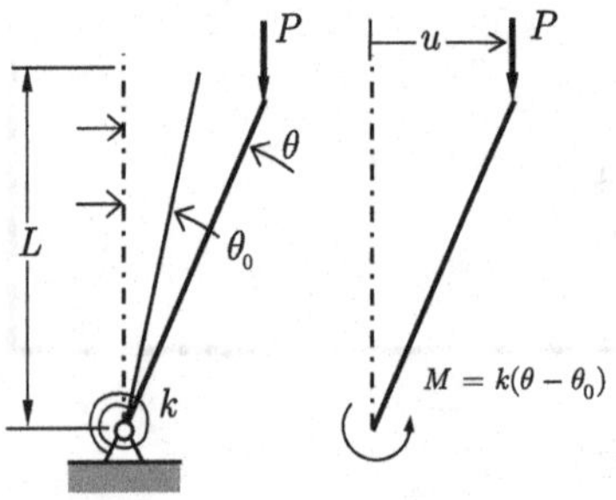


**FIGURE 1.15** Imperfect column.

# CHAPTER TWO

# MATHEMATICAL PRELIMINARIES

## 2.1 INTRODUCTION

In order to make this book sufficiently self-contained, it is necessary to include this chapter dealing with the mathematical tools that are needed to achieve a complete understanding of the topics discussed in the remaining chapters. Vector and tensor algebra is discussed, as is the important concept of the general directional derivative associated with the linearization of various nonlinear quantities that will appear throughout the book.

Readers, especially with engineering backgrounds, are often tempted to skip these mathematical preliminaries and move on directly to the main text. This temptation need not be resisted, as most readers will be able to follow most of the concepts presented even when they are unable to understand the details of the accompanying mathematical derivations. It is only when one needs to understand such derivations that this chapter may need to be consulted in detail. In this way, this chapter should, perhaps, be approached like an instruction manual, only to be referred to when absolutely necessary. The subjects have been presented without the excessive rigors of mathematical language and with a number of examples that should make the text more bearable.

## 2.2 VECTOR AND TENSOR ALGEBRA

Most quantities used in nonlinear continuum mechanics can only be described in terms of vectors or tensors. The purpose of this section, however, is not so much to give a rigorous mathematical description of tensor algebra, which can be found elsewhere, but to introduce some basic concepts and notation that will be used throughout the book. Most readers will have a degree of familiarity with the concepts described herein and may need to come back to this section only if clarification on the notation used further on is required. The topics will be presented in terms of Cartesian coordinate systems.

### 2.2.1 Vectors

Boldface italic characters are used to express points and vectors in a three-dimensional Cartesian space. For instance, $\boldsymbol{e}_1$, $\boldsymbol{e}_2$, and $\boldsymbol{e}_3$ denote the three unit base vectors shown in Figure 2.1, and any given vector $\boldsymbol{v}$ can be expressed as a linear combination of these vectors as

$$\boldsymbol{v} = \sum_{i=1}^{3} v_i\, \boldsymbol{e}_i. \tag{2.1}$$

In more mathematical texts, expressions of this kind are often written without the summation sign, $\sum$, as

$$\boldsymbol{v} = v_i\, \boldsymbol{e}_i. \tag{2.2}$$

Such expressions make use of the *Einstein* or *summation convention*, whereby the repetition of an index (such as the $i$ in the above equation) automatically implies its summation. For clarity, however, this convention will rarely be used in this text.

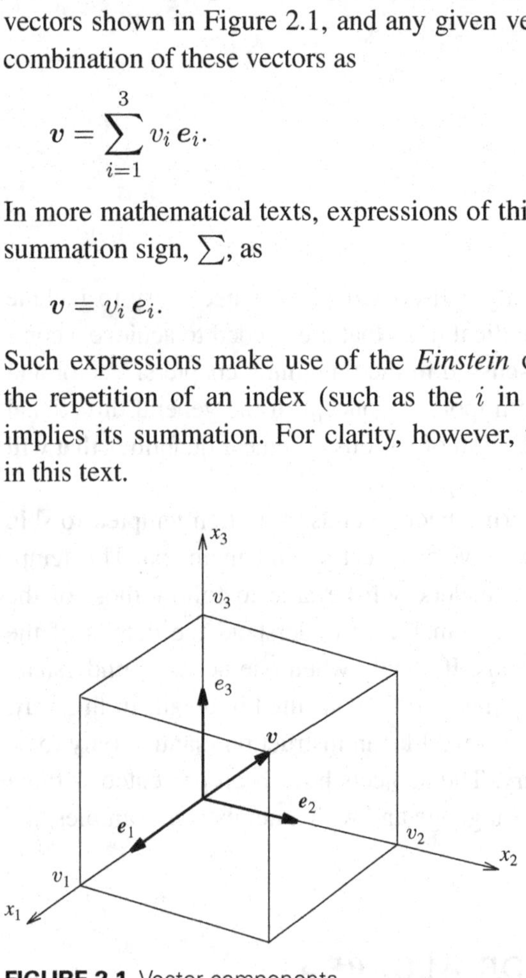


**FIGURE 2.1** Vector components.

The familiar *scalar* or *dot* product of two vectors is defined in such a way that the products of the Cartesian base vectors are given as

$$\boldsymbol{e}_i \cdot \boldsymbol{e}_j = \delta_{ij}, \tag{2.3}$$

where $\delta_{ij}$ is the Kronecker delta, defined as

$$\delta_{ij} = \begin{cases} 1 & i = j; \\ 0 & i \neq j. \end{cases} \tag{2.4}$$

Because the dot product is distributive with respect to addition, the well-known result that the scalar product of any two vectors $\boldsymbol{u}$ and $\boldsymbol{v}$ is given by the sum of the products of the components is shown by

$$\begin{aligned} \boldsymbol{u} \cdot \boldsymbol{v} &= \left( \sum_{i=1}^{3} u_i \, \boldsymbol{e}_i \right) \cdot \left( \sum_{j=1}^{3} v_j \, \boldsymbol{e}_j \right) \\ &= \sum_{i,j=1}^{3} u_i v_j \, (\boldsymbol{e}_i \cdot \boldsymbol{e}_j) \\ &= \sum_{i=1}^{3} u_i v_i = \boldsymbol{v} \cdot \boldsymbol{u}. \end{aligned} \tag{2.5}$$

An additional useful expression for the components of any given vector $\boldsymbol{v}$ follows from Equation (2.3) and the scalar product of Equation (2.1) by $\boldsymbol{e}_j$ to give

$$v_j = \boldsymbol{v} \cdot \boldsymbol{e}_j; \quad j = 1, 2, 3. \tag{2.6}$$

In engineering textbooks, vectors are often represented by single-column matrices containing their Cartesian components as

$$[\boldsymbol{v}] = \begin{bmatrix} v_1 \\ v_2 \\ v_3 \end{bmatrix}, \tag{2.7}$$

where the square bracket symbols [ ] have been used in order to distinguish the vector $\boldsymbol{v}$ itself from the single-column matrix $[\boldsymbol{v}]$ containing its Cartesian components. This distinction is somewhat superfluous unless more than one basis is being considered. For instance, in the new basis $\boldsymbol{e}'_1$, $\boldsymbol{e}'_2$, and $\boldsymbol{e}'_3$ shown in Figure 2.2, the same vector $\boldsymbol{v}$ will manifest itself with different components, in which case the following notation can be used:

$$[\boldsymbol{v}]' = \begin{bmatrix} v'_1 \\ v'_2 \\ v'_3 \end{bmatrix}. \tag{2.8}$$

It must be emphasized, however, that although the components of $\boldsymbol{v}$ are different in the two bases, the vector $\boldsymbol{v}$ itself remains unchanged; that is,

$$\boldsymbol{v} = \sum_{i=1}^{3} v_i \, \boldsymbol{e}_i = \sum_{i=1}^{3} v'_i \, \boldsymbol{e}'_i. \tag{2.9}$$

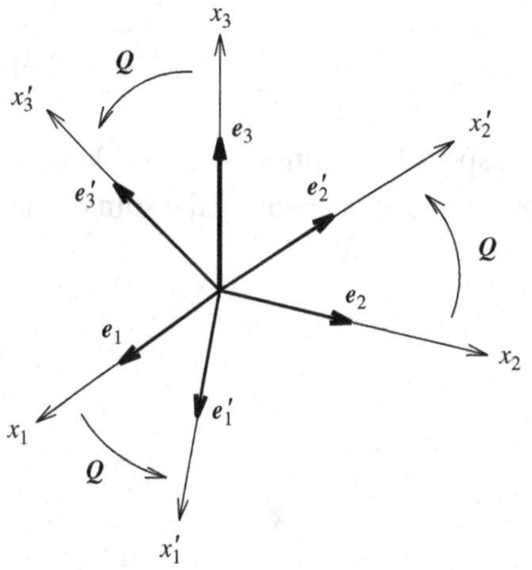


**FIGURE 2.2** Transformation of the Cartesian axes.

Furthermore, the above equation is the key to deriving a relationship between the two sets of components. For this purpose, let $Q_{ij}$ denote the dot products between the two bases as

$$Q_{ij} = \boldsymbol{e}_i \cdot \boldsymbol{e}'_j. \tag{2.10}$$

In fact, the definition of the dot product is such that $Q_{ij}$ is the cosine of the angle between $\boldsymbol{e}_i$ and $\boldsymbol{e}'_j$. Recalling Equation (2.6) for the components of a vector enables the new base vectors to be expressed in terms of the old, or vice versa, as

$$\boldsymbol{e}'_j = \sum_{i=1}^{3} (\boldsymbol{e}'_j \cdot \boldsymbol{e}_i)\, \boldsymbol{e}_i = \sum_{i=1}^{3} Q_{ij}\, \boldsymbol{e}_i; \tag{2.11a}$$

$$\boldsymbol{e}_i = \sum_{j=1}^{3} (\boldsymbol{e}_i \cdot \boldsymbol{e}'_j)\, \boldsymbol{e}'_j = \sum_{j=1}^{3} Q_{ij}\, \boldsymbol{e}'_j. \tag{2.11b}$$

Substituting for $\boldsymbol{e}_i$ in Equation (2.6) from Equation (2.11b) gives, after simple algebra, the old components of $\boldsymbol{v}$ in terms of the new components as

$$\begin{aligned} v_i &= \boldsymbol{v} \cdot \boldsymbol{e}_i \\ &= \boldsymbol{v} \cdot \sum_{j=1}^{3} Q_{ij} \boldsymbol{e}'_j \\ &= \sum_{j=1}^{3} Q_{ij} (\boldsymbol{v} \cdot \boldsymbol{e}'_j) = \sum_{j=1}^{3} Q_{ij} v'_j. \end{aligned} \tag{2.12a}$$

A similar derivation gives

$$v'_i = \sum_{j=1}^{3} Q_{ji} v_j. \tag{2.12b}$$

The above equations can be more easily expressed in matrix form with the help of the $3 \times 3$ matrix $[\boldsymbol{Q}]$ containing the angle cosines $Q_{ij}$ as

$$[\boldsymbol{v}] = [\boldsymbol{Q}][\boldsymbol{v}]'; \tag{2.13a}$$

$$[\boldsymbol{v}]' = [\boldsymbol{Q}]^T[\boldsymbol{v}]; \tag{2.13b}$$

where

$$[\boldsymbol{Q}] = \begin{bmatrix} \boldsymbol{e}_1 \cdot \boldsymbol{e}'_1 & \boldsymbol{e}_1 \cdot \boldsymbol{e}'_2 & \boldsymbol{e}_1 \cdot \boldsymbol{e}'_3 \\ \boldsymbol{e}_2 \cdot \boldsymbol{e}'_1 & \boldsymbol{e}_2 \cdot \boldsymbol{e}'_2 & \boldsymbol{e}_2 \cdot \boldsymbol{e}'_3 \\ \boldsymbol{e}_3 \cdot \boldsymbol{e}'_1 & \boldsymbol{e}_3 \cdot \boldsymbol{e}'_2 & \boldsymbol{e}_3 \cdot \boldsymbol{e}'_3 \end{bmatrix}. \tag{2.14}$$

As a precursor to the discussion of second-order tensors, it is worth emphasizing the coordinate-independent nature of vectors. For example, the vector equations $\boldsymbol{w} = \boldsymbol{u} + \boldsymbol{v}$ or $s = \boldsymbol{u} \cdot \boldsymbol{v}$ make sense without specific reference to the basis used to express the components of the vectors. Obviously, a vector will have different components when expressed in a different basis – but the vector remains unchanged.

**EXAMPLE 2.1: Vector product**

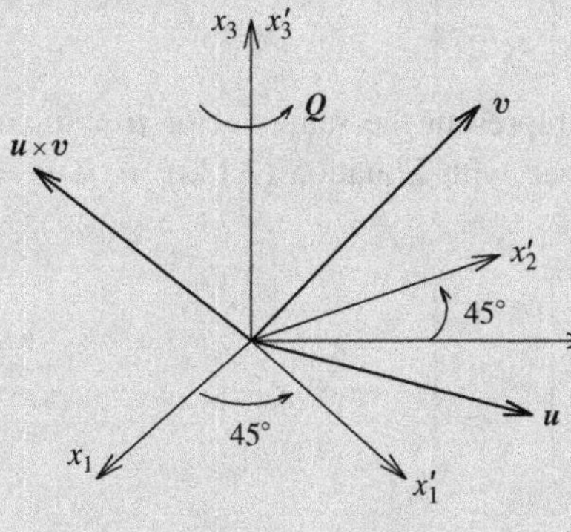


$$[\boldsymbol{Q}] = \frac{1}{\sqrt{2}} \begin{bmatrix} 1 & -1 & 0 \\ 1 & 1 & 0 \\ 0 & 0 & \sqrt{2} \end{bmatrix};$$

*(continued)*

**EXAMPLE 2.1:** *(cont.)*

$$[\boldsymbol{u}] = \begin{bmatrix} 1 \\ 2 \\ 0 \end{bmatrix};$$

$$[\boldsymbol{u}]' = [\boldsymbol{Q}]^T[\boldsymbol{u}] = \frac{1}{\sqrt{2}}\begin{bmatrix} 3 \\ 1 \\ 0 \end{bmatrix};$$

$$[\boldsymbol{v}] = \begin{bmatrix} 0 \\ 1 \\ 1 \end{bmatrix}; \qquad [\boldsymbol{v}]' = [\boldsymbol{Q}]^T[\boldsymbol{v}] = \frac{1}{\sqrt{2}}\begin{bmatrix} 1 \\ 1 \\ \sqrt{2} \end{bmatrix}.$$

As an example of the invariance of a vector under transformation, consider the two vectors $\boldsymbol{u}$ and $\boldsymbol{v}$ and the transformation $[\boldsymbol{Q}]$ shown above. The *vector* or *cross* product of $\boldsymbol{u}$ and $\boldsymbol{v}$ is a third vector $\boldsymbol{u} \times \boldsymbol{v}$ whose components in any base are given as

$$[\boldsymbol{u} \times \boldsymbol{v}] = \begin{bmatrix} u_2v_3 - u_3v_2 \\ u_3v_1 - u_1v_3 \\ u_1v_2 - u_2v_1 \end{bmatrix}.$$

We can apply this equation in both systems of coordinates and obtain a different set of components as

$$[\boldsymbol{u} \times \boldsymbol{v}] = \begin{bmatrix} 2 \\ -1 \\ 1 \end{bmatrix}; \quad [\boldsymbol{u} \times \boldsymbol{v}]' = \frac{1}{\sqrt{2}}\begin{bmatrix} 1 \\ -3 \\ \sqrt{2} \end{bmatrix}.$$

The fact that these two sets of components represent the same vector $\boldsymbol{u} \times \boldsymbol{v}$ can be verified by checking whether in accordance with Equation (2.13a) $[\boldsymbol{u} \times \boldsymbol{v}] = [\boldsymbol{Q}][\boldsymbol{u} \times \boldsymbol{v}]'$. This is clearly the case as

$$\begin{bmatrix} 2 \\ -1 \\ 1 \end{bmatrix} = \frac{1}{\sqrt{2}}\begin{bmatrix} 1 & -1 & 0 \\ 1 & 1 & 0 \\ 0 & 0 & \sqrt{2} \end{bmatrix}\frac{1}{\sqrt{2}}\begin{bmatrix} 1 \\ -3 \\ \sqrt{2} \end{bmatrix}.$$

### 2.2.2 Second-Order Tensors

A second-order tensor $\boldsymbol{S}$ is a linear mapping that associates a given vector $\boldsymbol{u}$ with a second vector $\boldsymbol{v}$ as

$$v = Su. \tag{2.15}$$

Italic boldface capitals will be used throughout this chapter to denote second-order tensors. Later on, however, the distinction between lower- and upper-case quantities will be needed for more important purposes and no explicit differentiation will be made between vectors and second-order tensors. The term "linear" in the above definition is used to imply that, given two arbitrary vectors $u_1$ and $u_2$ and arbitrary scalars $\alpha$ and $\beta$, then

$$S(\alpha u_1 + \beta u_2) = \alpha S u_1 + \beta S u_2. \tag{2.16}$$

Recognizing in Equation (2.15) that $u$ and $v$ are vectors and thus coordinate- independent, the tensor $S$ that operates on $u$ to give $v$ must also, in a similar sense, have a coordinate-independent nature. If the vectors $u$ and $v$ can be expressed in terms of components in a specific basis, namely, Equation (2.1), then it is to be expected that the tensor $S$ can somehow be expressed in terms of components in the same basis; this is shown below. Consequently, a tensor will have different components in different bases, but the tensor itself will remain unchanged.

The simple example of a second-order tensor that satisfies the above definition is the *identity tensor* $I$, which maps any given vector $u$ onto itself as

$$u = Iu. \tag{2.17}$$

Another example is the transformation tensor $Q$ shown in Figure 2.2, which rotates vectors in space in such a way that the standard Cartesian base vectors $e_1$, $e_2$, and $e_3$ become $e'_1$, $e'_2$, and $e'_3$, that is,

$$e'_i = Q e_i; \quad i = 1, 2, 3. \tag{2.18}$$

The relationship between this important tensor and the angle cosines $Q_{ij}$ introduced in the previous section will be elaborated below.

Simple operations such as the sum, product, and inverse of second-order tensors are defined in an obvious manner so that for any vector $u$

$$(S_1 + S_2)\, u = S_1 u + S_2 u; \tag{2.19a}$$
$$(S_1 S_2)\, u = S_1(S_2 u); \tag{2.19b}$$
$$S^{-1} S = I. \tag{2.19c}$$

Additionally, the transpose of a tensor $S$ is defined as the tensor $S^T$, which for any two vectors $u$ and $v$ satisfies

$$u \cdot Sv = v \cdot S^T u. \tag{2.20}$$

For example, the transpose of the identity tensor is again the identity $I$, because use of the above definition shows that for any pair of vectors $u$ and $v$, $I^T$ satisfies

$$\begin{aligned} \boldsymbol{v} \cdot \boldsymbol{I}^T \boldsymbol{u} &= \boldsymbol{u} \cdot \boldsymbol{I} \boldsymbol{v} \\ &= \boldsymbol{u} \cdot \boldsymbol{v} \\ &= \boldsymbol{v} \cdot \boldsymbol{u} \\ &= \boldsymbol{v} \cdot \boldsymbol{I} \boldsymbol{u}. \end{aligned} \tag{2.21}$$

A tensor $\boldsymbol{S}$ that, like the identity, satisfies $\boldsymbol{S}^T = \boldsymbol{S}$ is said to be *symmetric*; whereas a tensor for which $\boldsymbol{S}^T = -\boldsymbol{S}$ is said to be *skew*. As an example of a skew tensor consider the tensor $\boldsymbol{W}_{\boldsymbol{w}}$, associated with an arbitrary vector $\boldsymbol{w}$, defined in such a way that for any vector $\boldsymbol{u}$

$$\boldsymbol{W}_{\boldsymbol{w}} \boldsymbol{u} = \boldsymbol{w} \times \boldsymbol{u} \tag{2.22}$$

where $\times$ denotes the standard vector or cross product. Proof that $\boldsymbol{W}_{\boldsymbol{w}}$ is skew follows from the cyclic commutative property of the mixed vector product, which, for any $\boldsymbol{u}$ and $\boldsymbol{v}$, gives

$$\begin{aligned} \boldsymbol{v} \cdot \boldsymbol{W}_{\boldsymbol{w}}^T \boldsymbol{u} &= \boldsymbol{u} \cdot \boldsymbol{W}_{\boldsymbol{w}} \boldsymbol{v} \\ &= \boldsymbol{u} \cdot (\boldsymbol{w} \times \boldsymbol{v}) \\ &= -\boldsymbol{v} \cdot (\boldsymbol{w} \times \boldsymbol{u}) \\ &= -\boldsymbol{v} \cdot \boldsymbol{W}_{\boldsymbol{w}} \boldsymbol{u}. \end{aligned} \tag{2.23}$$

A final example of general interest is the transpose of the *transformation tensor* $\boldsymbol{Q}$. Applying the definition of the transpose tensor to the new and old base vectors and recalling the definition of $\boldsymbol{Q}$ given in Equation (2.18) gives

$$\begin{aligned} \boldsymbol{e}_j \cdot \boldsymbol{Q}^T \boldsymbol{e}'_i &= \boldsymbol{e}'_i \cdot \boldsymbol{Q} \boldsymbol{e}_j \\ &= \boldsymbol{e}'_i \cdot \boldsymbol{e}'_j \\ &= \boldsymbol{e}_j \cdot \boldsymbol{e}_i; \quad i, j = 1, 2, 3. \end{aligned} \tag{2.24}$$

Comparing the first and last expressions in this equation leads to

$$\boldsymbol{Q}^T \boldsymbol{e}'_i = \boldsymbol{e}_i; \qquad i = 1, 2, 3, \tag{2.25}$$

which implies that the transpose tensor $\boldsymbol{Q}^T$ rotates the new base vectors $\boldsymbol{e}'_i$ back to their original positions $\boldsymbol{e}_i$. Moreover, combining Equations (2.25) and (2.18) gives

$$\boldsymbol{Q}^T \boldsymbol{Q} = \boldsymbol{I}. \tag{2.26}$$

Tensors that satisfy this important property are said to be *orthogonal*. In fact, any arbitrary second-order tensor $\boldsymbol{A}$ can be expressed as the sum of a symmetric plus a skew tensor or as the product of an orthogonal times a symmetric tensor as

$$\boldsymbol{A} = \boldsymbol{S} + \boldsymbol{W}; \quad \boldsymbol{S}^T = \boldsymbol{S}, \quad \boldsymbol{W}^T = -\boldsymbol{W}; \tag{2.27a}$$

$$\boldsymbol{A} = \boldsymbol{Q}\boldsymbol{S}; \quad \boldsymbol{S}^T = \boldsymbol{S}, \quad \boldsymbol{Q}^T \boldsymbol{Q} = \boldsymbol{I}. \tag{2.27b}$$

The first expression is rather trivial and follows from taking $\boldsymbol{S} = (\boldsymbol{A} + \boldsymbol{A}^T)/2$ and $\boldsymbol{W} = (\boldsymbol{A} - \boldsymbol{A}^T)/2$; whereas the second, less obvious, equation is known as the *polar decomposition* and plays a crucial role in continuum mechanics. Many examples of symmetric and skew tensors will occur throughout the remaining continuum mechanics sections of this chapter.

Second-order tensors can often be derived from the *dyadic* or *tensor* product of two vectors $\boldsymbol{u}$ and $\boldsymbol{v}$ to give a tensor, denoted $\boldsymbol{u} \otimes \boldsymbol{v}$, which to any arbitrary third vector $\boldsymbol{w}$ assigns the following vector:

$$(\boldsymbol{u} \otimes \boldsymbol{v})\boldsymbol{w} = (\boldsymbol{w} \cdot \boldsymbol{v})\boldsymbol{u}, \tag{2.28}$$

where $(\boldsymbol{w} \cdot \boldsymbol{v})\boldsymbol{u}$ is obviously the projection in the $\boldsymbol{u}$ direction of the scalar component of $\boldsymbol{w}$ in the $\boldsymbol{v}$ direction. This seemingly bizarre definition of the tensor $\boldsymbol{u} \otimes \boldsymbol{v}$ transpires to make physical sense, particularly in the case of the stress tensor, from which, incidentally, the word tensor originates, from the association with a tensile stress.

The tensor product satisfies the following properties:

$$(\boldsymbol{u} \otimes \boldsymbol{v})^T = (\boldsymbol{v} \otimes \boldsymbol{u}); \tag{2.29a}$$

$$\boldsymbol{S}(\boldsymbol{u} \otimes \boldsymbol{v}) = (\boldsymbol{S}\boldsymbol{u} \otimes \boldsymbol{v}); \tag{2.29b}$$

$$(\boldsymbol{u} \otimes \boldsymbol{v})\boldsymbol{S} = (\boldsymbol{u} \otimes \boldsymbol{S}^T\boldsymbol{v}); \tag{2.29c}$$

$$\boldsymbol{u} \otimes (\boldsymbol{v}_1 + \boldsymbol{v}_2) = \boldsymbol{u} \otimes \boldsymbol{v}_1 + \boldsymbol{u} \otimes \boldsymbol{v}_2. \tag{2.29d}$$

**EXAMPLE 2.2: Proof of Equation (2.29b)**

Any of the properties of the dyadic product can be easily demonstrated using its definition, Equation (2.28). For example, we can prove (2.29b) by showing that for any vector $\boldsymbol{w}$ we have

$$\begin{aligned}\boldsymbol{S}(\boldsymbol{u} \otimes \boldsymbol{v})\boldsymbol{w} &= \boldsymbol{S}[(\boldsymbol{u} \otimes \boldsymbol{v})\boldsymbol{w}]\\ &= \boldsymbol{S}\boldsymbol{u}(\boldsymbol{v} \cdot \boldsymbol{w})\\ &= (\boldsymbol{S}\boldsymbol{u} \otimes \boldsymbol{v})\boldsymbol{w}.\end{aligned}$$

Now recall that a vector can be expressed in terms of a linear combination of the base vectors $\boldsymbol{e}_1$, $\boldsymbol{e}_2$, and $\boldsymbol{e}_3$ as shown in Equation (2.1). Hence it is not unreasonable to suggest that in a similar fashion a tensor could be expressed in terms of a linear combination of dyadic products of these base vectors. In particular, the nine tensors $\boldsymbol{e}_i \otimes \boldsymbol{e}_j$ for $i, j = 1, 2, 3$ obtained by the dyadic product of the three Cartesian base vectors form a basis on which any second-order tensor can be expressed. For instance, the identity tensor $\boldsymbol{I}$ can be written as

$$I = \sum_{i=1}^{3} e_i \otimes e_i \quad \text{or} \quad I = \sum_{i,j=1}^{3} \delta_{ij}\, e_i \otimes e_j. \tag{2.30a,b}$$

In order to verify these expressions simply note that when either equation is applied to a vector $u$, use of the definition (2.28) together with Equations (2.6) and (2.19a) gives

$$\begin{aligned} Iu &= \left( \sum_{i=1}^{3} e_i \otimes e_i \right) u \\ &= \sum_{i=1}^{3} (u \cdot e_i) e_i \\ &= \sum_{i=1}^{3} u_i\, e_i \\ &= u. \end{aligned} \tag{2.31}$$

In general, we can express any given tensor $S$ as a linear combination of $e_i \otimes e_j$ in terms of a set of nine components $S_{ij}$ as

$$S = \sum_{i,j=1}^{3} S_{ij}\, e_i \otimes e_j, \tag{2.32}$$

where the components $S_{ij}$ can be obtained in a manner similar to that used in Equation (2.6) for vectors as

$$S_{ij} = e_i \cdot S e_j. \tag{2.33}$$

Proof of this expression follows from Equations (2.32) and (2.28) as

$$\begin{aligned} e_i \cdot S e_j &= e_i \cdot \left( \sum_{k,l=1}^{3} S_{kl}\, e_k \otimes e_l \right) e_j \\ &= e_i \cdot \left( \sum_{k,l=1}^{3} S_{kl} (e_j \cdot e_l) e_k \right) \\ &= e_i \cdot \left( \sum_{k,l=1}^{3} S_{kl} \delta_{lj}\, e_k \right) \\ &= \sum_{k,l=1}^{3} S_{kl} \delta_{lj} \delta_{ik} \\ &= S_{ij}. \end{aligned} \tag{2.34}$$

For example, the transformation tensor $\boldsymbol{Q}$ can be expressed in terms of its components $Q_{ij}$ as

$$\boldsymbol{Q} = \sum_{i,j=1}^{3} Q_{ij}\, \boldsymbol{e}_i \otimes \boldsymbol{e}_j, \tag{2.35}$$

where, using Equation (2.33) together with Equation (2.18), the components $Q_{ij}$ can now be seen to coincide with the angle cosines introduced in the previous section as

$$Q_{ij} = \boldsymbol{e}_i \cdot \boldsymbol{Q}\boldsymbol{e}_j = \boldsymbol{e}_i \cdot \boldsymbol{e}'_j. \tag{2.36}$$

**EXAMPLE 2.3: Components of $\boldsymbol{u} \otimes \boldsymbol{v}$**

We can evaluate the components of the tensor product $\boldsymbol{u} \otimes \boldsymbol{v}$ in two different ways. First, direct use of Equation (2.33) and the definition (2.28) gives

$$\begin{aligned} (\boldsymbol{u} \otimes \boldsymbol{v})_{ij} &= \boldsymbol{e}_i \cdot (\boldsymbol{u} \otimes \boldsymbol{v})\boldsymbol{e}_j \\ &= (\boldsymbol{e}_i \cdot \boldsymbol{u})(\boldsymbol{v} \cdot \boldsymbol{e}_j) \\ &= u_i v_j. \end{aligned}$$

Alternatively, use of Equation (2.1) and property (2.29d) gives

$$\begin{aligned} (\boldsymbol{u} \otimes \boldsymbol{v}) &= \left(\sum_{i=1}^{3} u_i\, \boldsymbol{e}_i\right) \otimes \left(\sum_{j=1}^{3} v_j\, \boldsymbol{e}_j\right) \\ &= \sum_{i.j=1}^{3} u_i v_j\, \boldsymbol{e}_i \otimes \boldsymbol{e}_j, \end{aligned}$$

from which the same components of the dyadic product are immediately identified.

The components of any second-order tensor $\boldsymbol{S}$ can be arranged in the form of a $3 \times 3$ matrix as

$$[\boldsymbol{S}] = \begin{bmatrix} S_{11} & S_{12} & S_{13} \\ S_{21} & S_{22} & S_{23} \\ S_{31} & S_{32} & S_{33} \end{bmatrix}. \tag{2.37}$$

For instance, use of Equation (2.33) shows that the skew tensor $\boldsymbol{W}_w$ defined in Equation (2.22) can be written as

$$[\boldsymbol{W_w}] = \begin{bmatrix} 0 & -w_3 & w_2 \\ w_3 & 0 & -w_1 \\ -w_2 & w_1 & 0 \end{bmatrix}, \tag{2.38}$$

where $w_1$, $w_2$, and $w_3$ are the components of $\boldsymbol{w}$.

Tensorial expressions can now be duplicated in terms of matrix operations with the tensor components. For instance, it is simple to show that the basic Equation (2.15) becomes

$$[\boldsymbol{v}] = [\boldsymbol{S}][\boldsymbol{u}]. \tag{2.39}$$

Similar expressions apply for the sum, product, dyadic product, and transpose of tensors as

$$[\boldsymbol{S}_1 + \boldsymbol{S}_2] = [\boldsymbol{S}_1] + [\boldsymbol{S}_2]; \tag{2.40a}$$

$$[\boldsymbol{S}_1\boldsymbol{S}_2] = [\boldsymbol{S}_1][\boldsymbol{S}_2]; \tag{2.40b}$$

$$[\boldsymbol{S}^T] = [\boldsymbol{S}]^T; \tag{2.40c}$$

$$[\boldsymbol{u} \otimes \boldsymbol{v}] = [\boldsymbol{u}][\boldsymbol{v}]^T. \tag{2.40d}$$

The association between second-order tensors and $3 \times 3$ matrices is thus identical to that between vectors and column matrices. Similarly, the distinction between the tensor and its components is only useful when more than one basis is being considered at the same time. Under such circumstances, a single tensor $\boldsymbol{S}$ is expressed by two sets of components, $[\boldsymbol{S}]'$ or $S'_{ij}$ in the base $\boldsymbol{e}'_i \otimes \boldsymbol{e}'_j$ and $[\boldsymbol{S}]$ or $S_{ij}$ in the base $\boldsymbol{e}_i \otimes \boldsymbol{e}_j$, as

$$\boldsymbol{S} = \sum_{i,j=1}^{3} S_{ij}\, \boldsymbol{e}_i \otimes \boldsymbol{e}_j = \sum_{i,j=1}^{3} S'_{ij}\, \boldsymbol{e}'_i \otimes \boldsymbol{e}'_j. \tag{2.41}$$

Introducing the relationship between the new and old bases given by the angle cosines as shown in Equations (2.11a–b) and after simple algebra, a relationship between both sets of tensor components emerges as

$$[\boldsymbol{S}]' = [\boldsymbol{Q}]^T[\boldsymbol{S}][\boldsymbol{Q}] \quad \text{or} \quad S'_{ij} = \sum_{k,l=1}^{3} Q_{ki} S_{kl} Q_{lj}. \tag{2.42}$$

It is often necessary to use second-order tensors relating vectors expressed in *different* sets of Cartesian bases. This is illustrated in Example 2.4.

**EXAMPLE 2.4: Exploration of two-point second-order tensors**

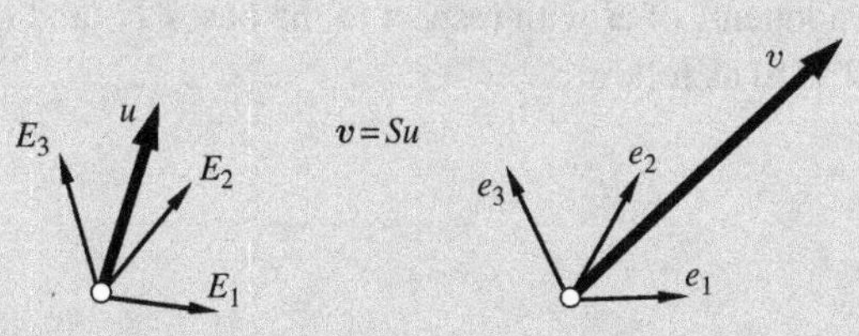


The figure above shows a vector $\boldsymbol{u}$ in Cartesian basis $\boldsymbol{E}_i$ which is mapped (transformed) into the vector $\boldsymbol{v}$ in a different Cartesian basis $\boldsymbol{e}_i$ as

$$\boldsymbol{v} = \boldsymbol{S}\boldsymbol{u} \quad \text{or} \quad \sum_{k=1}^{3} v_k \boldsymbol{e}_k = \boldsymbol{S} \sum_{j=1}^{3} u_j \, \boldsymbol{E}_j.$$

Multiplying through by $\boldsymbol{e}_i$ gives

$$\boldsymbol{e}_i \cdot \sum_{k=1}^{3} v_k \boldsymbol{e}_k = \boldsymbol{e}_i \cdot \boldsymbol{S} \sum_{j=1}^{3} u_j \, \boldsymbol{E}_j;$$

$$\sum_{k=1}^{3} v_k \delta_{ik} = \Big(\boldsymbol{S} \sum_{j=1}^{3} u_j \, \boldsymbol{E}_j\Big) \cdot \boldsymbol{e}_i = \sum_{j=1}^{3} (\boldsymbol{e}_i \cdot \boldsymbol{S}\boldsymbol{E}_j) \, u_j;$$

$$\text{alternatively,} \quad v_i = \sum_{j=1}^{3} S_{ij} u_j \quad \text{where} \quad S_{ij} = \boldsymbol{e}_i \cdot \boldsymbol{S}\boldsymbol{E}_j.$$

Clearly, the components $S_{ij}$ of the tensor $\boldsymbol{S}$ depend upon the bases in which the vectors $\boldsymbol{u}$ and $\boldsymbol{v}$ are expressed. It is instructive to explore a little further to discover how this dependency is expressed in terms of the tensor product $\otimes$ . The vector $\boldsymbol{v}$ can be written as

$$\boldsymbol{v} = \sum_{i,j=1}^{3} u_j S_{ij} \, \boldsymbol{e}_i = \sum_{i,j=1}^{3} S_{ij} \big(\boldsymbol{u} \cdot \boldsymbol{E}_j\big) \boldsymbol{e}_i;$$

recall (2.28) to give

$$\boldsymbol{v} = \sum_{i,j=1}^{3} S_{ij} \big(\boldsymbol{e}_i \otimes \boldsymbol{E}_j\big) \boldsymbol{u}.$$

Consequently,

$$\boldsymbol{S} = \sum_{i,j=1}^{3} S_{ij} \big(\boldsymbol{e}_i \otimes \boldsymbol{E}_j\big).$$

It is now possible to rotate just one set of bases and examine the effect this has on the components of the tensor $\boldsymbol{S}$. For example, if the bases $\boldsymbol{e}_i$ are rotated as $\boldsymbol{e}'_i = \boldsymbol{Q}\boldsymbol{e}_i$

*(continued)*

**EXAMPLE 2.4:** *(cont.)*

(see Equation (2.18)) then the components of $\boldsymbol{S}$ with respect to the bases $\boldsymbol{E}_i$ and $\boldsymbol{e}'_i$ can be obtained using Equation (2.11b) as follows:

$$\begin{aligned}\boldsymbol{S} &= \sum_{i,j=1}^{3} S_{ij}(\boldsymbol{e}_i \otimes \boldsymbol{E}_j) \\ &= \sum_{i,j,k=1}^{3} S_{ij}\, Q_{ik} \boldsymbol{e}'_k \otimes \boldsymbol{E}_j \\ &= \sum_{i,k=1}^{3} S'_{jk} \boldsymbol{e}'_k \otimes \boldsymbol{E}_j; \quad S'_{jk} = S_{ij}\, Q_{ik}.\end{aligned}$$

Using Equation (2.42) and assuming some elementary knowledge of a two-dimensional state of stresses, the invariant nature of the stress tensor is shown in Example 2.5.

**EXAMPLE 2.5: Tensor invariance**

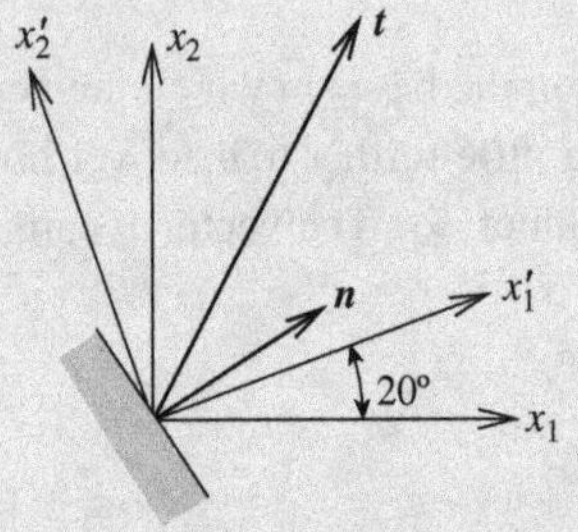


$$[\boldsymbol{t}] = \begin{bmatrix} 2.232 \\ 2.866 \end{bmatrix}; \quad [\boldsymbol{t}]' = \begin{bmatrix} 3.079 \\ 1.932 \end{bmatrix};$$

$$[\boldsymbol{n}] = \begin{bmatrix} 0.866 \\ 0.500 \end{bmatrix}; \quad [\boldsymbol{n}]' = \begin{bmatrix} 0.985 \\ 0.174 \end{bmatrix};$$

$$[\boldsymbol{\sigma}] = \begin{bmatrix} 2 & 1 \\ 1 & 4 \end{bmatrix}; \quad [\boldsymbol{\sigma}]' = \begin{bmatrix} 2.878 & 1.409 \\ 1.409 & 3.125 \end{bmatrix}.$$

*(continued)*

**EXAMPLE 2.5:** *(cont.)*

Consider a two-dimensional example of a stress tensor $\boldsymbol{\sigma}$ having Cartesian components $[\boldsymbol{\sigma}]$ and an associated unit normal vector $\boldsymbol{n}$ having components $[\boldsymbol{n}]$. The traction force $\boldsymbol{t}$ having components $[\boldsymbol{t}]$ on the surface normal to $\boldsymbol{n}$ is

$$[\boldsymbol{t}] = [\boldsymbol{\sigma}]\,[\boldsymbol{n}] \quad \text{or} \quad \begin{bmatrix} 2.232 \\ 2.866 \end{bmatrix} = \begin{bmatrix} 2 & 1 \\ 1 & 4 \end{bmatrix} \begin{bmatrix} 0.866 \\ 0.500 \end{bmatrix}.$$

Now rotate the Cartesian axes counterclockwise through an angle $\alpha = 20°$ in which the same stress tensor $\boldsymbol{\sigma}$ now has components $[\boldsymbol{\sigma}]'$ and the normal vector $\boldsymbol{n}$ components $[\boldsymbol{n}]'$. The traction vector $\boldsymbol{t}$ having components $[\boldsymbol{t}]'$ on the surface normal to $\boldsymbol{n}$ is

$$[\boldsymbol{t}]' = [\boldsymbol{\sigma}]'\,[\boldsymbol{n}]' \quad \text{or} \quad \begin{bmatrix} 3.079 \\ 1.932 \end{bmatrix} = \begin{bmatrix} 2.878 & 1.409 \\ 1.409 & 3.125 \end{bmatrix} \begin{bmatrix} 0.985 \\ 0.174 \end{bmatrix}.$$

The vectors $\boldsymbol{n}$ and $\boldsymbol{t}$ have remained unchanged even though their components have changed. Likewise the *stress tensor* $\boldsymbol{\sigma}$ has remained unchanged in that it *is that which operates on* $\boldsymbol{n}$ *to give* $\boldsymbol{t}$, even though its components have changed. Hence, the general expression $\boldsymbol{t} = \boldsymbol{\sigma}\boldsymbol{n}$ can justifiably be written.

Noting that from Equation (2.13) $[\boldsymbol{t}]' = [\boldsymbol{Q}]^T\,[\boldsymbol{t}]$ and $[\boldsymbol{n}]' = [\boldsymbol{Q}]^T\,[\boldsymbol{n}]$ it is easy to show that

$$[\boldsymbol{\sigma}]' = [\boldsymbol{Q}]^T\,[\boldsymbol{\sigma}]\,[\boldsymbol{Q}]; \quad [\boldsymbol{Q}] = \begin{bmatrix} \cos\alpha & -\sin\alpha \\ \sin\alpha & \cos\alpha \end{bmatrix}.$$

## 2.2.3 Vector and Tensor Invariants

The above sections have shown that when different bases are used, vectors and tensors manifest themselves via different sets of components. An interesting exception to this rule is the case of the identity tensor or its multiples $\alpha\boldsymbol{I}$. Using Equation (2.42) it is easy to show that the components of these tensors remain unchanged by the rotation of the axes as

$$\begin{aligned} [\alpha\boldsymbol{I}]' &= [\boldsymbol{Q}]^T[\alpha\boldsymbol{I}][\boldsymbol{Q}] \\ &= \alpha[\boldsymbol{Q}^T\boldsymbol{Q}] \\ &= [\alpha\boldsymbol{I}]. \end{aligned} \tag{2.43}$$

Tensors that satisfy this property are said to be *isotropic* and are used in continuum mechanics to describe materials that exhibit identical properties in all directions.

In general, however, the components of vectors and second-order tensors will change when the axes are rotated. Nevertheless, certain intrinsic magnitudes associated with them will remain invariant under such transformations. For example, the scalar product between two vectors $\boldsymbol{u}$ and $\boldsymbol{v}$, defined in Section 2.2.1 as

$$\boldsymbol{u} \cdot \boldsymbol{v} = \sum_{i=1}^{3} u_i v_i = [\boldsymbol{u}]^T [\boldsymbol{v}], \tag{2.44}$$

remains invariant when the components change as a result of a rotation $\boldsymbol{Q}$ of the axes. This is easily proved with the help of Equation (2.13a) and the orthogonality of $\boldsymbol{Q}$ as

$$\begin{aligned}\boldsymbol{u} \cdot \boldsymbol{v} &= [\boldsymbol{u}]^T [\boldsymbol{v}] \\ &= ([\boldsymbol{Q}][\boldsymbol{u}]')^T [\boldsymbol{Q}][\boldsymbol{v}]' \\ &= ([\boldsymbol{u}]')^T [\boldsymbol{Q}^T \boldsymbol{Q}][\boldsymbol{v}]' \\ &= ([\boldsymbol{u}]')^T [\boldsymbol{v}]'.\end{aligned} \tag{2.45}$$

Consequently, the modulus or magnitude of a vector, defined as

$$\|\boldsymbol{u}\| = \sqrt{\boldsymbol{u} \cdot \boldsymbol{u}}, \tag{2.46}$$

is an invariant, that is, an intrinsic physical property of the vector.

Insofar as tensors can be expressed in terms of dyadic products of base vectors it is reasonable to inquire whether invariant quantities can be associated with tensors. This is indeed the case, and the first of these magnitudes, denoted as $I_{\boldsymbol{S}}$, is given by the *trace*, which is defined as the sum of the diagonal components of the tensor $\boldsymbol{S}$ as

$$I_{\boldsymbol{S}} = \mathrm{tr}\boldsymbol{S} = \sum_{i=1}^{3} S_{ii}. \tag{2.47}$$

The invariance of this magnitude can be easily checked using Equation (2.41) in an indicial form and recalling the orthogonality of $\boldsymbol{Q}$ given by Equation (2.26), to give

$$\begin{aligned}\sum_{i=1}^{3} S'_{ii} &= \sum_{i,k,l=1}^{3} Q_{ki} S_{kl} Q_{li} \\ &= \sum_{k,l=1}^{3} S_{kl} \delta_{kl} \\ &= \sum_{k=1}^{3} S_{kk}.\end{aligned} \tag{2.48}$$

Some useful and simple-to-prove properties of the trace are

$$\mathrm{tr}(\boldsymbol{u} \otimes \boldsymbol{v}) = \boldsymbol{u} \cdot \boldsymbol{v}; \tag{2.49a}$$

$$\mathrm{tr}\boldsymbol{S}^T = \mathrm{tr}\boldsymbol{S}; \tag{2.49b}$$

$$\mathrm{tr}\boldsymbol{S}_1\boldsymbol{S}_2 = \mathrm{tr}\boldsymbol{S}_2\boldsymbol{S}_1. \tag{2.49c}$$

**EXAMPLE 2.6: Proof of Equation (2.49a)**

Property (2.49a) follows from the components of the dyadic product discussed in Example 2.3 and the definition of the scalar product as

$$\begin{aligned}\mathrm{tr}(\boldsymbol{u} \otimes \boldsymbol{v}) &= \sum_{i=1}^{3} (\boldsymbol{u} \otimes \boldsymbol{v})_{ii} \\ &= \sum_{i=1}^{3} u_i v_i \\ &= \boldsymbol{u} \cdot \boldsymbol{v}.\end{aligned}$$

Analogous to the scalar product of vectors, the *double product* or *double contraction* of two tensors $\boldsymbol{A}$ and $\boldsymbol{B}$ is an invariant magnitude defined in terms of the trace as

$$\boldsymbol{A} : \boldsymbol{B} = \mathrm{tr}(\boldsymbol{A}^T\boldsymbol{B}), \tag{2.50}$$

which recalling the properties of the trace can be variously written as

$$\boldsymbol{A} : \boldsymbol{B} = \mathrm{tr}(\boldsymbol{A}^T\boldsymbol{B}) = \mathrm{tr}(\boldsymbol{B}\boldsymbol{A}^T) = \mathrm{tr}(\boldsymbol{B}^T\boldsymbol{A}) = \mathrm{tr}(\boldsymbol{A}\boldsymbol{B}^T) = \sum_{i,j=1}^{3} A_{ij}B_{ij}. \tag{2.51}$$

Further useful properties of the double contraction are

$$\mathrm{tr}\boldsymbol{S} = \boldsymbol{I} : \boldsymbol{S}; \tag{2.52a}$$

$$\boldsymbol{S} : (\boldsymbol{u} \otimes \boldsymbol{v}) = \boldsymbol{u} \cdot \boldsymbol{S}\boldsymbol{v}; \tag{2.52b}$$

$$(\boldsymbol{u} \otimes \boldsymbol{v}) : (\boldsymbol{x} \otimes \boldsymbol{y}) = (\boldsymbol{u} \cdot \boldsymbol{x})(\boldsymbol{v} \cdot \boldsymbol{y}); \tag{2.52c}$$

$$\boldsymbol{S} : \boldsymbol{W} = 0 \quad \text{if} \quad \boldsymbol{S}^T{=}\boldsymbol{S} \quad \text{and} \quad \boldsymbol{W}^T = -\boldsymbol{W}. \tag{2.52d}$$

A second independent invariant of a tensor $\boldsymbol{S}$ can now be defined as*

$$II_{\boldsymbol{S}} = \boldsymbol{S} : \boldsymbol{S}. \tag{2.53}$$

* In other texts the following alternative definition is frequently used:

$$II_{\boldsymbol{S}} = \frac{1}{2}(I_{\boldsymbol{S}}^2 - \boldsymbol{S} : \boldsymbol{S}).$$

A third and final invariant of a second-order tensor is provided by its determinant, which is simply defined as the determinant of the matrix of components as

$$III_{\boldsymbol{S}} = \det \boldsymbol{S} = \det[\boldsymbol{S}]. \tag{2.54}$$

Proof of the invariance of the determinant follows easily from Equation (2.42) and the orthogonality of the transformation tensor $\boldsymbol{Q}$ as

$$\begin{aligned} \det\,[\boldsymbol{S}]' &= \det([\boldsymbol{Q}]^T[\boldsymbol{S}][\boldsymbol{Q}]) \\ &= \det[\boldsymbol{Q}]^T \det[\boldsymbol{S}] \det[\boldsymbol{Q}] \\ &= \det[\boldsymbol{Q}^T\boldsymbol{Q}] \det[\boldsymbol{S}] \\ &= \det[\boldsymbol{S}]. \end{aligned} \tag{2.55}$$

An alternative way in which invariant magnitudes of second-order tensors can be explored is by studying the existence of eigenvectors and eigenvalues. For a given tensor $\boldsymbol{S}$, a vector $\boldsymbol{n}$ is said to be an eigenvector with an associated eigenvalue $\lambda$ if

$$\boldsymbol{S}\boldsymbol{n} = \lambda \boldsymbol{n}. \tag{2.56}$$

Linear algebra theory shows that the eigenvalues $\lambda$ that satisfy the above equation are the roots of a third-degree polynomial obtained as

$$\det(\boldsymbol{S} - \lambda \boldsymbol{I}) = 0. \tag{2.57}$$

In general, however, the roots of such an equation can be imaginary and hence of little practical use. An important exception is the case of symmetric tensors, for which it is possible to prove not only that there exist three real roots of Equation (2.57), $\lambda_1$, $\lambda_2$, and $\lambda_3$, but also that the three corresponding eigenvectors $\boldsymbol{n}_1$, $\boldsymbol{n}_2$, and $\boldsymbol{n}_3$ are orthogonal, that is,

$$\boldsymbol{S}\boldsymbol{n}_i = \lambda_i \boldsymbol{n}_i; \quad i = 1, 2, 3; \tag{2.58a}$$

$$\boldsymbol{n}_i \cdot \boldsymbol{n}_j = \delta_{ij}; \quad i, j = 1, 2, 3. \tag{2.58b}$$

Consequently, the above unit eigenvectors can be used as an alternative Cartesian base in which Equations (2.33) and (2.58a) show that the off-diagonal components of $\boldsymbol{S}$ vanish whereas the three diagonal components are precisely the eigenvalues $\lambda_i$. Accordingly, the symmetric tensor $\boldsymbol{S}$ can be conveniently written as

$$\boldsymbol{S} = \sum_{i=1}^{3} \lambda_i \, \boldsymbol{n}_i \otimes \boldsymbol{n}_i. \tag{2.59}$$

The eigenvalues of a symmetric tensor are independent of the Cartesian axes chosen and therefore constitute an alternative set of invariants. Finally, relationships between the invariants $I_{\boldsymbol{S}}$, $II_{\boldsymbol{S}}$, and $III_{\boldsymbol{S}}$ and the eigenvalues can be derived by applying the definitions (2.47), (2.53), and (2.54) to Equation (2.59) to give

$$I_{\boldsymbol{S}} = \lambda_1 + \lambda_2 + \lambda_3; \tag{2.60a}$$

$$II_{\boldsymbol{S}} = \lambda_1^2 + \lambda_2^2 + \lambda_3^2; \tag{2.60b}$$

$$III_{\boldsymbol{S}} = \lambda_1\,\lambda_2\,\lambda_3. \tag{2.60c}$$

### 2.2.4 Higher-Order Tensors

It will be seen in Chapter 6 that the second-order stress and strain tensors are related via a fourth-order constitutive or material behavior tensor. In order to appreciate fourth-order tensors it is necessary to examine the intermediate third-order tensors. Inevitably, things are likely to get more difficult and, because an appreciation of this section is not immediately necessary, it is suggested that what follows be read just prior to Chapter 6.

Although there are several ways in which higher-order tensors can be defined, the procedure used here is a simple extension of the definition (2.15), employed for second-order tensors. In this way, a third-order tensor is defined as a linear map from an arbitrary vector $\boldsymbol{u}$ to a second-order tensor $\boldsymbol{S}$ as

$$\boldsymbol{\mathcal{A}}\boldsymbol{u} = \boldsymbol{S}. \tag{2.61}$$

In particular, the tensor product of three arbitrary vectors yields a third-order tensor $\boldsymbol{u}\otimes\boldsymbol{v}\otimes\boldsymbol{w}$ defined in a manner similar to Equation (2.28) so that any vector $\boldsymbol{x}$ is mapped to

$$(\boldsymbol{u}\otimes\boldsymbol{v}\otimes\boldsymbol{w})\boldsymbol{x} = (\boldsymbol{w}\cdot\boldsymbol{x})(\boldsymbol{u}\otimes\boldsymbol{v}). \tag{2.62}$$

This definition, combined with the fact that second-order tensors can be expressed as linear combinations of dyadic products, leads to the following expressions for the tensor product of a vector and a second-order tensor:

$$(\boldsymbol{S}\otimes\boldsymbol{v})\boldsymbol{u} = (\boldsymbol{u}\cdot\boldsymbol{v})\boldsymbol{S}; \tag{2.63a}$$

$$(\boldsymbol{v}\otimes\boldsymbol{S})\boldsymbol{u} = \boldsymbol{v}\otimes(\boldsymbol{S}\boldsymbol{u}). \tag{2.63b}$$

The 27 third-order tensors $\boldsymbol{e}_i\otimes\boldsymbol{e}_j\otimes\boldsymbol{e}_k$ for $i,j,k = 1,2,3$ which are obtained by taking the tensor products of the three Cartesian base vectors constitute a basis in which any third-order tensor $\boldsymbol{\mathcal{A}}$ can be expressed as

$$\boldsymbol{\mathcal{A}} = \sum_{i,j,k=1}^{3} \mathcal{A}_{ijk}\,\boldsymbol{e}_i\otimes\boldsymbol{e}_j\otimes\boldsymbol{e}_k, \tag{2.64}$$

where a manipulation similar to that employed in Equation (2.33) for second-order tensors will verify that the components of $\boldsymbol{\mathcal{A}}$ are

$$\mathcal{A}_{ijk} = (\boldsymbol{e}_i \otimes \boldsymbol{e}_j) : \boldsymbol{\mathcal{A}}\boldsymbol{e}_k. \tag{2.65}$$

An example of a third-order tensor of interest in continuum mechanics is the *alternating tensor* $\boldsymbol{\mathcal{E}}$, which is defined in such a way that any vector $\boldsymbol{w}$ is mapped to

$$\boldsymbol{\mathcal{E}}\boldsymbol{w} = -\boldsymbol{W}_{\boldsymbol{w}}\,, \tag{2.66}$$

where $\boldsymbol{W}_{\boldsymbol{w}}$ is the second-order tensor defined in Equation (2.22). The components of $\boldsymbol{\mathcal{E}}$ follow from Equations (2.22), (2.65), and (2.66) as

$$\begin{aligned}\mathcal{E}_{ijk} &= (\boldsymbol{e}_i \otimes \boldsymbol{e}_j) : (\boldsymbol{\mathcal{E}}\boldsymbol{e}_k)\\ &= -(\boldsymbol{e}_i \otimes \boldsymbol{e}_j) : \boldsymbol{W}_{\boldsymbol{e}_k}\\ &= -\boldsymbol{e}_i \cdot (\boldsymbol{W}_{\boldsymbol{e}_k}\boldsymbol{e}_j)\\ &= \boldsymbol{e}_i \cdot (\boldsymbol{e}_j \times \boldsymbol{e}_k).\end{aligned} \tag{2.67}$$

The result of the above triple product is 0 if any indices are repeated, 1 if the permutation $\{i, j, k\}$ is even, and $-1$ otherwise. Note that exactly the same components would be obtained if a different Cartesian base were used, that is, $\boldsymbol{\mathcal{E}}$ is an isotropic third-order tensor. Using these $\{0, 1, -1\}$ components, the tensor $\boldsymbol{\mathcal{E}}$ can be expressed as

$$\begin{aligned}\boldsymbol{\mathcal{E}} = {}& \boldsymbol{e}_1 \otimes \boldsymbol{e}_2 \otimes \boldsymbol{e}_3 + \boldsymbol{e}_3 \otimes \boldsymbol{e}_1 \otimes \boldsymbol{e}_2 + \boldsymbol{e}_2 \otimes \boldsymbol{e}_3 \otimes \boldsymbol{e}_1\\ &- \boldsymbol{e}_3 \otimes \boldsymbol{e}_2 \otimes \boldsymbol{e}_1 - \boldsymbol{e}_1 \otimes \boldsymbol{e}_3 \otimes \boldsymbol{e}_2 - \boldsymbol{e}_2 \otimes \boldsymbol{e}_1 \otimes \boldsymbol{e}_3.\end{aligned} \tag{2.68}$$

The double contraction of a third-order tensor and a second-order tensor is defined in such a way that for the dyadic product of any two vectors $\boldsymbol{u}$ and $\boldsymbol{v}$ a new vector is obtained as

$$\boldsymbol{\mathcal{A}} : (\boldsymbol{u} \otimes \boldsymbol{v}) = (\boldsymbol{\mathcal{A}}\boldsymbol{v})\boldsymbol{u}. \tag{2.69}$$

Note that the result of the double contraction is a vector and not a scalar. In fact, third-order tensors can be alternatively defined as linear maps from second-order tensors to vectors, as shown in the above double contraction. For example, applying the above definition to the alternating tensor gives

$$\begin{aligned}\boldsymbol{\mathcal{E}} : (\boldsymbol{u} \otimes \boldsymbol{v}) &= (\boldsymbol{\mathcal{E}}\boldsymbol{v})\boldsymbol{u}\\ &= -\boldsymbol{W}_{\boldsymbol{v}}\boldsymbol{u}\\ &= \boldsymbol{u} \times \boldsymbol{v}.\end{aligned} \tag{2.70}$$

This result, instead of Equation (2.66), could have been used to define the alternating tensor $\boldsymbol{\mathcal{E}}$.

In general, the double contraction of a third-order tensor by a second-order tensor $\boldsymbol{S}$ can be evaluated in terms of their respective components using the definition (2.69) together with Equations (2.32) and (2.65) to give

$$\boldsymbol{\mathcal{A}} : \boldsymbol{S} = \sum_{i,j,k=1}^{3} \mathcal{A}_{ijk} S_{jk}\, \boldsymbol{e}_i. \tag{2.71}$$

Additional properties of the double contraction are given below without proof:

$$(\boldsymbol{u} \otimes \boldsymbol{v} \otimes \boldsymbol{w}) : (\boldsymbol{x} \otimes \boldsymbol{y}) = (\boldsymbol{x} \cdot \boldsymbol{v})(\boldsymbol{y} \cdot \boldsymbol{w})\boldsymbol{u}; \tag{2.72a}$$
$$(\boldsymbol{u} \otimes \boldsymbol{S}) : \boldsymbol{T} = (\boldsymbol{S} : \boldsymbol{T})\boldsymbol{u}; \tag{2.72b}$$
$$(\boldsymbol{S} \otimes \boldsymbol{u}) : \boldsymbol{T} = \boldsymbol{S}\boldsymbol{T}\boldsymbol{u}. \tag{2.72c}$$

**EXAMPLE 2.7: Proof of Equation (2.71)**

In order to prove Equation (2.71), we first express both tensors $\boldsymbol{\mathcal{A}}$ and $\boldsymbol{S}$ in terms of their components as

$$\boldsymbol{\mathcal{A}} = \sum_{i,j,k=1}^{3} \mathcal{A}_{ijk}\, \boldsymbol{e}_i \otimes \boldsymbol{e}_j \otimes \boldsymbol{e}_k; \quad \boldsymbol{S} = \sum_{l,m=1}^{3} S_{lm}\, \boldsymbol{e}_l \otimes \boldsymbol{e}_m.$$

Taking the double contraction now gives

$$\begin{aligned}
\boldsymbol{\mathcal{A}} : \boldsymbol{S} &= \sum_{i,j,k,l,m=1}^{3} \mathcal{A}_{ijk} S_{lm} (\boldsymbol{e}_i \otimes \boldsymbol{e}_j \otimes \boldsymbol{e}_k) : (\boldsymbol{e}_l \otimes \boldsymbol{e}_m) \\
&= \sum_{i,j,k,l,m=1}^{3} \mathcal{A}_{ijk} S_{lm} [(\boldsymbol{e}_i \otimes \boldsymbol{e}_j \otimes \boldsymbol{e}_k)\boldsymbol{e}_m]\boldsymbol{e}_l \\
&= \sum_{i,j,k,l,m=1}^{3} \mathcal{A}_{ijk} S_{lm} (\boldsymbol{e}_k \cdot \boldsymbol{e}_m)(\boldsymbol{e}_j \cdot \boldsymbol{e}_l)\boldsymbol{e}_i \\
&= \sum_{i,j,k,l,m=1}^{3} \mathcal{A}_{ijk} S_{lm} \delta_{km} \delta_{jl}\, \boldsymbol{e}_i \\
&= \sum_{i,j,k=1}^{3} \mathcal{A}_{ijk} S_{jk}\, \boldsymbol{e}_i.
\end{aligned}$$

Tensors of any order can now be defined by recursive use of Equation (2.61). For instance, a fourth-order tensor $\boldsymbol{\mathcal{C}}$ is a linear map between an arbitrary vector $\boldsymbol{u}$ and a third-order tensor $\boldsymbol{\mathcal{A}}$ as

$$\boldsymbol{\mathcal{C}}\boldsymbol{u} = \boldsymbol{\mathcal{A}}. \tag{2.73}$$

Observe that no explicit notational distinction is made between third-, fourth-, or any higher-order tensors. Examples of fourth-order tensors are obtained by extending the definition of the tensor product of vectors in the obvious way to give

$$(\boldsymbol{u}_1 \otimes \boldsymbol{u}_2 \otimes \boldsymbol{u}_3 \otimes \boldsymbol{u}_4)\boldsymbol{v} = (\boldsymbol{v} \cdot \boldsymbol{u}_4)(\boldsymbol{u}_1 \otimes \boldsymbol{u}_2 \otimes \boldsymbol{u}_3). \tag{2.74}$$

Similar tensor products of a vector and a third-order tensor and second-order tensors are inferred from the above definition as

$$(\boldsymbol{\mathcal{A}} \otimes \boldsymbol{u})\boldsymbol{v} = (\boldsymbol{u} \cdot \boldsymbol{v})\boldsymbol{\mathcal{A}}; \tag{2.75a}$$

$$(\boldsymbol{u} \otimes \boldsymbol{\mathcal{A}})\boldsymbol{v} = \boldsymbol{u} \otimes (\boldsymbol{\mathcal{A}}\boldsymbol{v}); \tag{2.75b}$$

$$(\boldsymbol{T} \otimes \boldsymbol{S})\boldsymbol{v} = \boldsymbol{T} \otimes (\boldsymbol{S}\boldsymbol{v}). \tag{2.75c}$$

Additionally, the double contraction of a fourth- (or higher-) order tensor $\boldsymbol{\mathcal{C}}$ with a second-order tensor is again defined using Equation (2.69) to yield a second-order tensor as

$$\boldsymbol{\mathcal{C}} : (\boldsymbol{u} \otimes \boldsymbol{v}) = (\boldsymbol{\mathcal{C}}\boldsymbol{v})\boldsymbol{u}. \tag{2.76}$$

An illustration of this would be the crucially important constitutive relationship between a second-order stress tensor $\boldsymbol{\sigma}$ and a second-order strain tensor $\boldsymbol{\varepsilon}$, given as $\boldsymbol{\sigma} = \boldsymbol{\mathcal{C}} : \boldsymbol{\varepsilon}$, where $\boldsymbol{\mathcal{C}}$ would be a fourth-order elasticity tensor.

Properties similar to those listed in Equations (2.72a–c) are obtained as

$$(\boldsymbol{u}_1 \otimes \boldsymbol{u}_2 \otimes \boldsymbol{u}_3 \otimes \boldsymbol{u}_4) : (\boldsymbol{x} \otimes \boldsymbol{y}) = (\boldsymbol{x} \cdot \boldsymbol{u}_3)(\boldsymbol{y} \cdot \boldsymbol{u}_4)(\boldsymbol{u}_1 \otimes \boldsymbol{u}_2); \tag{2.77a}$$

$$(\boldsymbol{S}_1 \otimes \boldsymbol{S}_2) : \boldsymbol{T} = (\boldsymbol{S}_2 : \boldsymbol{T})\boldsymbol{S}_1; \tag{2.77b}$$

$$(\boldsymbol{\mathcal{A}} \otimes \boldsymbol{u}) : \boldsymbol{T} = \boldsymbol{\mathcal{A}}(\boldsymbol{T}\boldsymbol{u}); \tag{2.77c}$$

$$(\boldsymbol{u} \otimes \boldsymbol{\mathcal{A}}) : \boldsymbol{T} = \boldsymbol{u} \otimes (\boldsymbol{\mathcal{A}} : \boldsymbol{T}). \tag{2.77d}$$

Recalling that the double contraction of a fourth-order tensor with a second-order tensor gives a second-order tensor, fourth-order tensors can be also defined as linear mappings between second-order tensors. For instance, the fourth-order identity tensor $\boldsymbol{\mathcal{I}}$ and the transposition tensor $\tilde{\boldsymbol{\mathcal{I}}}$ are defined in such a way that any second-order tensor $\boldsymbol{S}$ is mapped onto itself and its transpose respectively as

$$\boldsymbol{\mathcal{I}} : \boldsymbol{S} = \boldsymbol{S}; \tag{2.78a}$$

$$\tilde{\boldsymbol{\mathcal{I}}} : \boldsymbol{S} = \boldsymbol{S}^T. \tag{2.78b}$$

Fourth-order tensors can be expressed as linear combinations of the 81 tensor products of the Cartesian base vectors $\boldsymbol{e}_i \otimes \boldsymbol{e}_j \otimes \boldsymbol{e}_k \otimes \boldsymbol{e}_l$ for $i, j, k, l = 1, 2, 3$ as

$$\boldsymbol{\mathcal{C}} = \sum_{i,j,k,l=1}^{3} \mathcal{C}_{ijkl}\, \boldsymbol{e}_i \otimes \boldsymbol{e}_j \otimes \boldsymbol{e}_k \otimes \boldsymbol{e}_l, \tag{2.79a}$$

where it can be proved that the components $\mathcal{C}_{ijkl}$ are given as

$$\mathcal{C}_{ijkl} = (\boldsymbol{e}_i \otimes \boldsymbol{e}_j) : \boldsymbol{\mathcal{C}} : (\boldsymbol{e}_k \otimes \boldsymbol{e}_l). \tag{2.79b}$$

For instance, the components of $\boldsymbol{\mathcal{I}}$ are obtained with the help of Equations (2.79a,b) and (2.52c) as

$$\begin{aligned}\mathcal{I}_{ijkl} &= (\boldsymbol{e}_i \otimes \boldsymbol{e}_j) : \boldsymbol{\mathcal{I}} : (\boldsymbol{e}_k \otimes \boldsymbol{e}_l)\\ &= (\boldsymbol{e}_i \otimes \boldsymbol{e}_j) : (\boldsymbol{e}_k \otimes \boldsymbol{e}_l)\\ &= (\boldsymbol{e}_i \cdot \boldsymbol{e}_k)(\boldsymbol{e}_j \cdot \boldsymbol{e}_l)\\ &= \delta_{ik}\delta_{jl};\end{aligned} \tag{2.80}$$

and, similarly, the components of $\tilde{\boldsymbol{\mathcal{I}}}$ are

$$\begin{aligned}\tilde{\mathcal{I}}_{ijkl} &= (\boldsymbol{e}_i \otimes \boldsymbol{e}_j) : \tilde{\boldsymbol{\mathcal{I}}} : (\boldsymbol{e}_k \otimes \boldsymbol{e}_l)\\ &= (\boldsymbol{e}_i \otimes \boldsymbol{e}_j) : (\boldsymbol{e}_l \otimes \boldsymbol{e}_k)\\ &= (\boldsymbol{e}_i \cdot \boldsymbol{e}_l)(\boldsymbol{e}_j \cdot \boldsymbol{e}_k)\\ &= \delta_{il}\delta_{jk}.\end{aligned} \tag{2.81}$$

Consequently, these two tensors can be expressed as

$$\boldsymbol{\mathcal{I}} = \sum_{i,j=1}^{3} \boldsymbol{e}_i \otimes \boldsymbol{e}_j \otimes \boldsymbol{e}_i \otimes \boldsymbol{e}_j; \tag{2.82a}$$

$$\tilde{\boldsymbol{\mathcal{I}}} = \sum_{i,j=1}^{3} \boldsymbol{e}_i \otimes \boldsymbol{e}_j \otimes \boldsymbol{e}_j \otimes \boldsymbol{e}_i. \tag{2.82b}$$

Observe that both these tensors are isotropic, as the same components would emerge from Equations (2.80) and (2.81) regardless of the Cartesian base being used. Isotropic fourth-order tensors are of particular interest in continuum mechanics because they will be used to describe the elasticity tensor of materials that exhibit equal properties in all directions. In fact, it is possible to prove that all fourth-order isotropic tensors are combinations of $\boldsymbol{\mathcal{I}}$, $\tilde{\boldsymbol{\mathcal{I}}}$ and the additional isotropic tensor $\boldsymbol{I} \otimes \boldsymbol{I}$ as

$$\boldsymbol{\mathcal{C}} = \alpha \boldsymbol{I} \otimes \boldsymbol{I} + \beta \boldsymbol{\mathcal{I}} + \gamma \tilde{\boldsymbol{\mathcal{I}}}. \tag{2.83}$$

Examples of such combinations are the tensors $\boldsymbol{\mathcal{S}}$, $\boldsymbol{\mathcal{W}}$, and $\boldsymbol{\mathcal{D}}$ defined as

$$\boldsymbol{\mathcal{S}} = \frac{1}{2}(\boldsymbol{\mathcal{I}} + \tilde{\boldsymbol{\mathcal{I}}}); \tag{2.84a}$$

$$\boldsymbol{\mathcal{W}} = \frac{1}{2}(\boldsymbol{\mathcal{I}} - \tilde{\boldsymbol{\mathcal{I}}}); \tag{2.84b}$$

$$\boldsymbol{\mathcal{D}} = \boldsymbol{\mathcal{I}} - \frac{1}{3}\boldsymbol{I} \otimes \boldsymbol{I}; \tag{2.84c}$$

which project a second-order tensor $\boldsymbol{S}$ onto its symmetric, skew, and deviatoric components as

$$\mathcal{S} : \boldsymbol{S} = \frac{1}{2}(\boldsymbol{S} + \boldsymbol{S}^T); \tag{2.85a}$$

$$\mathcal{W} : \boldsymbol{S} = \frac{1}{2}(\boldsymbol{S} - \boldsymbol{S}^T); \tag{2.85b}$$

$$\mathcal{D} : \boldsymbol{S} = \boldsymbol{S}' = \boldsymbol{S} - \frac{1}{3}(\mathrm{tr}\boldsymbol{S})\boldsymbol{I}. \tag{2.85c}$$

Finally, the definition of symmetric second-order tensors given in the previous section can be extended to fourth-order tensors. In this way a fourth-order tensor $\mathcal{C}$ is said to be symmetric if, for any arbitrary second-order tensors $\boldsymbol{S}$ and $\boldsymbol{T}$, the following expression is satisfied:

$$\boldsymbol{S} : \mathcal{C} : \boldsymbol{T} = \boldsymbol{T} : \mathcal{C} : \boldsymbol{S}. \tag{2.86}$$

For example, it is easy to show that $\boldsymbol{I} \otimes \boldsymbol{I}$, $\mathcal{I}$, and $\tilde{\mathcal{I}}$ are symmetric fourth-order tensors and consequently combinations of these three tensors such as $\mathcal{S}$, $\mathcal{W}$, and $\mathcal{D}$ are also symmetric.

**EXAMPLE 2.8: Elasticity tensor**

In linear isotropic elasticity the stress tensor $\boldsymbol{\sigma}$ is related to the small strain tensor $\boldsymbol{\varepsilon}$ by the Lamé material coefficients $\lambda$ and $\mu$ as

$$\boldsymbol{\sigma} = \lambda(\mathrm{tr}\boldsymbol{\varepsilon})\boldsymbol{I} + 2\mu\boldsymbol{\varepsilon}.$$

This equation can be rewritten in terms of the fourth-order elasticity tensor $\mathcal{C}$ as

$$\boldsymbol{\sigma} = \mathcal{C} : \boldsymbol{\varepsilon}; \quad \mathcal{C} = \lambda\boldsymbol{I} \otimes \boldsymbol{I} + \mu(\mathcal{I} + \tilde{\mathcal{I}}); \; \mathcal{C}_{ijkl} = \lambda\delta_{ij}\delta_{kl} + \mu(\delta_{ik}\delta_{jl} + \delta_{il}\delta_{jk}).$$

Alternatively, the above relationship can be inverted to give the strain expressed in terms of the stress tensor. To achieve this, note first that taking the trace of the above stress–strain equation gives

$$\mathrm{tr}\boldsymbol{\sigma} = (3\lambda + 2\mu)\mathrm{tr}\boldsymbol{\varepsilon},$$

and consequently $\boldsymbol{\varepsilon}$ can be written as

$$\boldsymbol{\varepsilon} = \frac{1}{2\mu}\boldsymbol{\sigma} - \frac{\lambda\mathrm{tr}\boldsymbol{\sigma}}{2\mu(3\lambda + 2\mu)}\boldsymbol{I},$$

or in terms of the Young's modulus $E$ and Poisson's ratio $\nu$ as

$$\boldsymbol{\varepsilon} = \frac{1}{E}[(1+\nu)\boldsymbol{\sigma} - \nu(\mathrm{tr}\boldsymbol{\sigma})\boldsymbol{I}]; \quad E = \frac{\mu(3\lambda + 2\mu)}{\lambda + \mu}; \quad \nu = \frac{\lambda}{2\lambda + 2\mu}.$$

Hence the inverse elasticity tensor can be defined as

$$\boldsymbol{\varepsilon} = \mathcal{C}^{-1} : \boldsymbol{\sigma}; \quad \mathcal{C}^{-1} = -\frac{\nu}{E}\boldsymbol{I} \otimes \boldsymbol{I} + \frac{1+\nu}{E}\mathcal{I}.$$

## 2.3 LINEARIZATION AND THE DIRECTIONAL DERIVATIVE

Nonlinear problems in continuum mechanics are invariably solved by linearizing the nonlinear equations and iteratively solving the resulting linear equations until a solution to the nonlinear problem is found. The Newton–Raphson method is the most popular example of such a technique. Correct linearization of the nonlinear equations is fundamental for the success of such techniques. In this section we will consolidate the concept of the directional derivative introduced in Chapter 1. The familiar Newton–Raphson scheme will be used as the initial vehicle for exploring the ideas that will eventually be generalized.

### 2.3.1 One Degree of Freedom

Consider the one-degree-of-freedom nonlinear equation shown in Figure 2.3:

$$f(x) = 0. \tag{2.87}$$

Given an initial guess of the solution, $x_0$, the function $f(x)$ can be expressed in the neighborhood of $x_0$ using a Taylor's series as

$$f(x) = f(x_0) + \left.\frac{df}{dx}\right|_{x_0} (x - x_0) + \frac{1}{2}\left.\frac{d^2 f}{dx^2}\right|_{x_0} (x - x_0)^2 + \cdots . \tag{2.88}$$

If the increment in $x$ is expressed as $u = (x - x_0)$ then (2.88) can be rewritten as

$$f(x_0 + u) = f(x_0) + \left.\frac{df}{dx}\right|_{x_0} u + \frac{1}{2}\left.\frac{d^2 f}{dx^2}\right|_{x_0} u^2 + \cdots . \tag{2.89}$$

To establish the Newton–Raphson procedure for this single-degree-of-freedom case, (2.89) is linearized by truncating the Taylor's expression to give

$$f(x_0 + u) \approx f(x_0) + \left.\frac{df}{dx}\right|_{x_0} u. \tag{2.90}$$

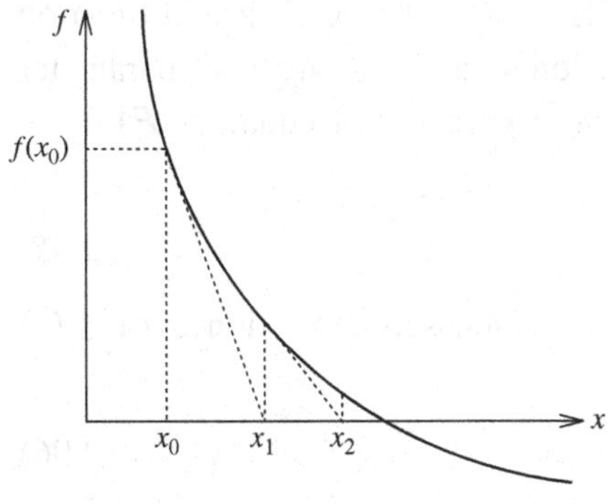


**FIGURE 2.3** One-degree-of-freedom nonlinear problem, $f(x) = 0$.

This is clearly a linear function in $u$, and the term $u(df/dx)|_{x_0}$ is called the linearized increment in $f(x)$ at $x_0$ with respect to $u$. This is generally expressed as

$$Df(x_0)[u] = \left.\frac{df}{dx}\right|_{x_0} u \approx f(x_0 + u) - f(x_0). \tag{2.91}$$

The symbol $Df(x_0)[u]$ denotes a derivative, formed at $x_0$, that operates in some linear manner (not necessarily multiplicative as here) on $u$.

Using Equation (2.90) the Newton–Raphson iterative procedure is set up by requiring the function $f(x_k + u)$ to vanish, thus giving a linear equation in $u$ as

$$f(x_k) + Df(x_k)[u] = 0, \tag{2.92}$$

from which the new iterative value $x_{k+1}$, illustrated in Figure 2.3, is obtained as

$$u = \left[ -\left.\frac{df}{dx}\right|_{x_k} \right]^{-1} f(x_k); \quad x_{k+1} = x_k + u. \tag{2.93}$$

This simple one-degree-of-freedom case will now be generalized in order to further develop the concept of the directional derivative.

### 2.3.2 General Solution to a Nonlinear Problem

Consider a set of general nonlinear equations given as

$$\mathcal{F}(\mathbf{x}) = \mathbf{0}, \tag{2.94}$$

where the function $\mathcal{F}(\mathbf{x})$ can represent anything from a system of nonlinear algebraic equations to more complex cases such as nonlinear differential equations where the unknowns $\mathbf{x}$ could be sets of *functions*. Consequently, $\mathbf{x}$ represents a list of unknown variables or functions.

Consider an initial guess $\mathbf{x}_0$ and a general change or increment $\mathbf{u}$ that, it is hoped, will generate $\mathbf{x} = \mathbf{x}_0 + \mathbf{u}$ closer to the solution of Equation (2.94). In order to replicate the example given in Section 2.3.1 and because, in general, it is not immediately obvious how to express the derivative of a complicated function $\mathcal{F}$ with respect to what could also be a function $\mathbf{x}$, a single artificial parameter $\epsilon$ is introduced that enables a nonlinear function $\mathbf{F}$ in $\epsilon$ (not equal to $\mathcal{F}$) to be established as

$$\mathbf{F}(\epsilon) = \mathcal{F}(\mathbf{x}_0 + \epsilon\mathbf{u}). \tag{2.95}$$

For example, in the one-degree-of-freedom case discussed in Section 2.3.1, $\mathbf{F}(\epsilon)$ becomes

$$F(\epsilon) = f(x_0 + \epsilon u). \tag{2.96}$$

This is illustrated for the one-degree-of-freedom case in Figure 2.4.

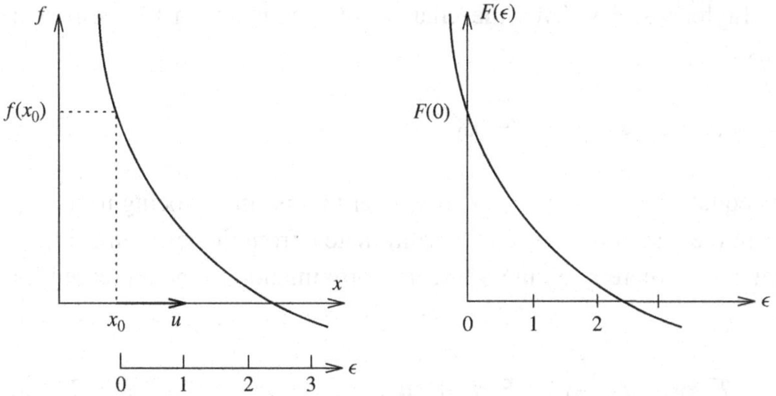


**FIGURE 2.4** Single-degree-of-freedom nonlinear problem, $f(x) = 0$ and $F(\epsilon) = 0$.

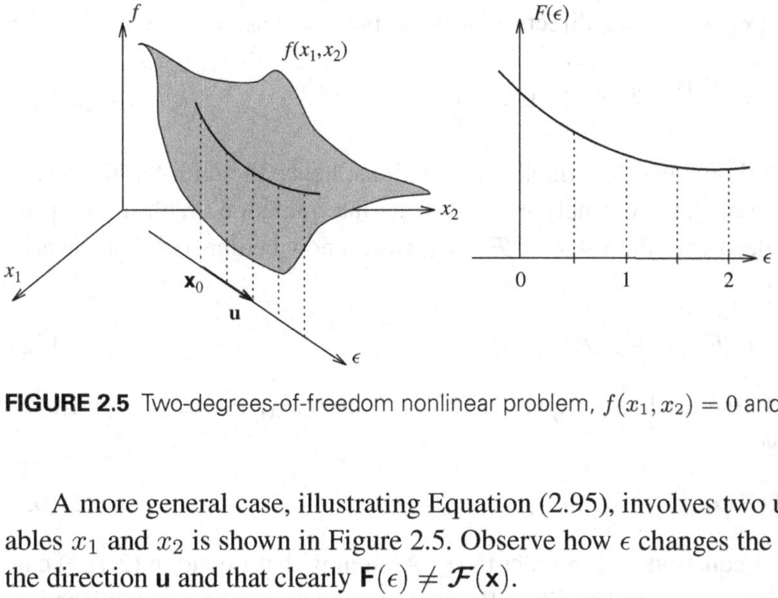


**FIGURE 2.5** Two-degrees-of-freedom nonlinear problem, $f(x_1, x_2) = 0$ and $F(\epsilon) = 0$.

A more general case, illustrating Equation (2.95), involves two unknown variables $x_1$ and $x_2$ is shown in Figure 2.5. Observe how $\epsilon$ changes the function $\boldsymbol{\mathcal{F}}$ in the direction $\mathbf{u}$ and that clearly $\mathbf{F}(\epsilon) \neq \boldsymbol{\mathcal{F}}(\mathbf{x})$.

In order to develop the Newton–Raphson method together with the associated linearized equations, a Taylor's series expansion of the nonlinear function $\mathbf{F}(\epsilon)$ about $\epsilon = 0$, corresponding to $\mathbf{x} = \mathbf{x}_0$, gives

$$\mathbf{F}(\epsilon) = \mathbf{F}(0) + \left.\frac{d\mathbf{F}}{d\epsilon}\right|_{\epsilon=0} \epsilon + \frac{1}{2}\left.\frac{d^2\mathbf{F}}{d\epsilon^2}\right|_{\epsilon=0} \epsilon^2 + \cdots . \tag{2.97}$$

Introducing the definition of $\mathbf{F}$ given in Equation (2.95) into the above Taylor's series yields

$$\boldsymbol{\mathcal{F}}(\mathbf{x}_0 + \epsilon\mathbf{u}) = \boldsymbol{\mathcal{F}}(\mathbf{x}_0) + \epsilon\left.\frac{d}{d\epsilon}\right|_{\epsilon=0} \boldsymbol{\mathcal{F}}(\mathbf{x}_0 + \epsilon\mathbf{u}) + \frac{\epsilon^2}{2}\left.\frac{d^2}{d\epsilon^2}\right|_{\epsilon=0} \boldsymbol{\mathcal{F}}(\mathbf{x}_0 + \epsilon\mathbf{u}) + \cdots . \tag{2.98}$$

Truncating this Taylor's series gives the change, or increment, in the nonlinear function $\mathcal{F}(\mathbf{x})$ as

$$\mathcal{F}(\mathbf{x}_0 + \epsilon\mathbf{u}) - \mathcal{F}(\mathbf{x}_0) \approx \epsilon \frac{d}{d\epsilon}\bigg|_{\epsilon=0} \mathcal{F}(\mathbf{x}_0 + \epsilon\mathbf{u}). \tag{2.99}$$

Note that in this equation $\epsilon$ is an artificial parameter that is simply being used as a vehicle to perform the derivative. In order to eliminate $\epsilon$ from the left-hand side of this equation, let $\epsilon = 1$, thereby giving a linear approximation to the increment of $\mathcal{F}(\mathbf{x})$ as

$$\mathcal{F}(\mathbf{x}_0 + \mathbf{u}) - \mathcal{F}(\mathbf{x}_0) \approx 1 \frac{d}{d\epsilon}\bigg|_{\epsilon=0} \mathcal{F}(\mathbf{x}_0 + \epsilon\mathbf{u}), \tag{2.100}$$

where the term on the right-hand side of the above equation is the directional derivative of $\mathcal{F}(\mathbf{x})$ at $\mathbf{x}_0$ in the direction of $\mathbf{u}$ and is written as

$$D\mathcal{F}(\mathbf{x}_0)[\mathbf{u}] = \frac{d}{d\epsilon}\bigg|_{\epsilon=0} \mathcal{F}(\mathbf{x}_0 + \epsilon\mathbf{u}). \tag{2.101}$$

Note that $\mathbf{u}$ could be a list of variables or functions; hence the term "in the direction" is, at the moment, extremely general in its interpretation. With the help of the directional derivative the value of $\mathcal{F}(\mathbf{x}_0 + \mathbf{u})$ can now be linearized or linearly approximated as

$$\mathcal{F}(\mathbf{x}_0 + \mathbf{u}) \approx \mathcal{F}(\mathbf{x}_0) + D\mathcal{F}(\mathbf{x}_0)[\mathbf{u}]. \tag{2.102}$$

Returning to the nonlinear Equation (2.94), setting $\mathcal{F}(\mathbf{x}_0 + \mathbf{u}) = \mathbf{0}$ in Equation (2.102) gives

$$\mathcal{F}(\mathbf{x}_0) + D\mathcal{F}(\mathbf{x}_0)[\mathbf{u}] = \mathbf{0}, \tag{2.103}$$

which is a linear equation with respect to $\mathbf{u}$.† Assuming that Equation (2.103) can be solved for $\mathbf{u}$, then a general Newton–Raphson procedure can be re-established as

$$D\mathcal{F}(\mathbf{x}_k)[\mathbf{u}] = -\mathcal{F}(\mathbf{x}_k)\ ; \quad \mathbf{x}_{k+1} = \mathbf{x}_k + \mathbf{u}. \tag{2.104}$$

### 2.3.3 Properties of the Directional Derivative

The directional derivative defined above satisfies the usual properties of the derivative. These are listed for completeness below:

† The term $D\mathcal{F}(\mathbf{x}_0)[\mathbf{u}]$ is linear with respect to $\mathbf{u}$ in the sense that, for any $\mathbf{u}_1$ and $\mathbf{u}_2$,

$$D\mathcal{F}(\mathbf{x}_0)[\mathbf{u}_1 + \mathbf{u}_2] = D\mathcal{F}(\mathbf{x}_0)[\mathbf{u}_1] + D\mathcal{F}(\mathbf{x}_0)[\mathbf{u}_2].$$

(a) If $\mathcal{F}(\mathbf{x}) = \mathcal{F}_1(\mathbf{x}) + \mathcal{F}_2(\mathbf{x})$ then

$$D\mathcal{F}(\mathbf{x}_0)[\mathbf{u}] = D\mathcal{F}_1(\mathbf{x}_0)[\mathbf{u}] + D\mathcal{F}_2(\mathbf{x}_0)[\mathbf{u}]. \tag{2.105a}$$

(b) The product rule: if $\mathcal{F}(\mathbf{x}) = \mathcal{F}_1(\mathbf{x}) \cdot \mathcal{F}_2(\mathbf{x})$, where "$\cdot$" means any type of product, then

$$D\mathcal{F}(\mathbf{x}_0)[\mathbf{u}] = D\mathcal{F}_1(\mathbf{x}_0)[\mathbf{u}] \cdot \mathcal{F}_2(\mathbf{x}_0) + \mathcal{F}_1(\mathbf{x}_0) \cdot D\mathcal{F}_2(\mathbf{x}_0)[\mathbf{u}]. \tag{2.105b}$$

(c) The chain rule: if $\mathcal{F}(\mathbf{x}) = \mathcal{F}_1(\mathcal{F}_2(\mathbf{x}))$ then

$$D\mathcal{F}(\mathbf{x}_0)[\mathbf{u}] = D\mathcal{F}_1(\mathcal{F}_2(\mathbf{x}_0))[D\mathcal{F}_2(\mathbf{x}_0)[\mathbf{u}]]. \tag{2.105c}$$

**EXAMPLE 2.9: Interpretation of Equation (2.105c)**

The chain rule Equation (2.105c) is not easy to interpret. In an attempt to clarify the meaning of this equation consider the approximation statement

$$\mathcal{F}(\mathbf{x}_0 + \mathbf{u}) \approx \mathcal{F}(\mathbf{x}_0) + D\mathcal{F}(\mathbf{x}_0)[\mathbf{u}].$$

In the case where $\mathcal{F}(\mathbf{x}) = \mathcal{F}_1(\mathcal{F}_2(\mathbf{x}))$ the left-hand side becomes

$$\mathcal{F}(\mathbf{x}_0 + \mathbf{u}) = \mathcal{F}_1(\mathcal{F}_2(\mathbf{x}_0 + \mathbf{u})),$$

and using the linearization of $\mathcal{F}_2$ this can be written as

$$\mathcal{F}(\mathbf{x}_0 + \mathbf{u}) \approx \mathcal{F}_1(\mathcal{F}_2(\mathbf{x}_0) + D\mathcal{F}_2(\mathbf{x}_0)[\mathbf{u}]).$$

Now linearizing $\mathcal{F}_1$ at $\mathcal{F}_2(\mathbf{x}_0)$ in the direction of the increment $D\mathcal{F}_2(\mathbf{x}_0)[\mathbf{u}]$ gives

$$\mathcal{F}(\mathbf{x}_0 + \mathbf{u}) \approx \mathcal{F}_1(\mathcal{F}_2(\mathbf{x}_0)) + D\mathcal{F}_1(\mathcal{F}_2(\mathbf{x}_0))[D\mathcal{F}_2(\mathbf{x}_0)[\mathbf{u}]].$$

Comparing the first and last equations gives Equation (2.105c).

### 2.3.4 Examples of Linearization

**Algebraic systems of equations.** Consider a set of nonlinear algebraic equations $\mathbf{f}(\mathbf{x}) = [f_1, f_2, \ldots, f_n]^T$ with unknowns $\mathbf{x} = [x_1, x_2, \ldots, x_n]^T$ as

$$\begin{aligned} f_1(x_1, x_2, \ldots, x_n) &= 0; \\ f_2(x_1, x_2, \ldots, x_n) &= 0; \\ &\vdots \\ f_n(x_1, x_2, \ldots, x_n) &= 0. \end{aligned} \tag{2.106}$$

The directional derivative of $\mathbf{f}(\mathbf{x})$ at a solution estimate $\mathbf{x}_0$ in the general direction $\mathbf{u} = [u_1, u_2, \ldots, u_n]^T$ is given by (2.101) as

$$D\mathbf{f}(\mathbf{x}_0)[\mathbf{u}] = \frac{d}{d\epsilon}\bigg|_{\epsilon=0} \mathbf{f}(\mathbf{x}_0 + \epsilon\mathbf{u}). \tag{2.107}$$

The above expression can be evaluated using the standard chain rule for the partial derivatives of a function of several variables as

$$\begin{aligned}
D\mathbf{f}(\mathbf{x}_0)[\mathbf{u}] &= \frac{d}{d\epsilon}\bigg|_{\epsilon=0} \mathbf{f}(\mathbf{x}_0 + \epsilon\mathbf{u}) \\
&= \sum_{i=1}^{n} \frac{\partial \mathbf{f}}{\partial x_i}\bigg|_{x_i=x_{0,i}} \frac{d(x_{0,i}+\epsilon u_i)}{d\epsilon}\bigg|_{\epsilon=0} \\
&= \sum_{i=1}^{n} u_i \frac{\partial \mathbf{f}}{\partial x_i}\bigg|_{x_i=x_{0,i}} \\
&= \mathbf{K}(\mathbf{x}_0)\mathbf{u},
\end{aligned} \tag{2.108}$$

where the tangent matrix $\mathbf{K}$ is

$$\mathbf{K} = \begin{bmatrix} \frac{\partial f_1}{\partial x_1} & \frac{\partial f_1}{\partial x_2} & \cdots & \frac{\partial f_1}{\partial x_n} \\ \frac{\partial f_2}{\partial x_1} & \frac{\partial f_2}{\partial x_2} & \cdots & \frac{\partial f_2}{\partial x_n} \\ \vdots & \vdots & \ddots & \vdots \\ \frac{\partial f_n}{\partial x_1} & \frac{\partial f_n}{\partial x_2} & \cdots & \frac{\partial f_n}{\partial x_n} \end{bmatrix}. \tag{2.109}$$

Consequently, the Newton–Raphson iterative scheme becomes

$$\mathbf{K}(\mathbf{x}_k)\,\mathbf{u} = -\mathbf{f}(\mathbf{x}_k); \quad \mathbf{x}_{k+1} = \mathbf{x}_k + \mathbf{u}. \tag{2.110}$$

**EXAMPLE 2.10: Linearization of $A^2 = AA$**

Expanding $A^2 = AA$ gives

$$\begin{aligned}
DA^2[U] &= \frac{d}{d\epsilon}\bigg|_{\epsilon=0} (\boldsymbol{A}+\epsilon\boldsymbol{U})(\boldsymbol{A}+\epsilon\boldsymbol{U}) \\
&= \frac{d}{d\epsilon}\bigg|_{\epsilon=0} (\boldsymbol{AA}+\epsilon\boldsymbol{UA}+\epsilon\boldsymbol{AU}+\epsilon^2\boldsymbol{UU}) \\
&= (\boldsymbol{UA}+\boldsymbol{AU}+2\epsilon\boldsymbol{UU})\Big|_{\epsilon=0} \\
&= \boldsymbol{UA}+\boldsymbol{AU}.
\end{aligned}$$

Alternatively the product rule given in Equation (2.105b) can be employed to give

$$\begin{aligned}
DA^2[U] &= DA(\boldsymbol{U})\boldsymbol{A} + \boldsymbol{A}DA(\boldsymbol{U}) \\
&= \frac{d}{d\epsilon}\bigg|_{\epsilon=0} (\boldsymbol{A}+\epsilon\boldsymbol{U})\boldsymbol{A} + \boldsymbol{A}\frac{d}{d\epsilon}\bigg|_{\epsilon=0} (\boldsymbol{A}+\epsilon\boldsymbol{U}) \\
&= \boldsymbol{UA}+\boldsymbol{AU}.
\end{aligned}$$

**Function minimization.** The directional derivative given in (2.101) need not necessarily be associated with the Newton–Raphson method and can equally be applied to other purposes. An interesting application is the minimization of a functional, which is a familiar problem that often arises in continuum or structural mechanics. For example, consider the total potential energy for a simply supported beam under the action of a uniformly distributed load $q(x)$ given as (Figure 2.6)

$$\mathcal{V}(w(x)) = \frac{1}{2}\int_0^l EI\left(\frac{d^2w(x)}{dx^2}\right)^2 dx - \int_0^l q(x)w(x)\,dx, \tag{2.111}$$

where $w(x)$ is the lateral deflection (which satisfies the boundary conditions a priori), $E$ is Young's modulus, $I$ is the second moment of area, and $l$ is the length of the beam. A functional such as $\mathcal{V}$ is said to be stationary at point $w_0(x)$ when the directional derivative of $\mathcal{V}$ vanishes for any arbitrary increment $u(x)$ in $w_0(x)$. Consequently, the equilibrium position $w_0(x)$ satisfies

$$D\mathcal{V}(w_0(x))[u(x)] = \frac{d}{d\epsilon}\bigg|_{\epsilon=0} \mathcal{V}(w_0(x) + \epsilon u(x)) = 0 \tag{2.112}$$

for any function $u(x)$ compatible with the boundary conditions. Note that $w_0(x)$ is the unknown function in the problem and is not to be confused with a Newton–Raphson iterative estimate of the solution. Substituting for $\mathcal{V}$ in (2.112) from (2.111) gives

$$\begin{aligned} D\mathcal{V}(w_0(x))[u(x)] &= \frac{d}{d\epsilon}\bigg|_{\epsilon=0} \frac{1}{2}\int_0^l EI\left[\frac{d^2(w_0(x)+\epsilon u(x))}{dx^2}\right]^2 dx \\ &\quad - \frac{d}{d\epsilon}\bigg|_{\epsilon=0}\int_0^l q(x)(w_0(x)+\epsilon u(x))dx = 0. \end{aligned} \tag{2.113}$$

Hence

$$D\mathcal{V}(w_0(x))[u(x)] = \int_0^l EI\frac{d^2w_0(x)}{dx^2}\frac{d^2u(x)}{dx^2}\,dx - \int_0^l q(x)u(x)\,dx = 0. \tag{2.114}$$

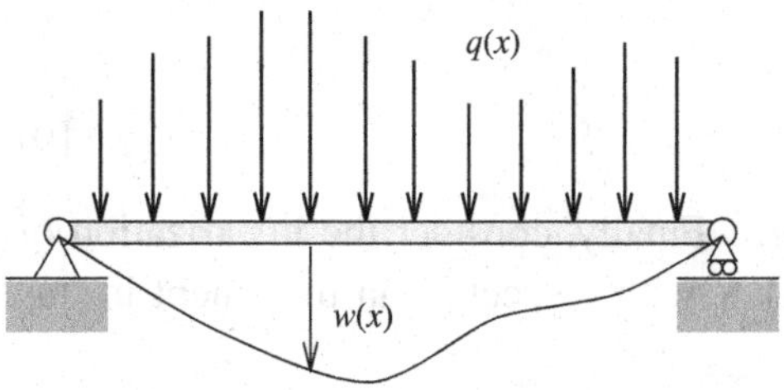


**FIGURE 2.6** Simply supported beam.

If $u(x)$ is considered to be the virtual displacement $\delta u(x)$ then the above equation is easily recognized as the virtual work equation for the beam, which is an alternative expression of equilibrium.

**Linearization of the determinant of a tensor.** This example further illustrates the generality of the concept of the linearization obtained using the directional derivative. Consider the linearization of the determinant det $\boldsymbol{S}$ of the second-order tensor $\boldsymbol{S}$ (or square matrix) with respect to an increment in this tensor $\boldsymbol{U}$ as

$$\det(\boldsymbol{S}+\boldsymbol{U}) \approx \det \boldsymbol{S} + D\det(\boldsymbol{S})[\boldsymbol{U}], \tag{2.115}$$

where the directional derivative of the determinant can be found by direct application of Equation (2.111) as

$$\begin{aligned} D\det(\boldsymbol{S})[\boldsymbol{U}] &= \frac{d}{d\epsilon}\bigg|_{\epsilon=0} \det(\boldsymbol{S}+\epsilon\boldsymbol{U}) \\ &= \frac{d}{d\epsilon}\bigg|_{\epsilon=0} \det[\boldsymbol{S}(\boldsymbol{I}+\epsilon\boldsymbol{S}^{-1}\boldsymbol{U})] \\ &= \det \boldsymbol{S}\, \frac{d}{d\epsilon}\bigg|_{\epsilon=0} \det(\boldsymbol{I}+\epsilon\boldsymbol{S}^{-1}\boldsymbol{U}). \end{aligned} \tag{2.116}$$

In order to proceed, note that the characteristic equation of a matrix $\boldsymbol{B}$ with eigenvalues $\lambda_1^{\boldsymbol{B}}$, $\lambda_2^{\boldsymbol{B}}$, and $\lambda_3^{\boldsymbol{B}}$ is

$$\det(\boldsymbol{B}-\lambda\boldsymbol{I}) = \left(\lambda_1^{\boldsymbol{B}}-\lambda\right)\left(\lambda_2^{\boldsymbol{B}}-\lambda\right)\left(\lambda_3^{\boldsymbol{B}}-\lambda\right). \tag{2.117}$$

Using this equation with $\lambda = -1$ and $\boldsymbol{B} = \epsilon\boldsymbol{S}^{-1}\boldsymbol{U}$ gives

$$D\det(\boldsymbol{S})[\boldsymbol{U}] = \det \boldsymbol{S}\,\frac{d}{d\epsilon}\bigg|_{\epsilon=0} \left(1+\epsilon\lambda_1^{\boldsymbol{S}^{-1}\boldsymbol{U}}\right)\left(1+\epsilon\lambda_2^{\boldsymbol{S}^{-1}\boldsymbol{U}}\right)\left(1+\epsilon\lambda_3^{\boldsymbol{S}^{-1}\boldsymbol{U}}\right), \tag{2.118}$$

where $\lambda_1^{\boldsymbol{S}^{-1}\boldsymbol{U}}$, $\lambda_2^{\boldsymbol{S}^{-1}\boldsymbol{U}}$, and $\lambda_3^{\boldsymbol{S}^{-1}\boldsymbol{U}}$ are the eigenvalues of $\boldsymbol{S}^{-1}\boldsymbol{U}$. Using the standard product rule of differentiation in (2.118) and recalling the definition and properties of the trace of a tensor introduced in Section 2.2.3 gives the directional derivative of the determinant of a tensor as

$$\begin{aligned} D\det(\boldsymbol{S})[\boldsymbol{U}] &= \det \boldsymbol{S}\left(\lambda_1^{\boldsymbol{S}^{-1}\boldsymbol{U}}+\lambda_2^{\boldsymbol{S}^{-1}\boldsymbol{U}}+\lambda_3^{\boldsymbol{S}^{-1}\boldsymbol{U}}\right) \\ &= \det \boldsymbol{S}\,\mathrm{tr}(\boldsymbol{S}^{-1}\boldsymbol{U}) \\ &= \det \boldsymbol{S}\,(\boldsymbol{S}^{-T} : \boldsymbol{U}). \end{aligned} \tag{2.119}$$

**Linearization of the inverse of a tensor.** Finally, consider the linearization of the inverse of a tensor (or square matrix) $\boldsymbol{S}$ with respect to an increment in this matrix $\boldsymbol{U}$ as

$$(\boldsymbol{S}+\boldsymbol{U})^{-1} \approx \boldsymbol{S}^{-1} + D(\boldsymbol{S}^{-1})[\boldsymbol{U}], \tag{2.120}$$

where the directional derivative is given as

$$D(\boldsymbol{S}^{-1})[\boldsymbol{U}] = \frac{d}{d\epsilon}\bigg|_{\epsilon=0} (\boldsymbol{S} + \epsilon \boldsymbol{U})^{-1}. \tag{2.121}$$

Clearly, the evaluation of this derivative is far from obvious. A simple procedure to evaluate this linearization, however, emerges from the product rule property given in Equation (2.105b). For this purpose, note first that the linearization of $\boldsymbol{I}$, the identity tensor, is the null tensor $\boldsymbol{0}$ because $\boldsymbol{I}$ is independent of the increment $\boldsymbol{U}$, that is,

$$D(\boldsymbol{S}^{-1}\boldsymbol{S})[\boldsymbol{U}] = D(\boldsymbol{I})[\boldsymbol{U}] = \boldsymbol{0}. \tag{2.122}$$

Consequently, using the product rule gives

$$D(\boldsymbol{S}^{-1})[\boldsymbol{U}]\boldsymbol{S} + \boldsymbol{S}^{-1}D(\boldsymbol{S})[\boldsymbol{U}] = \boldsymbol{0}, \tag{2.123}$$

which, after some simple algebra, leads to

$$D(\boldsymbol{S}^{-1})[\boldsymbol{U}] = -\boldsymbol{S}^{-1}\boldsymbol{U}\boldsymbol{S}^{-1}. \tag{2.124}$$

**EXAMPLE 2.11: Linearization of** $\det(\boldsymbol{S}^{-1})$

An interesting application of the chain rule Equation (2.105c) is obtained by combining the linearizations of the determinant and the inverse of a tensor into the linearization of the functional $\det \boldsymbol{S}^{-1}$. First, the directional derivative of this functional can be obtained directly by noting that $\det(\boldsymbol{S}^{-1}) = 1/\det \boldsymbol{S}$ and using Equation (2.119) to give

$$\begin{aligned}
D\det(\boldsymbol{S}^{-1})[\boldsymbol{U}] &= \frac{d}{d\epsilon}\bigg|_{\epsilon=0} \det(\boldsymbol{S}+\epsilon\boldsymbol{U})^{-1} \\
&= \frac{d}{d\epsilon}\bigg|_{\epsilon=0} \frac{1}{\det(\boldsymbol{S}+\epsilon\boldsymbol{U})} \\
&= \frac{-1}{(\det \boldsymbol{S})^2}\frac{d}{d\epsilon}\bigg|_{\epsilon=0} \det(\boldsymbol{S}+\epsilon\boldsymbol{U}) \\
&= -\det(\boldsymbol{S}^{-1})(\boldsymbol{S}^{-T} : \boldsymbol{U}).
\end{aligned}$$

An alternative route can be followed to reach the same result by using the chain rule Equation (2.105c) and both Equations (2.119) and (2.124) to give

$$\begin{aligned}
D\det(\boldsymbol{S}^{-1})[\boldsymbol{U}] &= \det(\boldsymbol{S}^{-1})(\boldsymbol{S}^{T} : D\boldsymbol{S}^{-1}[\boldsymbol{U}]) \\
&= -\det(\boldsymbol{S}^{-1})(\boldsymbol{S}^{T} : (\boldsymbol{S}^{-1}\boldsymbol{U}\boldsymbol{S}^{-1})) \\
&= -\det(\boldsymbol{S}^{-1})(\boldsymbol{S}^{-T} : \boldsymbol{U}).
\end{aligned}$$

## 2.4 TENSOR ANALYSIS

Section 2.2 dealt with constant vectors and tensors. In contrast, such items in continuum mechanics invariably change from point to point throughout a problem domain. The resulting magnitudes are known as *fields* in a three-dimensional Cartesian space and can be of a scalar, vector, or tensor nature. Examples of scalar fields are the temperature or density of a body. Alternatively, the velocity of the body particles would constitute a vector field and the stresses a tensor field. The study of these quantities requires operations such as differentiation and integration, which are the subject of tensor analysis.

### 2.4.1 The Gradient and Divergence Operators

Consider first a scalar field, that is, a function $f(\boldsymbol{x})$ that varies throughout a three-dimensional space. At a given point $\boldsymbol{x}_0$, the change in $f$ in the direction of an arbitrary incremental vector $\boldsymbol{u}$ is given by a vector $\boldsymbol{\nabla} f(\boldsymbol{x}_0)$ known as the *gradient* of $f$ at $\boldsymbol{x}_0$, which is defined in terms of the directional derivative as

$$\boldsymbol{\nabla} f(\boldsymbol{x}_0) \cdot \boldsymbol{u} = Df(\boldsymbol{x}_0)[\boldsymbol{u}]. \tag{2.125}$$

The components of the gradient can be obtained by using the definition of the directional derivative (2.101) in the above equation to give

$$\begin{aligned}\boldsymbol{\nabla} f(\boldsymbol{x}_0) \cdot \boldsymbol{u} &= \left.\frac{d}{d\epsilon}\right|_{\epsilon=0} f(\boldsymbol{x}_0 + \epsilon \boldsymbol{u}) \\ &= \sum_{i=1}^{3} \left.\frac{\partial f}{\partial x_i}\right|_{\epsilon=0} \left.\frac{d(x_{0,i} + \epsilon u_i)}{d\epsilon}\right|_{\epsilon=0} \\ &= \sum_{i=1}^{3} u_i \left.\frac{\partial f}{\partial x_i}\right|_{x_i = x_{0,i}}.\end{aligned} \tag{2.126}$$

Hence the components of the gradient are the partial derivatives of the function $f$ in each of the three spatial direction as

$$\boldsymbol{\nabla} f = \sum_{i=1}^{3} \frac{\partial f}{\partial x_i} \boldsymbol{e}_i. \tag{2.127}$$

For obvious reasons, the following alternative notation is frequently used:

$$\boldsymbol{\nabla} f = \frac{\partial f}{\partial \boldsymbol{x}}. \tag{2.128}$$

The gradient of a vector field $\boldsymbol{v}$ at a point $\boldsymbol{x}_0$ is a second-order tensor $\boldsymbol{\nabla}\boldsymbol{v}(\boldsymbol{x}_0)$ that maps an arbitrary vector $\boldsymbol{u}$ into the directional derivative of $\boldsymbol{v}$ at $\boldsymbol{x}_0$ in the direction of $\boldsymbol{u}$ as

$$\boldsymbol{\nabla v}(\boldsymbol{x}_0)\,\boldsymbol{u} = D\boldsymbol{v}(\boldsymbol{x}_0)[\boldsymbol{u}]. \tag{2.129}$$

A procedure identical to that employed in Equation (2.126) shows that the components of this gradient tensor are simply the partial derivatives of the vector components, thereby leading to the following expression and useful alternative notation:

$$\boldsymbol{\nabla v} = \sum_{i,j=1}^{3} \frac{\partial v_i}{\partial x_j} \boldsymbol{e}_i \otimes \boldsymbol{e}_j; \quad \boldsymbol{\nabla v} = \frac{\partial \boldsymbol{v}}{\partial \boldsymbol{x}}. \tag{2.130}$$

The trace of the gradient of a vector field defines the *divergence* of such a field as a scalar, div $\boldsymbol{v}$, which can be variously written as

$$\operatorname{div} \boldsymbol{v} = \operatorname{tr} \boldsymbol{\nabla v} = \boldsymbol{\nabla v} : \boldsymbol{I} = \sum_{i=1}^{3} \frac{\partial v_i}{\partial x_i}. \tag{2.131}$$

Similarly to Equation (2.129), the gradient of a second-order tensor $\boldsymbol{S}$ at $\boldsymbol{x}_0$ is a third-order tensor $\boldsymbol{\nabla S}(\boldsymbol{x}_0)$, which maps an arbitrary vector $\boldsymbol{u}$ to the directional derivative of $\boldsymbol{S}$ at $\boldsymbol{x}_0$ as

$$\boldsymbol{\nabla S}(\boldsymbol{x}_0)\,\boldsymbol{u} = D\boldsymbol{S}(\boldsymbol{x}_0)[\boldsymbol{u}]. \tag{2.132}$$

Moreover, the components of $\boldsymbol{\nabla S}$ are again the partial derivatives of the components of $\boldsymbol{S}$ and consequently

$$\boldsymbol{\nabla S} = \sum_{i,j,k=1}^{3} \frac{\partial S_{ij}}{\partial x_k} \boldsymbol{e}_i \otimes \boldsymbol{e}_j \otimes \boldsymbol{e}_k; \quad \boldsymbol{\nabla S} = \frac{\partial \boldsymbol{S}}{\partial \boldsymbol{x}}. \tag{2.133}$$

Additionally, the divergence of a second-order tensor $\boldsymbol{S}$ is the vector div $\boldsymbol{S}$, which results from the double contraction of the gradient $\boldsymbol{\nabla S}$ with the identity tensor as

$$\operatorname{div} \boldsymbol{S} = \boldsymbol{\nabla S} : \boldsymbol{I} = \sum_{i,j=1}^{3} \frac{\partial S_{ij}}{\partial x_j} \boldsymbol{e}_i. \tag{2.134}$$

Finally, the following useful properties of the gradient and divergence are a result of the product rule:

$$\boldsymbol{\nabla}(f\boldsymbol{v}) = f\boldsymbol{\nabla v} + \boldsymbol{v} \otimes \boldsymbol{\nabla} f; \tag{2.135a}$$

$$\operatorname{div}(f\boldsymbol{v}) = f \operatorname{div} \boldsymbol{v} + \boldsymbol{v} \cdot \boldsymbol{\nabla} f; \tag{2.135b}$$

$$\boldsymbol{\nabla}(\boldsymbol{v} \cdot \boldsymbol{w}) = (\boldsymbol{\nabla v})^T \boldsymbol{w} + (\boldsymbol{\nabla w})^T \boldsymbol{v}; \tag{2.135c}$$

$$\operatorname{div}(\boldsymbol{v} \otimes \boldsymbol{w}) = \boldsymbol{v} \operatorname{div} \boldsymbol{w} + (\boldsymbol{\nabla v})\boldsymbol{w}; \tag{2.135d}$$

$$\operatorname{div}(\boldsymbol{S}^T \boldsymbol{v}) = \boldsymbol{S} : \boldsymbol{\nabla v} + \boldsymbol{v} \cdot \operatorname{div} \boldsymbol{S}; \tag{2.135e}$$

$$\operatorname{div}(f\boldsymbol{S}) = f \operatorname{div} \boldsymbol{S} + \boldsymbol{S}\boldsymbol{\nabla} f; \tag{2.135f}$$

$$\boldsymbol{\nabla}(f\boldsymbol{S}) = f\boldsymbol{\nabla S} + \boldsymbol{S} \otimes \boldsymbol{\nabla} f. \tag{2.135g}$$

**EXAMPLE 2.12: Proof of Equation (2.135e)**

Any one of Equations (2.135a–g) can be easily proved in component form with the help of the product rule. For example, using Equations (2.131) and (2.134) gives (2.135e) as

$$\begin{aligned}\operatorname{div}(\boldsymbol{S}^T\boldsymbol{v}) &= \sum_{i,j=1}^{3} \frac{\partial}{\partial x_j}(S_{ij}v_i)\\ &= \sum_{i,j=1}^{3} S_{ij}\frac{\partial v_i}{\partial x_j} + v_i\frac{\partial S_{ij}}{\partial x_j}\\ &= \boldsymbol{S} : \boldsymbol{\nabla}\boldsymbol{v} + \boldsymbol{v}\cdot\operatorname{div}\boldsymbol{S}.\end{aligned}$$

## 2.4.2 Integration Theorems

Many derivations in continuum mechanics are dependent upon the ability to relate the integration of fields over general volumes to the integration over the boundary of such volumes. For this purpose, consider a volume $V$ with boundary surface $\partial V$ and let $\boldsymbol{n}$ be the unit normal to this surface as shown in Figure 2.7. All integration theorems can be derived from a basic equation giving the integral of the gradient of a scalar field $f$ as

$$\int_V \boldsymbol{\nabla} f \, dV = \int_{\partial V} f\boldsymbol{n}\, dA. \tag{2.136}$$

Proof of this equation can be found in any standard text on calculus.

Expressions similar to Equation (2.136) can be obtained for any given vector or tensor field $\boldsymbol{v}$ by simply using Equation (2.136) on each of the components of $\boldsymbol{v}$ to give

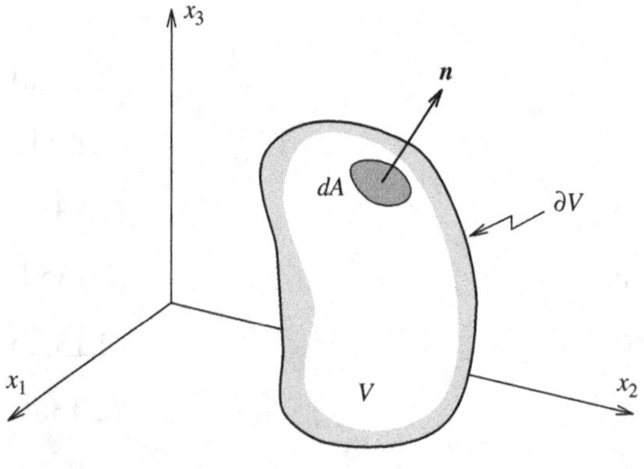


**FIGURE 2.7** General volume and element of area.

$$\int_V \nabla \boldsymbol{v}\, dV = \int_{\partial V} \boldsymbol{v} \otimes \boldsymbol{n}\, dA. \tag{2.137}$$

A more familiar expression is obtained by taking the trace of the above equation to give the Gauss or divergence theorem for a vector field $\boldsymbol{v}$ as

$$\int_V \operatorname{div} \boldsymbol{v}\, dV = \int_{\partial V} \boldsymbol{v} \cdot \boldsymbol{n}\, dA. \tag{2.138}$$

Similarly, taking the trace of Equation (2.137) when $\boldsymbol{v}$ is replaced by a second-order tensor $\boldsymbol{S}$ and noting that, as a result of Equation (2.72c), $(\boldsymbol{S} \otimes \boldsymbol{n}) : \boldsymbol{I} = \boldsymbol{S}\boldsymbol{n}$, gives

$$\int_V \operatorname{div} \boldsymbol{S}\, dV = \int_{\partial V} \boldsymbol{S}\boldsymbol{n}\, dA. \tag{2.139}$$

**EXAMPLE 2.13: Volume of a three-dimensional body**

The volume of a three-dimensional body is evaluated by the integral

$$V = \int_V dV.$$

Using Equation (2.138) it is possible and often useful to rewrite this volume in terms of an area integral. For this purpose note first that the divergence of the function $\boldsymbol{v}(\boldsymbol{x}) = \boldsymbol{x}/3$ is 1 and therefore Equation (2.138) gives

$$V = \frac{1}{3} \int_{\partial V} \boldsymbol{x} \cdot \boldsymbol{n}\, dA.$$

## Exercises

1. The second-order tensor $\boldsymbol{P}$ maps any vector $\boldsymbol{u}$ to its projection on a plane passing through the origin and with unit normal $\boldsymbol{a}$. Show that

$$P_{ij} = \delta_{ij} - a_i a_j; \quad \boldsymbol{P} = \boldsymbol{I} - \boldsymbol{a} \otimes \boldsymbol{a}.$$

   Show that the invariants of $\boldsymbol{P}$ are $I_{\boldsymbol{P}} = II_{\boldsymbol{P}} = 2$, $III_{\boldsymbol{P}} = 0$, and find the eigenvalues and eigenvectors of $\boldsymbol{P}$.
2. Using a procedure similar to that employed in Equations (2.41) and (2.42), obtain transformation equations for the components of third- and fourth-order tensors in two sets of bases $\boldsymbol{e}_i$ and $\boldsymbol{e}'_i$ that are related by the three-dimensional transformation tensor $\boldsymbol{Q}$ with components $Q_{ij} = \boldsymbol{e}_i \cdot \boldsymbol{e}'_j$.

3. If $L$ and $l$ are initial and current lengths, respectively, of an axial rod, the associated engineering, logarithmic, Green, and Almansi strains given in Section 1.3.1 are

$$\varepsilon_E(l) = \frac{l - L}{L}; \quad \varepsilon_L(l) = \ln \frac{l}{L}; \quad \varepsilon_G(l) = \frac{l^2 - L^2}{2L^2};$$
$$\varepsilon_A(l) = \frac{l^2 - L^2}{2l^2}.$$

Find the directional derivatives $D\varepsilon_E(l)[u]$, $D\varepsilon_L(l)[u]$, $D\varepsilon_G(l)[u]$, and $D\varepsilon_A(l)[u]$, where $u$ is a small increment in the length $l$.

4. Consider a functional $I$ that when applied to the function $y(x)$ gives the integral

$$I(y(x)) = \int_a^b f(x, y, y')\, dx,$$

where $f$ is a general expression involving $x$, $y(x)$, and the derivative $y'(x) = dy/dx$. Show that the function $y(x)$ that renders the above functional stationary and satisfies the boundary conditions $y(a) = y_a$ and $y(b) = y_b$ is the solution of the following Euler–Lagrange differential equation:

$$\frac{d}{dx}\left(\frac{\partial f}{\partial y'}\right) - \frac{\partial f}{\partial y} = 0.$$

5. Prove Equations (2.135a–g) following the procedure shown in Example 2.11.
6. Show that the volume of a closed three-dimensional body $V$ is variously given as

$$V = \int_{\partial V} x\, n_x\, dA = \int_{\partial V} y\, n_y\, dA = \int_{\partial V} z\, n_z\, dA,$$

where $n_x, n_y$, and $n_z$ are the $x, y$, and $z$ components of the unit normal $\boldsymbol{n}$.

7. A scalar field $\Phi(\boldsymbol{x}) = x_1^2 + 3x_2x_3$ describes some physical quantity (e.g. total potential energy). Show that the directional derivative of $\Phi$ in the direction $\boldsymbol{u} = \frac{1}{\sqrt{3}}(1, 1, 1)^T$ at the position $\boldsymbol{x} = (2, -1, 0)^T$ is $\frac{1}{\sqrt{3}}$.
8. Given the second-order tensor $\boldsymbol{A}$, obtain the directional derivative of the expression $\boldsymbol{A}^3 = \boldsymbol{AAA}$ in the direction of an arbitrary increment $\boldsymbol{U}$ of $\boldsymbol{A}$.
9. If the second-order tensors $\boldsymbol{S} = \boldsymbol{S}^T$ and $\boldsymbol{W}^T = -\boldsymbol{W}$, show that $\boldsymbol{S} : \boldsymbol{W} = 0$ (i.e. $\text{tr}(\boldsymbol{SW}) = 0$).

## CHAPTER THREE

# ANALYSIS OF THREE-DIMENSIONAL TRUSS STRUCTURES

## 3.1 INTRODUCTION

This chapter considers the uniaxial (one-dimensional) large displacement, large strain, rate-independent elasto-plastic behavior applicable to structural analysis of pin-jointed trusses. The motivation is to expand and reinforce previous material and to introduce some topics that will reappear later when elasto-plastic behavior of continua is considered. For example, various nonlinear geometrical descriptors will be linearized, providing further examples of the use of the directional derivative.

Formulations start with the kinematic description of the motion in three-dimensional space of a truss member (axial rod) that undergoes large displacements and rotations, leading to large or small strain that causes stress which may reach the limit or yield stress of the material. For simplicity, it will be assumed that the strain in the truss member is uniform. Consequently, the fundamental measure of deformation in the axial rod is the stretch $\lambda = l/L$, which is the ratio of the deformed length to the undeformed length; see Figure 3.1.

The internal forces in the truss are easily determined from simple strength-of-material considerations involving the true (or Cauchy) stress $\sigma$, defined for a truss as the internal axial force $T$ divided by the deformed cross-sectional area $a$. However, for large deformation the elasto-plastic behavior is best characterized using an alternative stress known as the *Kirchhoff stress* $\tau$ defined as $\sigma v/V$ see Figure 3.1. In preparation for Chapter 6 it will be shown how the Kirchhoff stress can be derived from a hyperelastic energy function involving the natural logarithm of the stretch.

Global equilibrium equations are derived from simple joint equilibrium equations. Since these are nonlinear with respect to the position of the rod, a Newton–Raphson solution procedure is adopted which requires the linearization of the equilibrium equations. This is achieved using the directional derivative to yield the tangent matrix necessary for the solution. The tangent matrix contains the material tangent modulus, which is the derivative of the stress with respect to the strain,

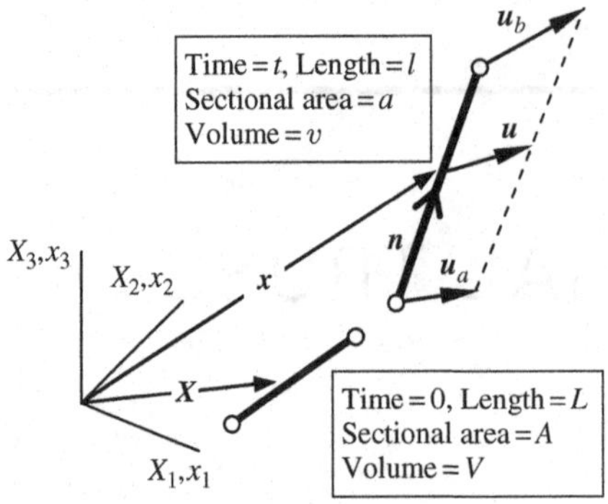


**FIGURE 3.1** Three-dimensional axial truss member – kinematics.

which for elastic behavior will emerge as a Young's modulus-like constant. However, for inelastic (plastic) cases detailed consideration of elasto-plastic material behavior is necessary.

In anticipation of Chapters 5 and 8, the concept of Total Potential Energy is introduced as an alternative way of obtaining the equilibrium equations. This involves a discussion of the stationary condition of the Total Potential Energy obtained as a variational statement which can be interpreted as the Principle of Virtual Work.

Plasticity (behavior at the elastic limit) is dealt with in the context of a simple one-dimensional yield criterion with hardening and associated flow rule for the plastic strain rate.

The equilibrium equations require the evaluation of the stress, and insofar as the stress is either lower or at the current (in the case of hardening) elastic limit, the stress only depends upon the elastic strain. A strain in excess of that required to reach the yield stress results, after complete removal of the stress, in a nonrecoverable permanent (plastic) strain which remains locked into the material. In such a situation the elastic strain measures the difference between the total strain and the plastic strain. However, given a current state of deformation of a truss member, it is not possible to directly determine the relative proportions of elastic to plastic strain that comprise the observed total strain. This can only be accomplished by tracking the development of the plastic strain throughout the history of the deformation. Such a material is called a *path-dependent material*. Consequently, central to any numerical algorithm is the need to integrate the plastic strain rate over time in order to calculate the accumulated plastic strain and thence the elastic strain. This must be accomplished under the constraint that the stress does not exceed the current yield stress. It will be shown that this can be conveniently expressed as a return-mapping algorithm from a guessed or trial stress. Although the calculation of the accumulated plastic strain introduces the notion of plastic strain rate, nevertheless material behavior can remain rate-independent in the sense that a given deformation history will lead to a specific state of stress irrespective of the speed at which this path is followed. This implies that for rate-independent materials the time variable used is purely notional. Consideration of the elasto-plastic material behavior is completed by the formulation of the material tangent modulus required for the numerical solution.

The resulting discretized equilibrium equations and tangent stiffness matrix are included in the FLagSHyP program that is used to provide a number of examples illustrating an interesting range of nonlinear structural behavior.

The subject matter covered in this chapter is considered in greater detail in Simo and Hughes (1997) in the context of small strains.

## 3.2 KINEMATICS

The description of the motion of the axial rod shown in Figure 3.1 is deliberately set up to provide an introduction to the more general concepts of large deformation kinematics to be encountered in Chapter 4 and large deformation elasto-plastic behavior presented in Chapter 7.

Figure 3.1 shows the motion of the rod with respect to Cartesian axes. To correspond to later developments, upper-case letters are used to describe the initial undeformed position $\boldsymbol{X}$, length $L$, cross-sectional area $A$, and volume $V$ at time $t = 0$. Lower-case letters are used to describe the current deformed postion $\boldsymbol{x}$, length $l$, cross-sectional area $a$, and volume $v$ at time $t$. Both upper- and lower-case axes are labeled in Figure 3.1 because, in principle, it is possible to use different Cartesian axes to describe the initial and deformed (current) positions; however, here the assumption is made that they coalesce.

The current (deformed) length of the rod and the unit vector $\boldsymbol{n}$ can be found in terms of the current end coordinates $\boldsymbol{x}_a$ and $\boldsymbol{x}_b$ as

$$l = \{(\boldsymbol{x}_b - \boldsymbol{x}_a) \cdot (\boldsymbol{x}_b - \boldsymbol{x}_a)\}^{1/2}; \quad \text{unit vector } \boldsymbol{n} = \frac{1}{l}(\boldsymbol{x}_b - \boldsymbol{x}_a). \tag{3.1a,b}$$

In the current position the rod may undergo an incremental displacement $\boldsymbol{u}$ which at the end nodes takes values $\boldsymbol{u}_a$ and $\boldsymbol{u}_b$; see Figure 3.1. Note that the end displacements are not the displacements from the initial position but *incremental* displacements that the current position may experience in the progress of its motion.

Under the assumption that the rod undergoes large uniform deformation and strain, the basic quantity from which a strain measure will be derived is the stretch $\lambda$, defined as the ratio of the current length $l$ to the initial length $L$, while the associated volume ratio $J$ is the ratio of the current to the initial volume; these are given as

$$\lambda = \frac{l}{L}; \; J = \frac{v}{V}. \tag{3.2a,b}$$

If the rod elongates by $dl$ from the current length $l$, then the instantaneous engineering strain is $d\varepsilon = dl/l$. Summing all $d\varepsilon$ as the rod elongates from the initial length $L$ to current length $l$ provides a definition of the logarithmic strain $\varepsilon$ as

$$\varepsilon = \ln \lambda = \int_L^l \frac{dl}{l}; \qquad d\varepsilon = \frac{dl}{l}. \tag{3.3a,b}$$

If the deformation normal to the axis of the rod is the same in all directions (which is the case for transversely isotropic materials) then it is shown in Example 3.1 that the volume ratio $J$ can be related to the stretch in terms of Poisson's ratio $\nu$ as

$$J = \lambda^{(1-2\nu)} \quad \text{or} \quad v = V\left(\frac{l}{L}\right)^{1-2\nu}. \tag{3.4a,b}$$

**EXAMPLE 3.1: Proof of Equation (3.4a,b)**

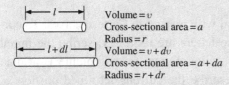

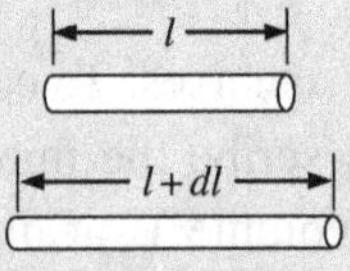


Assuming the usual relationship between axial and radial strain given by Poisson's ratio $\nu$, the ratio between the initial volume $V$ and current volume $v$ can be determined as follows:

Current volume $v = al;$

Current cross-sectional area $a = \pi r^2;$

Instantaneous axial strain $d\varepsilon = \dfrac{dl}{l};$

Instantaneous radial strain $d\varepsilon_r = \dfrac{dr}{r} = -\nu \dfrac{dl}{l};$

Instantaneous volume change $dv = adl + lda;$

where $da = 2\pi r dr = 2a\dfrac{dr}{r} = -2a\nu\dfrac{dl}{l}.$

Hence $dv = adl - 2a\nu dl = (1 - 2\nu)adl$

or $\dfrac{dv}{v} = (1 - 2\nu)\dfrac{dl}{l}.$

*(continued)*

**EXAMPLE 3.1:** *(cont.)*

Assuming Poisson's ratio is constant during deformation,

integration yields
$$\int_V^v \frac{dv}{v} = (1-2\nu)\int_L^l \frac{dl}{l}$$
$$\ln\frac{v}{V} = (1-2\nu)\ln\frac{l}{L}$$
or
$$J = \frac{v}{V} = \left(\frac{l}{L}\right)^{(1-2\nu)} = \lambda^{(1-2\nu)}.$$

### 3.2.1 Linearization of Geometrical Descriptors

In order to develop the tangent stiffness matrix required for the Newton–Raphson solution of the nonlinear equilibrium equations, it is necessary to linearize kinematic quantities. This will be accomplished using the directional derivative discussed in Chapters 1 and 2, determined with respect to the incremental end displacements $\boldsymbol{u}_a$ and $\boldsymbol{u}_b$.

Employing Equation (3.1a,b), the directional derivative of the current length vector, $l\boldsymbol{n} = (\boldsymbol{x}_b - \boldsymbol{x}_a)$, is determined as

$$D(\boldsymbol{x}_b - \boldsymbol{x}_a)[\boldsymbol{u}] = \frac{d}{d\epsilon}\bigg|_{\epsilon=0} (\boldsymbol{x}_b + \epsilon\boldsymbol{u}_b - \boldsymbol{x}_a - \epsilon\boldsymbol{u}_a) = (\boldsymbol{u}_b - \boldsymbol{u}_a). \tag{3.5}$$

The directional derivative of the current length $l$ is most easily derived by noting that

$$Dl^2(\boldsymbol{x})[\boldsymbol{u}] = 2l Dl(\boldsymbol{x})[\boldsymbol{u}]. \tag{3.6}$$

Consequently,

$$Dl(\boldsymbol{x})[\boldsymbol{u}] = \frac{1}{2l} Dl^2(\boldsymbol{x})[\boldsymbol{u}]; \tag{3.7}$$

substituting for $l^2$ from Equation (3.1a,b) gives

$$\begin{aligned} Dl^2(\boldsymbol{x})[\boldsymbol{u}] &= \frac{d}{d\epsilon}\bigg|_{\epsilon=0} \{(\boldsymbol{x}_b + \epsilon\boldsymbol{u}_b - \boldsymbol{x}_a - \epsilon\boldsymbol{u}_a)\cdot(\boldsymbol{x}_b + \epsilon\boldsymbol{u}_b - \boldsymbol{x}_a - \epsilon\boldsymbol{u}_a)\} \\ &= 2(\boldsymbol{x}_b - \boldsymbol{x}_a)\cdot(\boldsymbol{u}_b - \boldsymbol{u}_a), \end{aligned} \tag{3.8}$$

from which

$$Dl(\boldsymbol{x})[\boldsymbol{u}] = \boldsymbol{n}\cdot(\boldsymbol{u}_b - \boldsymbol{u}_a). \tag{3.9}$$

Similarly, it is useful to note for later derivations that

$$Dl^{-1}(\boldsymbol{x})[\boldsymbol{u}] = -l^{-2}Dl(\boldsymbol{x})[\boldsymbol{u}] = -l^{-2}\boldsymbol{n}\cdot(\boldsymbol{u}_b - \boldsymbol{u}_a). \tag{3.10}$$

Recalling the definition of the logarithmic strain in Equation (3.3a,b), $\varepsilon = \ln \lambda$, enables its directional derivative to be found as

$$D\varepsilon(\boldsymbol{x})[\boldsymbol{u}] = D(\ln l(\boldsymbol{x}) - \ln L)[\boldsymbol{u}] = \frac{1}{l}Dl(\boldsymbol{x})[\boldsymbol{u}] = \frac{1}{l}\boldsymbol{n}\cdot(\boldsymbol{u}_b - \boldsymbol{u}_a). \tag{3.11}$$

## 3.3 INTERNAL FORCES AND HYPERELASTIC CONSTITUTIVE EQUATIONS

Internal truss forces $\boldsymbol{T}_a$ and $\boldsymbol{T}_b$ (see Figure 3.2) are easily determined in terms of the true (or Cauchy) stress, the current cross-sectional area $a$ and the unit vector $\boldsymbol{n}$ as

$$\boldsymbol{T}_a = -\sigma a\boldsymbol{n}; \quad \boldsymbol{T}_b = +\sigma a\boldsymbol{n}. \tag{3.12a,b}$$

Before considering the tangent stiffness matrix which, as seen in Chapter 1, relates changes in equilibriating forces to corresponding changes in position, it is necessary to develop a constitutive equation suitable for large strain applications. A constitutive equation for a specified material relates stress to a strain measure and is necessary to enable the calculation of the internal force components of the equilibriating forces. Consequently, the change in the stress due to a change in strain is essential for the development of the tangent stiffness.

As a precursor to Chapter 6 concerning hyperelastic constitutive equations, a start is made by introducing the *stored strain energy*, $\Psi$, per unit initial volume developed as the rod deforms from initial to current position. At any time $t$ the absolute value of the current axial force $T(l)$ acting *on* the rod is a function of the

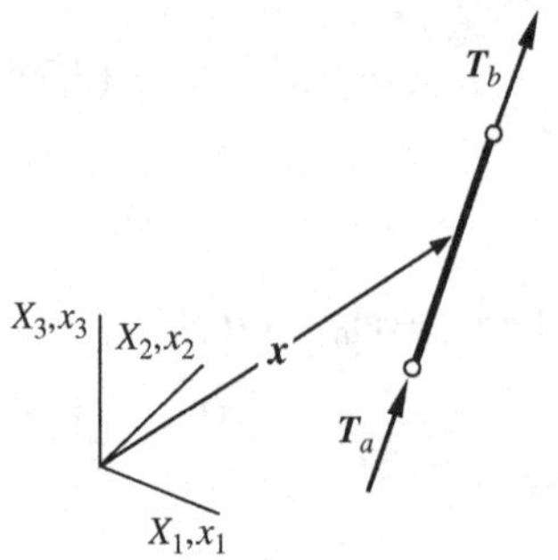


**FIGURE 3.2** Three-dimensional axial truss member – forces.

current length $l$. If the rod elongates by a small amount $dl$, then the work done is simply $T(l)\,dl$ and the total work $\Psi$ done per unit initial volume $V$ is

$$\Psi = \frac{1}{V}\int_L^l T(l)\,dl = \int_L^l \frac{\sigma a}{V}\,dl. \tag{3.13}$$

Since the current volume $v = al$, and recalling Equation (3.3a,b) for $d\varepsilon$, the strain energy $\Psi$ can be rewritten as

$$\Psi = \int_L^l \tau(l)d\varepsilon; \quad \tau = J\sigma; \tag{3.14}$$

where $\tau$ is called the *Kirchhoff stress* and $J$ is the volume ratio given in Equation (3.2a,b) as $J = v/V$. As discussed above, $d\varepsilon$ is the instantaneous engineering strain and, insofar as $\tau$ times $d\varepsilon$ gives work done (or strain energy developed) per unit initial volume, $\tau$ and $d\varepsilon$ are said to be *work conjugate* to one another with respect to the initial configuration. In other words, if the increment in strain is $d\varepsilon$ then $\tau$ must be used to obtain the correct work done. In Chapter 5 other important examples of conjugate stress and strain measures will emerge.

As yet a material parameter relating stress to strain has not been introduced and, in lieu of anything better at the moment, a Young's modulus-like constant $E$ is defined as the gradient of a measured $\tau$ versus the logarithmic strain $\varepsilon$ as

$$E = \frac{d\tau}{d\varepsilon}; \quad d\tau = E d\varepsilon. \tag{3.15}$$

Assuming for simplicity that this gradient $E$ is not dependent upon the strain, then a relationship from which the Kirchhoff stress $\tau$ can be determined as a function of the logarithm of the stretch $\lambda = l/L$ can be found as

$$\int_0^\tau d\tau = E\int_L^l \frac{dl}{l}; \quad \tau = E\,\ln\lambda. \tag{3.16}$$

This, it transpires, is the particular axial rod case of a general logarithmic stretch-based constitutive equation to be discussed in detail in Chapter 6. Again the logarithmic strain, $\ln\lambda$, emerges naturally as the integral of the instantaneous engineering strain. Substitution of $\tau$ from Equation (3.16) into Equation (3.14) enables the strain energy per unit initial volume to be found as

$$\Psi = E\int_L^l \ln\left(\frac{l}{L}\right)\frac{dl}{l} = \frac{1}{2}E\,(\ln\lambda)^2. \tag{3.17}$$

It is obvious that the Kirchhoff stress $\tau$ can also be found as

$$\tau = \frac{d\Psi}{d\,(\ln\lambda)}. \tag{3.18}$$

The strain energy per unit initial volume, $\Psi$, given by Equation (3.17) is a function of the initial length $L$ and the current length $l$ and is consequently independent of the path taken by the rod as it moves from $L$ to $l$. Such a material is called *hyperelastic*. Equation (3.18) is often used as a definition of a hyperelastic material. If the rod is incompressible, Poisson's ratio $\nu = 0.5$ and $J = 1$, and the constitutive equation used in Chapter 1, $\sigma = E \ln \lambda$, is recovered.

Returning to the internal truss forces $\boldsymbol{T}_a$ and $\boldsymbol{T}_b$, substitution of $\sigma = \tau/J$ into Equation (3.12a,b) gives, after some rearrangement,

$$\boldsymbol{T}_b = \frac{VE}{l} \ln\left(\frac{l}{L}\right) \boldsymbol{n}; \quad \boldsymbol{T}_a = -\boldsymbol{T}_b. \tag{3.19}$$

## 3.4 NONLINEAR EQUILIBRIUM EQUATIONS AND THE NEWTON–RAPHSON SOLUTION

### 3.4.1 Equilibrium Equations

The truss joint (node) equilibrium equations are established with respect to the current position by assembling the typical internal forces $\boldsymbol{T}_a$ and external forces $\boldsymbol{F}_a$ at all nodes ($a = 1, \ldots, N$) in the truss. This assembly is performed in the standard structural manner by adding contributions from all truss members (elements) meeting at the typical node $a$; see Figure 3.3. This is expressed in terms of the *residual* or *out-of-balance* nodal force $\boldsymbol{R}_a$ as the balance between the internal and the external forces as

$$\boldsymbol{R}_a = \sum_{\substack{e=1 \\ e \ni a}}^{m_a} \boldsymbol{T}_a^e - \boldsymbol{F}_a = \boldsymbol{0}, \tag{3.20}$$

where the slightly unusual symbol $e \ni a$ is used here to denote the elements $e$ that meet at node $a$, and $m_a$ is the total number of such elements. Since the internal

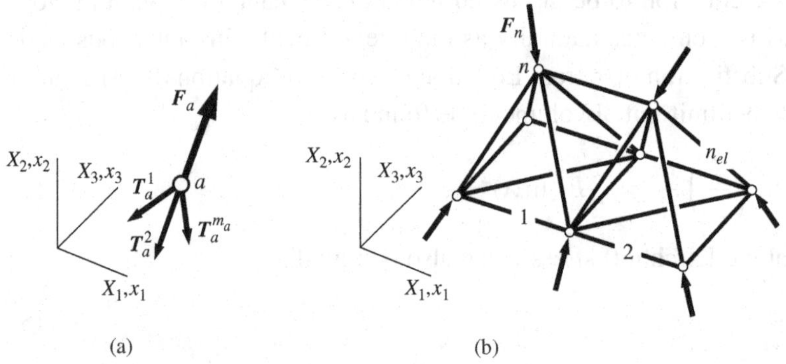


**FIGURE 3.3** Three-dimensional truss – equilibrium: (a) Joint forces; (b) Three-dimensional truss.

joint (nodal) force given by Equation (3.19) is a function of the current member (element) length $l$ and unit vector $\boldsymbol{n}$, the above equilibrium equations are nonlinear functions of the node positions. Consequently, the equilibrium equations need to be solved using the Newton–Raphson procedure given in Section 1.4.2. Equation (3.20) can be written in compact form in terms of the out-of-balance residual forces $\mathbf{R}(\mathbf{x})$, which are a function of the current nodal positions $\mathbf{x}$, giving

$$\mathbf{R}(\mathbf{x}) = \mathbf{T}(\mathbf{x}) - \mathbf{F};\ \mathbf{x} = \begin{bmatrix} \boldsymbol{x}_1 \\ \boldsymbol{x}_2 \\ \vdots \\ \boldsymbol{x}_N \end{bmatrix};\ \mathbf{T} = \begin{bmatrix} \boldsymbol{T}_1 \\ \boldsymbol{T}_2 \\ \vdots \\ \boldsymbol{T}_N \end{bmatrix};\ \mathbf{F} = \begin{bmatrix} \boldsymbol{F}_1 \\ \boldsymbol{F}_2 \\ \vdots \\ \boldsymbol{F}_N \end{bmatrix}. \tag{3.21a,b,c,d}$$

Observe that in the above equilibrium equation the external forces $\mathbf{F}$ are not a function of the current nodal positions $\mathbf{x}$. Generally this is not the case; for example, in the simple cantilever example given in Figure 1.1, the force $F$ could remain at right angles to the beam, in which case the force vector $\boldsymbol{F}$ is a function of the angle $\theta$. Another example is the inflation of a balloon, where the pressure changes in magnitude *and* direction as the balloon inflates.

### 3.4.2 Newton–Raphson Procedure

The Newton–Raphson procedure can be summarized by recalling Equations (1.36) and (1.37a,b) and Box 3.1 in which a linear set of equations involving the assembled tangent matrix $\mathbf{K}$, the incremental displacements $\mathbf{u}$, and the out-of-balance or residual forces $\mathbf{R}$ are written for an iterative step $k$ as

$$\mathbf{K}(\mathbf{x}_k)\mathbf{u} = -\mathbf{R}(\mathbf{x}_k);\ \mathbf{x}_{k+1} = \mathbf{x}_k + \mathbf{u}\ ;\ \mathbf{K}(\mathbf{x}_k)\mathbf{u} = D\mathbf{R}(\mathbf{x}_k)[\mathbf{u}]. \tag{3.22a,b,c}$$

Box 1.1, which shows the Newton–Raphson algorithm, will be revised later in this chapter to include elasto-plastic considerations.

The tangent matrix $\mathbf{K}$ is assembled in the usual finite element manner from individual element stiffness contributions $\mathbf{K}^{(e)}$, where

$$\mathbf{K}^{(e)}(\mathbf{x}_k^{(e)}) = \begin{bmatrix} \boldsymbol{K}_{aa}^{(e)} & \boldsymbol{K}_{ab}^{(e)} \\ \boldsymbol{K}_{ba}^{(e)} & \boldsymbol{K}_{bb}^{(e)} \end{bmatrix};\ \mathbf{x}_k^{(e)} = \begin{bmatrix} \boldsymbol{x}_a \\ \boldsymbol{x}_b \end{bmatrix}_k. \tag{3.23a,b}$$

For the case where external forces are not dependent upon the deformation, the element stiffness contribution to $\mathbf{K}(\mathbf{x}_k)\mathbf{u}$ in Equation (3.22a,b,c) is the directional derivative of $\mathbf{T}^{(e)}$, that is,

$$\mathbf{K}^{(e)}(\mathbf{x}_k^{(e)})\mathbf{u}^{(e)} = D\mathbf{T}^{(e)}(\mathbf{x}^{(e)})[\mathbf{u}] \; ; \; \mathbf{T}^{(e)} = \begin{bmatrix} \boldsymbol{T}_a \\ \boldsymbol{T}_b \end{bmatrix} \; ; \; \mathbf{u}_k^{(e)} = \begin{bmatrix} \boldsymbol{u}_a \\ \boldsymbol{u}_b \end{bmatrix}_k . \tag{3.24a,b,c}$$

Consequently in the following section attention is focused on the directional derivative $D\mathbf{T}^{(e)}(\mathbf{x}^{(e)})[\mathbf{u}^{(e)}]$ for a typical element $(e)$.

### 3.4.3 Tangent Elastic Stiffness Matrix

The directional derivative $D\mathbf{T}^{(e)}(\mathbf{x}^{(e)})[\mathbf{u}^{(e)}]$, which yields the tangent stiffness, results from the linearization with respect to the current position $\boldsymbol{x}$ of the nonlinear function $\boldsymbol{T}_b = -\boldsymbol{T}_a$, where $a$ and $b$ are the two nodes comprising element $(e)$. In addition, this reveals again the tangent modulus $E$ as the derivative of the Kirchhoff stress $\tau$ with respect to the logarithmic strain $\varepsilon$. Later, Section 3.6.6 will reconsider the tangent modulus to account for elasto-plastic behavior.

From Equations (3.12a,b) and (3.14) and Equations (3.3a,b) the internal force $\boldsymbol{T}_b^{(e)}$ and the logarithmic strain can be expressed as

$$\boldsymbol{T}_b^{(e)} = \tau \frac{V}{l}\boldsymbol{n}; \quad \varepsilon = \ln \frac{l}{L}. \tag{3.25a,b}$$

The directional derivative makes use of Equations (3.5), (3.9), (3.10), and (3.11) and will be displayed in detail as an example of the use of the directional derivative. From Equations (3.25a,b)* above,

$$D\boldsymbol{T}_b(\boldsymbol{x})[\boldsymbol{u}]$$

$$= D\tau(\boldsymbol{x})[\boldsymbol{u}]\frac{V}{l}\boldsymbol{n} + \tau D\left(\frac{V}{l}\right)[\boldsymbol{u}]\boldsymbol{n} + \tau\frac{V}{l}D\boldsymbol{n}(\boldsymbol{x})[\boldsymbol{u}] \tag{3.26a}$$

$$= \frac{d\tau}{d\varepsilon}D\varepsilon(\boldsymbol{x})[\boldsymbol{u}]\frac{V}{l}\boldsymbol{n} + \tau D\left(\frac{V}{l}\right)[\boldsymbol{u}]\boldsymbol{n} + \tau\frac{V}{l}D\boldsymbol{n}(\boldsymbol{x})[\boldsymbol{u}] \tag{3.26b}$$

$$= \frac{d\tau}{d\varepsilon}\frac{1}{l}\boldsymbol{n}\cdot(\boldsymbol{u}_b - \boldsymbol{u}_a)\frac{V}{l}\boldsymbol{n} + \tau V D\left(l^{-1}\right)[\boldsymbol{u}]\boldsymbol{n}$$
$$+ \tau\frac{V}{l}\left(D\left(l^{-1}\right)[\boldsymbol{u}](\boldsymbol{x}_b - \boldsymbol{x}_a) + \frac{1}{l}D(\boldsymbol{x}_b - \boldsymbol{x}_a)[\boldsymbol{u}]\right) \tag{3.26c}$$

$$= \frac{d\tau}{d\varepsilon}\frac{V}{l^2}\boldsymbol{n}\cdot(\boldsymbol{u}_b - \boldsymbol{u}_a)\boldsymbol{n} + 2\tau V D\left(l^{-1}\right)[\boldsymbol{u}]\boldsymbol{n} + \frac{\tau V}{l^2}(\boldsymbol{u}_b - \boldsymbol{u}_a) \tag{3.26d}$$

$$= \frac{d\tau}{d\varepsilon}\frac{V}{l^2}\boldsymbol{n}\cdot(\boldsymbol{u}_b - \boldsymbol{u}_a)\boldsymbol{n} - \frac{2}{l^2}\tau V\boldsymbol{n}\cdot(\boldsymbol{u}_b - \boldsymbol{u}_a)\boldsymbol{n} + \frac{\tau V}{l^2}(\boldsymbol{u}_b - \boldsymbol{u}_a) \tag{3.26e}$$

* The element superscript $(e)$ is omitted for clarity.

$$= \left(\frac{d\tau}{d\varepsilon}\frac{V}{l^2} - \frac{2\tau V}{l^2}\right) \boldsymbol{n} \cdot (\boldsymbol{u}_b - \boldsymbol{u}_a)\boldsymbol{n} + \frac{\tau V}{l^2}(\boldsymbol{u}_b - \boldsymbol{u}_a) \tag{3.26f}$$

$$= \left(\frac{V}{v}\frac{d\tau}{d\varepsilon}\frac{a}{l} - \frac{2\sigma a}{l}\right) \boldsymbol{n} \cdot (\boldsymbol{u}_b - \boldsymbol{u}_a)\boldsymbol{n} + \frac{\sigma a}{l}(\boldsymbol{u}_b - \boldsymbol{u}_a) \tag{3.26g}$$

$$= \left(\frac{V}{v}\frac{d\tau}{d\varepsilon}\frac{a}{l} - \frac{2\sigma a}{l}\right) (\boldsymbol{n} \otimes \boldsymbol{n})_{3\times 3}\,(\boldsymbol{u}_b - \boldsymbol{u}_a) + \frac{\sigma a}{l}\boldsymbol{I}_{3\times 3}\,(\boldsymbol{u}_b - \boldsymbol{u}_a). \tag{3.26h}$$

$$D\boldsymbol{T}_a(\boldsymbol{x})[\boldsymbol{u}] = -D\boldsymbol{T}_b(\boldsymbol{x})[\boldsymbol{u}].$$

Rearrangement in matrix form yields

$$D\mathbf{T}^{(e)}(\mathbf{x}^{(e)})[\mathbf{u}^{(e)}] = \mathbf{K}^{(e)}\mathbf{u}^{(e)} = \begin{bmatrix} \boldsymbol{K}_{aa}^{(e)} & \boldsymbol{K}_{ab}^{(e)} \\ \boldsymbol{K}_{ba}^{(e)} & \boldsymbol{K}_{bb}^{(e)} \end{bmatrix} \begin{bmatrix} \boldsymbol{u}_a \\ \boldsymbol{u}_b \end{bmatrix}, \tag{3.27a,b}$$

where, in terms of a scalar stiffness term $k$ and the axial force $T = \sigma a$, the component tangent stiffness matrices are

$$\boldsymbol{K}_{aa}^{(e)} = \boldsymbol{K}_{bb}^{(e)} = k\boldsymbol{n} \otimes \boldsymbol{n} + \frac{T}{l}\boldsymbol{I}_{3\times 3}; \quad \boldsymbol{K}_{ab}^{(e)} = \boldsymbol{K}_{ba}^{(e)} = -\boldsymbol{K}_{bb}^{(e)}. \tag{3.28a,b}$$

The scalar stiffness term $k$ requires evaluation of the *elastic tangent modulus*, which from Equations (3.3a,b) and (3.15) is simply $d\tau/d\varepsilon = E$. Hence the scalar stiffness can be expressed in alternative ways as

$$k = \left(\frac{V}{v}\frac{d\tau}{d\varepsilon}\frac{a}{l} - \frac{2T}{l}\right) = \left(\frac{V}{v}\frac{Ea}{l} - \frac{2T}{l}\right) = \frac{V}{l^2}(E - 2\tau)\;; \quad T = \sigma a. \tag{3.29a,b}$$

**EXAMPLE 3.2: Truss load correction matrix**

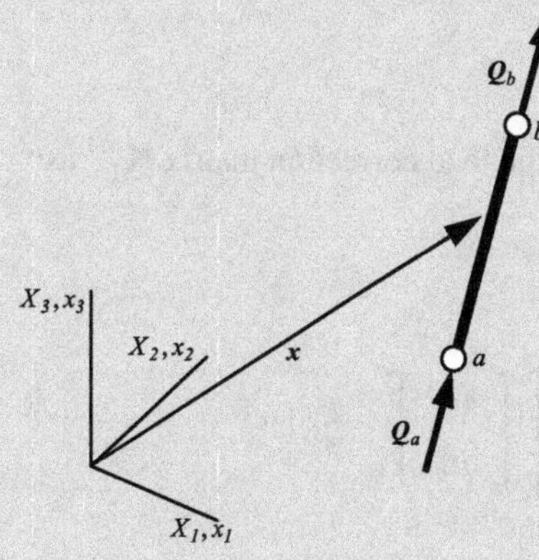


*(continued)*

**EXAMPLE 3.2:** *(cont.)*

This is an example of a single truss element, with load $\boldsymbol{Q}_a = \boldsymbol{0}$ and deformation-dependent load $\boldsymbol{Q}_b$ given as

$$\boldsymbol{Q}_b = Q_b \boldsymbol{n},$$

where $\boldsymbol{n}$ is given by Equation (3.1a,b) as $\boldsymbol{n} = \frac{1}{l}(\boldsymbol{x}_b - \boldsymbol{x}_a)$. During deformation the magnitude of this nodal force remains unchanged but the direction changes since $\boldsymbol{n}$ is a function of the nodal positions. Consequently there will be a change in nodal forces due to deformation and, as a function of the incremental nodal displacements $\boldsymbol{u}$, these changes are given by the *load correction matrix* $\mathbf{K}_L$, alternatively called the *load stiffness matrix* .

The directional derivative of $\boldsymbol{Q}_b$, which gives the change in $\boldsymbol{Q}_b$ due to the incremental displacements $\boldsymbol{u}$, can now be found in a similar manner to Equation (3.26h) as

$$\begin{aligned}
&D\boldsymbol{Q}_b(\boldsymbol{x})[\boldsymbol{u}] \\
&= Q_b D\, l^{-1}(\boldsymbol{x}_b - \boldsymbol{x}_a)[\boldsymbol{u}] \\
&= Q_b\big(Dl^{-1}[\boldsymbol{u}](\boldsymbol{x}_b - \boldsymbol{x}_a) + l^{-1}D(\boldsymbol{x}_b - \boldsymbol{x}_a)[\boldsymbol{u}]\big)\big) \\
&= Q_b\big(-l^{-2}\boldsymbol{n}\cdot(\boldsymbol{u}_b - \boldsymbol{u}_a)(\boldsymbol{x}_b - \boldsymbol{x}_a) + l^{-1}(\boldsymbol{u}_b - \boldsymbol{u}_a)\big) \\
&= Q_b\big(-l^{-1}\boldsymbol{n}\cdot(\boldsymbol{u}_b - \boldsymbol{u}_a)\boldsymbol{n} + l^{-1}(\boldsymbol{u}_b - \boldsymbol{u}_a)\big) \\
&= \frac{Q_b}{l}\big((\boldsymbol{n}\otimes\boldsymbol{n})_{3\times 3}(\boldsymbol{u}_a - \boldsymbol{u}_b) - \boldsymbol{I}_{3\times 3}(\boldsymbol{u}_a - \boldsymbol{u}_b)\big).
\end{aligned}$$

$D\boldsymbol{Q}_b(\boldsymbol{x})[\boldsymbol{u}]$ can now be expressed as

$$D\boldsymbol{Q}_b(\boldsymbol{x})[\boldsymbol{u}] = (\boldsymbol{K}_L^{(e)})_{ba}\boldsymbol{u}_a + (\boldsymbol{K}_L^{(e)})_{bb}\boldsymbol{u}_b.$$

Writing

$$\boldsymbol{Q} = \begin{bmatrix} \boldsymbol{Q}_a \\ \boldsymbol{Q}_b \end{bmatrix}, \text{ where } \boldsymbol{Q}_a = \boldsymbol{0}$$

allows $D\boldsymbol{Q}(\boldsymbol{x})[\boldsymbol{u}]$ to be expressed in terms of the load correction matrix $\mathbf{K}_L^{(e)}$ as

$$D\boldsymbol{Q}(\boldsymbol{x})[\boldsymbol{u}] = \mathbf{K}_L^{(e)}\mathbf{u}^{(e)},$$

where

$$\mathbf{K}_L^{(e)}\mathbf{u}^{(e)} = \frac{Q_b}{l}\begin{bmatrix} \boldsymbol{0} & \boldsymbol{0} \\ (\boldsymbol{K}_L^{(e)})_{ba} & (\boldsymbol{K}_L^{(e)})_{bb} \end{bmatrix}\begin{bmatrix} \boldsymbol{u}_a \\ \boldsymbol{u}_b \end{bmatrix}$$

and

$$(\boldsymbol{K}_L^{(e)})_{ba} = \boldsymbol{n}\otimes\boldsymbol{n} - \boldsymbol{I}_{3\times 3};\ (\boldsymbol{K}_L^{(e)})_{bb} = -(\boldsymbol{K}_L^{(e)})_{ba}.$$

## 3.5 TOTAL POTENTIAL ENERGY

The nodal equilibrium Equation (3.20) was found by simply considering the equilibrium between the internal and external forces acting at a node. Although perfectly adequate in the context of truss structures such a formulation is not suitable when considering the finite element analysis of three-dimensional solids. An approach based on the concept of total potential† energy (TPE) and leading to the principle of virtual work as an alternative expression of equilibrium is extremely convenient when dealing with solid structures comprising assemblages of three-dimensional finite elements often subject to constraints such as incompressibility or other constitutive requirements. This is because equilibrium, either pointwise in the case of a continuum or nodal in the case of the truss, can be expressed in integral form (or summation, for the case of the truss) which enables, when necessary, approximations to equilibrium to be formulated.

It will be shown below that the TPE can be found as a function of the displacements of the solid, and the equilibrium equations are then discovered by considering the *stationary condition* of the TPE. The stationary condition means that if the solid, under load, is subject to a small variation of displacement the change in the TPE is zero, diagrammatically shown in Figure 3.4. The small variation in displacement could be thought of as being real or indeed imaginary; in the latter case it is called a *virtual displacement* $\delta \boldsymbol{u}$.

Although not explained in this text, equilibrium does not imply stability; in Figure 3.4 position (a) is stable while (b) is unstable.

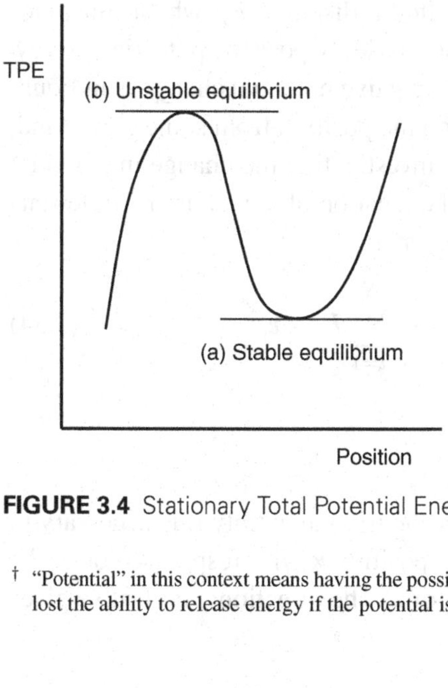


**FIGURE 3.4** Stationary Total Potential Energy.

† "Potential" in this context means having the possibility of releasing energy if the potential is positive or having lost the ability to release energy if the potential is negative.

Total Potential Energy comprises internal *strain energy* and external potential energy due to the applied forces. The strain energy per unit initial volume is given by Equation (3.31a,b,c) as

$$\Psi(\lambda) = \frac{1}{2}E\,(\ln\lambda)^2 \quad ; \lambda = \frac{l}{L} \quad ; l^2 = (\boldsymbol{x}_b - \boldsymbol{x}_a)\cdot(\boldsymbol{x}_b - \boldsymbol{x}_a). \qquad (3.31\text{a,b,c})$$

The total strain energy in the truss system is given by summing the above equation over all members as

$$\Pi_{\text{int}}(\mathbf{x}) = \sum_{e=1}^{m} \Psi(\lambda_e)V_e, \qquad (3.32)$$

where $\mathbf{x}$, given by Equation (3.21a,b,c,d), is the list of current nodal positions and $m$ is the number of elements (truss members).

In addition to the strain energy the external forces acting at all nodes $N$ of the truss also have a potential energy

$$\Pi_{\text{ext}}(\mathbf{x}) = -\sum_{a=1}^{N} \boldsymbol{F}_a \cdot \boldsymbol{x}_a, \qquad (3.33)$$

where $\boldsymbol{F}_a$ and $\boldsymbol{x}_a$ are the force and position of node $a$ respectively. The negative sign is employed above to accommodate the fact that work done against a force increases potential energy, while work done by the force decreases potential energy. As an example consider the weight of a particle of mass $m$ in a gravitational field of magnitude $g$. If this is downward, the corresponding force vector is $\boldsymbol{F} = [0\ 0\ -mg]^T$. Allow the force to move from the origin to $\boldsymbol{x} = [x_1\, x_2\, x_3]^T$, giving the scalar product $-\boldsymbol{F}\cdot\boldsymbol{x} = mg\,x_3$. This is the well-known expression for the potential energy gained by weight $mg$ lifted a distance $x_3$, which illustrates that negative work done (work against a force) yields positive potential energy. Conversely positive work done is manifest as negative potential energy as in Equation (3.33). Potential energy is dependent upon the position from some datum and, insofar as equilibrium will be determined by investigating the change in potential energy as the position of the force changes, the location of the datum is irrelevant.

The Total Potential Energy can now be given as

$$\Pi(\mathbf{x}) = \Pi_{\text{int}}(\mathbf{x}) + \Pi_{\text{ext}}(\mathbf{x}) = \sum_{e=1}^{m} \Psi(\lambda_e)V_e - \sum_{a=1}^{N} \boldsymbol{F}_a \cdot \boldsymbol{x}_a. \qquad (3.34)$$

### 3.5.1 Principle of Virtual Work

To establish the Principle of Virtual Work for the truss assembly it is necessary to consider the directional derivative of $\Pi(\mathbf{x})$ at position $\mathbf{x}$ with respect to (or in the direction of) an imaginary arbitrary variation of $\mathbf{x}$. The directional derivative gives the change in the total potential energy due to a variation (change) of the position $\mathbf{x}$ of the nodes of the truss. It would be usual to denote this nodal variation by $\mathbf{u}$ as in Chapters 1 and 2. However, $\mathbf{u}$ is reserved for use as the Newton–Raphson real increment in $\mathbf{x}$ required to achieve equilibrium. To avoid confusion we will use the notation $\delta\mathbf{u}$ to denote the variation of the nodal positions $\mathbf{x}$ when considering the variation of the total potential energy. It will be shown below that if, at some nodal position $\mathbf{x}$ the variation in the total potential energy is zero (called a stationary condition) then the truss is in equilibrium. From Equation (3.32) the variation of internal total potential energy is

$$D\Pi_{\text{int}}(\mathbf{x})[\delta\mathbf{u}] = \sum_{e=1}^{m} V_e\, D\Psi(\lambda_e)[\delta\boldsymbol{u}_e];\ \delta\boldsymbol{u}_e = \begin{bmatrix} \delta\boldsymbol{u}_a^e \\ \delta\boldsymbol{u}_b^e \end{bmatrix}. \tag{3.35a,b}$$

For truss member $e$ and using Equations (3.31a,b,c) and (3.9),

$$D\Psi(\lambda_e)[\delta\boldsymbol{u}_e] = \frac{d\Psi}{d\ln\lambda_e} D\ln\lambda_e[\delta\boldsymbol{u}^e] \tag{3.36a}$$

$$= \tau_e \frac{1}{\lambda_e} D\lambda_e[\delta\boldsymbol{u}^e] \tag{3.36b}$$

$$= \tau_e \frac{L_e}{l_e}\frac{1}{L_e} Dl_e[\delta\boldsymbol{u}^e] \tag{3.36c}$$

$$= \tau_e \frac{1}{l_e}\boldsymbol{n}\cdot(\delta\boldsymbol{u}_b^e - \delta\boldsymbol{u}_a^e). \tag{3.36d}$$

Recalling the explanation following Equation (3.14), the Kirchhoff stress is $\tau = J\sigma$ and $J = v/V$, enabling the directional derivative of the total strain energy of Equation (3.35a,b) to be written as

$$D\Pi_{\text{int}}(\mathbf{x})[\delta\mathbf{u}] = \sum_{e=1}^{m} J\sigma_e \frac{V_e}{l_e}\boldsymbol{n}\cdot(\delta\boldsymbol{u}_b^e - \delta\boldsymbol{u}_a^e) \tag{3.37a}$$

$$= \sum_{e=1}^{m} \sigma_e \frac{v_e}{V_e}\frac{V_e}{l_e}\boldsymbol{n}\cdot(\delta\boldsymbol{u}_b^e - \delta\boldsymbol{u}_a^e) \tag{3.37b}$$

$$= \sum_{e=1}^{m} \sigma_e a_e \boldsymbol{n}\cdot(\delta\boldsymbol{u}_b^e - \delta\boldsymbol{u}_a^e). \tag{3.37c}$$

From Equation (3.12a,b) the internal truss force is $\boldsymbol{T}_e = \sigma a_e \boldsymbol{n}$; consequently the directional derivation (or variation) of $\Pi_{\text{int}}(\mathbf{x})$ with respect to $\delta\mathbf{u}$ simply becomes

$$D\Pi_{\text{int}}(\mathbf{x})[\delta\mathbf{u}] = \sum_{e=1}^{m} \boldsymbol{T}_e\cdot(\delta\boldsymbol{u}_b^e - \delta\boldsymbol{u}_a^e). \tag{3.38}$$

Observe that $D\Pi_{\text{int}}(\mathbf{x})[\delta\mathbf{u}]$ for a truss member $m$ depends (obviously) on the variation of position at either end of the member. Attention now shifts to the contribution to the direction derivative of the strain energy due to all truss members connected to the variation $\delta\boldsymbol{u}_a$ of the position of a typical node $a$; see Figure 3.3. This gives

$$D\Pi_{\text{int}}(\mathbf{x})[\delta\boldsymbol{u}_a] = \sum_{e=1}^{m} \boldsymbol{T}_e \cdot (\delta\boldsymbol{u}_{\textcircled{b}}^{e} - \delta\boldsymbol{u}_{\textcircled{a}}^{e}) = \sum_{e=1}^{m} (\boldsymbol{T}_{\textcircled{a}}^{e}\delta\boldsymbol{u}_{\textcircled{a}}^{e} + \boldsymbol{T}_{\textcircled{b}}^{e}\delta\boldsymbol{u}_{\textcircled{b}}^{e}) \tag{3.39a}$$

$$= \sum_{\substack{e=1\\ e\ni a}}^{m_a} \boldsymbol{T}_{\textcircled{a}}^{e} \cdot \delta\boldsymbol{u}_a \quad ; \; \boldsymbol{T}_{\textcircled{a}}^{e} = -\boldsymbol{T}_e \; ; \; \boldsymbol{T}_{\textcircled{b}}^{e} = \boldsymbol{T}_e; \tag{3.39b}$$

see Equation (3.20) for an explanation of the notation. The directional derivative of the external potential energy due to a variation $\delta\boldsymbol{u}_a$ of position at node $a$ is

$$D\Pi_{\text{ext}}(\mathbf{x})[\delta\boldsymbol{u}_a] = \boldsymbol{F}_a \cdot \delta\boldsymbol{u}_a. \tag{3.40}$$

Using Equations (3.39) and (3.40), the directional derivative of the total potential energy due to the variation $\delta\mathbf{u}$ of position of all nodes $N$ is equated to zero to yield the stationary condition of the total potential energy as

$$D\Pi(\mathbf{x})[\delta\mathbf{u}] = \sum_{a=1}^{N} \left[ \sum_{\substack{e=1\\ e\ni a}}^{m_a} \boldsymbol{T}_{\textcircled{a}}^{e} - \boldsymbol{F}_a \right] \cdot \delta\boldsymbol{u}_a = 0. \tag{3.41}$$

This important equation implies that the work done by the internal forces equals that done by the external forces during the imposition of variation of displacement $\delta\mathbf{u}$. This variation, while arbitrary, is not zero; hence

$$\sum_{\substack{e=1\\ e\ni a}}^{m_a} \boldsymbol{T}_{\textcircled{a}}^{e} - \boldsymbol{F}_a = 0 \tag{3.42}$$

must be zero. This equation demonstrates that the stationary condition of the total potential energy is equivalent to the nodal equilibrium equation derived from statics given by Equation ( 3.20).

The small variation $\delta\boldsymbol{u}_a$ was introduced above without any precise qualification regarding magnitude. Indeed it transpires from Equation (3.41) that the magnitude of $\delta\boldsymbol{u}_a$ is of no consequence to the statement of equilibrium that is derived from Equation (3.41). Nevertheless, when establishing continuum nonlinear equilibrium equations using energy methods the issue of the magnitude of the variation of strain resulting from the variation of position is often resolved by considering $\delta\boldsymbol{u}_a$ to be the product of a virtual (imaginary) velocity $\delta\boldsymbol{v}_a$ and a time increment $\Delta t$, that is,

$$\delta\boldsymbol{u}_a = \delta\boldsymbol{v}_a \, \Delta t. \tag{3.43}$$

As a consequence of being instantaneous $\delta\boldsymbol{v}_a$ is not restricted to small magnitudes.

Insofar as $\delta\boldsymbol{v}_a$ is quantitatively a virtual displacement occurring during time interval $\Delta t$, the stationary condition given by Equation (3.41) can be expressed as the *Principle of Virtual Work*‡ by employing $\delta\boldsymbol{v}_a$ instead of $\delta\boldsymbol{u}_a$, etc., to give

$$D\Pi(\mathbf{x})[\delta\mathbf{v}] = \sum_{a=1}^{N}\Bigg[\sum_{\substack{e=1\\ e\ni a}}^{m_a}\boldsymbol{T}_a^e - \boldsymbol{F}_a\Bigg]\cdot\delta\boldsymbol{v}_a = 0, \tag{3.44}$$

where the column vector containing all nodal virtual velocities is denoted as $\delta\mathbf{v}$. Observe that the summation terms in the above equation represent virtual work done by the forces $\boldsymbol{T}_a^e$ and $\boldsymbol{F}_a$ displacing by $\delta\boldsymbol{v}_a$. Consequently Equation (3.44) can alternatively be written in terms of the virtual work $\delta W$ as

$$D\Pi(\mathbf{x})[\delta\mathbf{v}] = \delta W(\mathbf{x}, \delta\mathbf{v}). \tag{3.45}$$

Using the Principle of Virtual Work either directly or as a stationary condition of the Total Potential Energy will allow equilibrium to be established for more general mechanical systems such as continuum solids, considered later in this book.

## 3.6 ELASTO-PLASTIC BEHAVIOR

This section extends the previous formulation to include elasto-plastic material behavior of the axial rod. The kinematics of the rod are reconsidered in order to distinguish between elastic and permanent plastic deformation.

Although only axial behavior is considered, it may be helpful to recall what happens when a metal wire is bent. If the bending moment is small, then upon release the wire will return to its original configuration. This is called *elastic behavior*. If the bending moment is re-applied and continues to increase, a maximum moment, dependent upon the yield stress of the material, will be reached and further deformation can be achieved with little or no increase in the moment. This further deformation is *plastic deformation*. If the bending moment is now released for a second time, the wire will partially elastically return to its original shape but not completely; that is, it sustains *permanent plastic deformation*. If upon reaching the maximum moment a further small increase in moment is required to maintain the increasing deformation, then the material exhibits *hardening behavior*, which is a gradual increase in the yield stress. In order to reconstitute the original configuration of the wire, it would be necessary to reverse this process by applying a negative bending moment to produce reverse yielding such that upon removal of the moment the partial elastic deformation would return the wire to its original shape. This scenario obviously occurs for purely axial deformation but the various configurations are less apparent.

‡ Sometimes more precisely called the Principle of Virtual Power.

A description of the overall elasto-plastic deformation will be followed by a reasonably detailed discussion of the elasto-plastic material behavior necessary for the subsequent numerical implementation.

### 3.6.1 Multiplicative Decomposition of the Stretch

Assume that the force in the axial rod with current stretch $\lambda = l/L$ is such that the Kirchhoff stress $\tau$ has reached the maximum capable of being sustained by the material, that is, the yield stress $\tau_y$. This could imply that the stretch is far greater than that necessary to initially attain the yield stress. Now consider that the axial rod is completely unloaded, as discussed above; as a result, it will change length but not fully recover the original length $L$. In fact, it will partially recover elastically to an unloaded length $l_p$ as the load is removed. This will remain as the permanent length of the rod unless further loading takes place. Referring to Figure 3.5, the elastic stretch $\lambda_e$ and the plastic stretch $\lambda_p$ are defined as

$$\lambda_e = \frac{l}{l_p}; \quad \lambda_p = \frac{l_p}{L}; \tag{3.46a,b}$$

from which the *multiplicative decomposition* of the stretch $\lambda$ into elastic and plastic components is

$$\lambda = \lambda_e \lambda_p. \tag{3.47}$$

The unloaded configuration having length $l_p$ can be viewed as a new reference state from which the rod is elastically stretched by $\lambda_e$ to its current configuration $\boldsymbol{x}$ with length $l$. This alternative reference state is often referred to as the inelastic or plastic reference configuration.

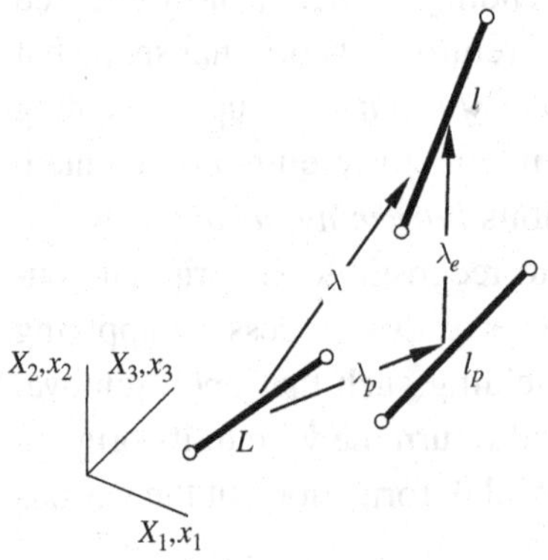


**FIGURE 3.5** Decomposition of the stretch.

By taking the natural logarithm of the stretch given in Equation (3.47) an *additive decomposition* of the logarithmic strain into elastic and plastic components is achieved as

$$\ln \lambda = \ln \lambda_e + \ln \lambda_p. \tag{3.48}$$

This corresponds to the additive decomposition of strain employed in small strain elasto-plastic formulations. For convenience logarithmic strains are notationally defined as

$$\varepsilon = \ln \lambda; \quad \varepsilon_e = \ln \lambda_e; \quad \varepsilon_p = \ln \lambda_p; \quad \varepsilon = \varepsilon_e + \varepsilon_p. \tag{3.49a,b,c,d}$$

Note that in small strain theory Equation $(3.49\text{a,b,c,d})_d$ is taken as the initial kinematic assumption.

### 3.6.2 Rate-Independent Plasticity

The elasto-plastic material is most easily introduced by observing the behavior of the one-dimensional rheological model shown in Figure 3.6(a). The Kirchhoff stress $\tau$, discussed in Section 3.3, is retained as the work conjugate stress to the logarithmic strain $\varepsilon$. The model, as illustrated, is not entirely representative of the behavior but is adequate as an introduction provided the load is monotonically increasing.[§]

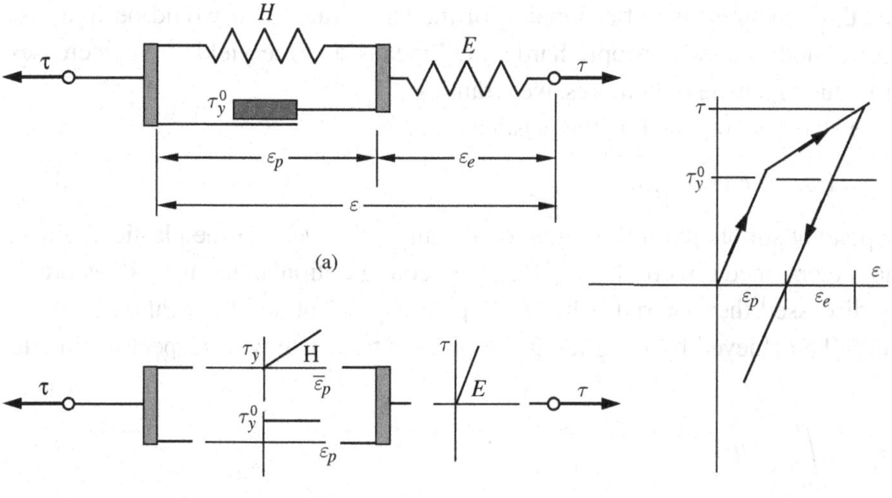


**FIGURE 3.6** One-dimensional elasto-plastic behavior: (a) Rheological model; (b) Component stress–strain behavior; (c) Combined stress–strain behavior.

§ For the model to be complete it is necessary to include the caveat that upon any reversal of load the yield stress $\tau_y$ in the frictional slip device must be re-initialized to that developed by hardening prior to the reversal of the load.

The model has three components (see Figure 3.6(b)): a hyperelastic spring in series with a frictional slip device which is in parallel with an additional linear hardening elastic spring. The hyperelastic spring represents the nonlinear elastic behavior of Section 3.3. The frictional slip device determines the onset of plastic behavior insofar as slip is not initiated until the stress reaches the initial yield stress $\tau_y^0$. Thereafter the hardening spring introduces an increase in the yield stress as a function of post-yielding deformation which is characterized by the amount of slip in the friction device. The total strain $\varepsilon$ is the addition of the elastic recoverable strain $\varepsilon_e$ (in the hyperelastic spring) and the permanent plastic strain $\varepsilon_p$ (in the frictional slip device) as given by the additive decomposition of the logarithmic stretch in Equation (3.48). The stress increases from zero until the initial yield stress $\tau_y^0$ is reached. In the absence of the hardening spring, further increases in strain can only be accommodated by irreversible sliding of the frictional device at constant stress. If the hardening spring is present, further increase in stress can be tolerated as the effective yield stress $\tau_y(\varepsilon_p)$ now becomes a function of the plastic strain $\varepsilon_p$. Upon removal of the stress, elastic strain $\varepsilon_e$ is recovered, leaving a permanent plastic strain $\varepsilon_p$. Observe that the permanent plastic strain is not equal to the additional strain occurring after the first onset of yielding, since this additional strain has an elastic component. Furthermore, note that the stress can never be higher than the current yield stress, but can be lower if unloading occurs. The combined action of the model shown in Figure 3.6(c) is the sum of the individual components shown in Figure 3.6(b). To reiterate, the elasto-plastic behavior described is called *rate-independent finite strain plasticity with isotropic hardening*. "Rate-independent" means that the stress is not a function of the strain rate (as it would be in a viscoplastic model) and "isotropic hardening" means that the yield stress increases equally due to tensile or compressive straining.

The stress is governed by the elastic spring as

$$\tau = E\varepsilon_e = E(\varepsilon - \varepsilon_p). \tag{3.50}$$

The apparent simplicity of this equation disguises the fact that the plastic strain $\varepsilon_p$ cannot be obtained directly from the current configuration alone. Indeed, as previously discussed the material behavior is path-dependent and the evaluation of $\varepsilon_p$ can only be achieved by integrating the plastic strain rate with respect to time to give

$$\varepsilon_p = \int_0^t \dot{\varepsilon}_p \, dt. \tag{3.51}$$

**Remark 3.1**: In the present one-dimensional case it was simple to define the plastic strain $\varepsilon_p$ and, as shown below, its rate $\dot{\varepsilon}_p$. However, in the general three-dimensional context it is not obvious how physically meaningful plastic strain measures and corresponding rates can be defined. Nevertheless, such difficulties can be circumvented by expressing plastic strain rates in terms of rates of elastic strain at constant overall deformation. This can be illustrated for the truss member case as follows. From Equation (3.49a,b,c,d)$_d$ the elastic strain is $\varepsilon_e = \varepsilon - \varepsilon_p$. Differentiating this expression with respect to time with $\varepsilon$ remaining constant gives

$$\left.\frac{d\varepsilon_e}{dt}\right|_{\epsilon=\text{const}} = -\dot{\varepsilon}_p. \tag{3.52}$$

This equation can be expressed in terms of the elastic stretch $\lambda_e$ to yield

$$\dot{\varepsilon}_p = -\left.\frac{d\ln\lambda_e}{dt}\right|_{\ln\lambda=\text{const}} \tag{3.53a}$$

$$= -\frac{1}{\lambda_e}\left.\frac{d\lambda_e}{dt}\right|_{\lambda=\text{const}} \tag{3.53b}$$

$$= -\frac{1}{2}\left.\frac{d\lambda_e^2}{dt}\right|_{\lambda=\text{const}} \lambda_e^{-2}. \tag{3.53c}$$

This last term is directly analogous to equations to be used in Chapter 7; see Remark 7.3.

For the simple one-dimensional case under consideration, the plastic strain rate can obviously be expressed as

$$\dot{\varepsilon}_p = |\dot{\varepsilon}_p| \text{ sign}(\tau); \quad \text{sign}(\tau) = \begin{cases} +1 & \text{if} \quad \tau > 0 \\ -1 & \text{if} \quad \tau < 0 \end{cases}. \tag{3.54}$$

This, it transpires, is the particular one-dimensional equivalent of the general *flow rule*¶

$$\dot{\varepsilon}_p = \dot{\gamma}\frac{\partial f}{\partial \tau}, \tag{3.55}$$

where $\dot{\gamma}$, a proportionality factor, is called the *consistency parameter* or *plastic multiplier*. The function $f(\tau, \bar{\varepsilon}_p)$, which determines the maximum attainable stress, is called the *yield condition*; for the simple model under consideration, this is given by

$$f(\tau, \bar{\varepsilon}_p) = |\tau| - (\tau_y^0 + H\bar{\varepsilon}_p) \leq 0; \quad \bar{\varepsilon}_p \geq 0; \tag{3.56}$$

where $\bar{\varepsilon}_p$ is the *hardening parameter* and $H$ is a property of the material called the *plastic modulus*; see Figure 3.6(b). If $f(\tau, \bar{\varepsilon}_p) < 0$, the deformation is elastic; if $f(\tau, \bar{\varepsilon}_p) = 0$, further deformation may be either elastic or plastic depending upon the subsequent loading.

¶ As discussed in Chapter 7, this is a consequence of the maximum plastic dissipation principle.

At its simplest the hardening parameter $\bar{\varepsilon}_p$ is defined as the accumulated absolute plastic strain occurring over time, that is,

$$\bar{\varepsilon}_p = \int_0^t \dot{\bar{\varepsilon}}_p \, dt; \quad \dot{\bar{\varepsilon}}_p = |\dot{\varepsilon}_p| \,. \tag{3.57a,b}$$

For the one-dimensional case a simple derivation and comparison with Equation (3.54) shows that

$$\frac{\partial f}{\partial \tau} = \text{sign}(\tau); \quad \dot{\gamma} = |\dot{\varepsilon}_p| = \dot{\bar{\varepsilon}}_p. \tag{3.58a,b}$$

> **Remark 3.2**: For our particular plasticity model it so happens that $\dot{\gamma} = |\dot{\varepsilon}_p| = \dot{\bar{\varepsilon}}_p$. However, this is not always the case for other possible choices of yield function. Clearly we could progress our development using either $|\dot{\varepsilon}_p|$ or $\dot{\bar{\varepsilon}}_p$ in place of $\dot{\gamma}$; nevertheless, the use of $\dot{\gamma}$ is preferred in order to align with classical plasticity theory. It is worth emphasizing that it is precisely $\dot{\gamma} = |\dot{\varepsilon}_p|$ that is needed for the evaluation of the plastic strain rate given by Equation (3.54).

The term $(\tau_y^0 + H\bar{\varepsilon}_p)$ in the yield condition of Equation (3.56) is the effective yield stress dependent upon the accumulated plastic strain $\bar{\varepsilon}_p$. As explained above, plastic straining, that is, $\dot{\varepsilon}_p \neq 0$, will only occur when the stress equals the yield stress, that is, $f(\tau, \bar{\varepsilon}_p) = 0$. If $f(\tau, \bar{\varepsilon}_p) < 0$ then the response is elastic and both $\dot{\gamma}$ and $\dot{\varepsilon}_p$ are zero. These statements can be combined into the *loading/unloading conditions*** as

$$\dot{\gamma} \geq 0; \quad f(\tau, \bar{\varepsilon}_p) \leq 0; \quad \dot{\gamma} f(\tau, \bar{\varepsilon}_p) = 0. \tag{3.59a,b,c}$$

The first two of the above equations are self-evident, while the last implies that if $\dot{\gamma} > 0$ then $f(\tau, \bar{\varepsilon}_p) = 0$ and if $f(\tau, \bar{\varepsilon}_p) < 0$ then $\dot{\gamma} = 0$. Put simply, the loading/unloading conditions determine whether the stress is "at or less than" the current effective yield stress or, alternatively, whether the behavior is plastic or elastic. If the stress is such that $f(\tau, \bar{\varepsilon}_p) = 0$ then $\dot{\varepsilon}_p \neq 0$ and must be such that $f(\tau, \bar{\varepsilon}_p)$ remains equal to zero, that is, $\frac{d}{dt} f(\tau, \bar{\varepsilon}_p) = 0$; in other words, the stress must remain at the yield stress value, taking into account any increase due to hardening. This requirement is known as the *consistency condition*,†† expressed as

$$\dot{\gamma} \dot{f}(\tau, \bar{\varepsilon}_p) = 0. \tag{3.60}$$

Expanding $\dot{f}(\tau, \bar{\varepsilon}_p)$ with the help of Equations (3.57a,b) and (3.58a,b) gives

$$\dot{f}(\tau, \bar{\varepsilon}_p) = \frac{\partial f}{\partial \tau} \dot{\tau} + \frac{\partial f}{\partial \bar{\varepsilon}_p} \dot{\gamma} = 0; \quad \text{if} \quad f(\tau, \bar{\varepsilon}_p) = 0. \tag{3.61}$$

** Often called the Kuhn–Tucker conditions in recognition of affinities with mathematical programming.

†† A more descriptive, but less used, term is *persistency condition*, in which case $\dot{\gamma}$ would be called the persistency parameter.

With the help of this equation it is now possible to evaluate $\dot{\gamma}$ and hence, from Equations (3.54) and (3.58a,b), $\dot{\bar{\varepsilon}}_p$, enabling the determination of the plastic strain $\varepsilon_p$ from Equation (3.51). Using the yield condition given by Equation (3.56), the derivatives of the yield function in Equation (3.61) are found as

$$\frac{\partial f}{\partial \tau} = \text{sign}(\tau); \quad \frac{\partial f}{\partial \bar{\varepsilon}_p} = -H. \tag{3.62a,b}$$

Substituting these expressions into Equation (3.61) and using the time derivative of Equation (3.50) as $\dot{\tau} = E(\dot{\varepsilon} - \dot{\varepsilon}_p)$ gives, after some simple manipulation, the plastic multiplier $\dot{\gamma}$ as

$$\dot{\gamma} = \left(\frac{E}{E+H}\right) \text{sign}(\tau)\,\dot{\varepsilon}; \quad \text{if} \quad f(\tau, \bar{\varepsilon}_p) = 0. \tag{3.63}$$

This enables the plastic strain rate of Equation (3.54) to be found as

$$\dot{\varepsilon}_p = \left(\frac{E}{E+H}\right) \dot{\varepsilon}; \quad \text{if} \quad f(\tau, \bar{\varepsilon}_p) = 0. \tag{3.64}$$

The above rate could now be integrated in time according to Equation (3.51) to provide the plastic strain $\varepsilon_p$ and thus enable the Kirchhoff stress $\tau$ to be obtained from Equation (3.50) as $\tau = E(\varepsilon - \varepsilon_p)$. Unfortunately, in a computational setting the time integration implied in Equation (3.51) can only be performed approximately from a finite sequence of values determined at different time steps. The inaccuracies in this process will lead to a stress that may not satisfy the yield condition. This can be resolved by taking into account the incremental nature of the computational process to derive an algorithm that ensures consistency with the yield condition at each incremental time step. Such a procedure is known as a *return-mapping algorithm*, which will be considered in the following sections.

### 3.6.3 Incremental Kinematics

An incremental framework suitable for elasto-plastic computation is established by considering discrete time steps $\Delta t$ during which the rod moves from position $n$ at time $t$ to position $n+1$ at time $t+\Delta t$. The lengths of the rod at times $t$ and $t+\Delta t$ are $l_n$ and $l_{n+1}$ respectively; see Figure 3.7.

Furthermore, at each incremental step an unloaded configuration of the rod can be defined at times $t$ and $t+\Delta t$ as $l_{p,n}$ and $l_{p,n+1}$ respectively. Of course, if no plastic straining takes place during the increment then $l_{p,n+1} = l_{p,n}$.

In the following development it is assumed that the deformation is strain-driven; that is, forces are acting that produce a known stretch $\lambda_{n+1}$ at time $t+\Delta t$. The problem is then to determine whether this stretch results in the rod developing an increment in plastic strain and, if so, what stress satisfies the yield condition

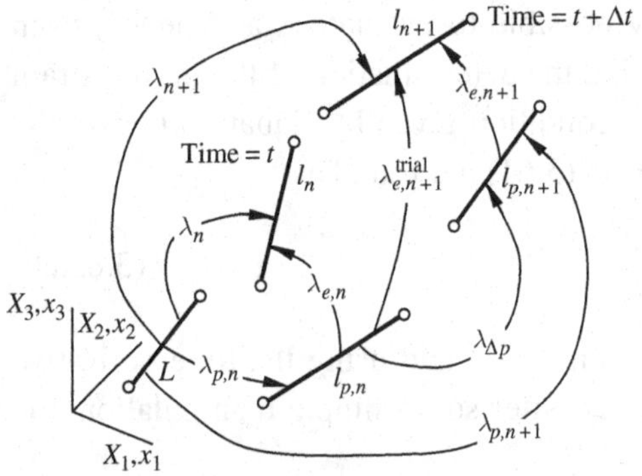


**FIGURE 3.7** Incremental kinematics.

given by Equation (3.56). It then remains to determine whether this stress, together with all other stresses in rods making up the truss, is in equilibrium with the current applied forces. For a given load the Newton–Raphson iteration adjusts the deformation pattern so that the constitutive equations and equilibrium are satisfied.

Assuming, as a trial in lieu of better information, that the deformation over time step $\Delta t$ is purely elastic, the stretch $\lambda_{n+1}$ can be expressed as a multiplicative decomposition involving any already existing permanent length $l_{p,n}$ and a so-called *trial elastic stretch* $\lambda^{\text{trial}}_{e,n+1}$ (see Figure 3.7) as

$$\lambda_{n+1} = \lambda^{\text{trial}}_{e,n+1}\,\lambda_{p,n}. \tag{3.65}$$

Note that if no plastic strain has occurred then $\lambda_{p,n} = 1$. Insofar as plastic deformation is incompressible, the volume ratio $J$ given by Equation (3.4a,b) is always calculated using the elastic component of the stretch as

$$J = \left(\lambda^{\text{trial}}_{e,n+1}\right)^{(1-2\nu)}. \tag{3.66}$$

From Equation (3.65) the trial elastic stretch is

$$\lambda^{\text{trial}}_{e,n+1} = \lambda_{n+1}\,\lambda^{-1}_{p,n} = \frac{l_{n+1}}{l_{p,n}}, \tag{3.67}$$

enabling an additive decomposition of the trial elastic logarithmic strain to be established as

$$\varepsilon^{\text{trial}}_{e,n+1} = \varepsilon_{n+1} - \varepsilon_{p,n}; \quad \varepsilon_{n+1} = \ln\frac{l_{n+1}}{L}; \quad \varepsilon_{p,n} = \ln\frac{l_{p,n}}{L}. \tag{3.68a,b,c}$$

A trial state of stress $\tau^{\text{trial}}_{n+1}$ will be calculated in Section 3.6.5 using the above trial elastic strain. If this trial stress satisfies the yield condition, that is, $f(\tau^{\text{trial}}_{n+1}, \bar{\varepsilon}_{p,n}) \leq 0$, given in Equation (3.56), no further plastic strain will occur during the increment. However, should this not be the case, that is, $f(\tau^{\text{trial}}_{n+1}, \bar{\varepsilon}_{p,n}) > 0$, then plastic strain will take place during increment $n$ to $n + 1$ and a different unloaded length $l_{p,n+1}$ will emerge; see Figure 3.7. Comparing unloaded lengths at steps $n$ and $n+1$ enables an incremental plastic stretch $\lambda_{\Delta p}$ and its corresponding logarithmic strain $\Delta\varepsilon_p$ to be defined as

$$\lambda_{\Delta p} = \frac{l_{p,n+1}}{l_{p,n}}; \quad \Delta\varepsilon_p = \ln \lambda_{\Delta p} = \ln l_{p,n+1} - \ln l_{p,n}. \tag{3.69a,b}$$

The incremental plastic stretch permits an alternative description of the trial elastic stretch as

$$\lambda_{e,n+1}^{\text{trial}} = \lambda_{e,n+1}\lambda_{\Delta p}; \quad \lambda_{e,n+1} = \frac{l_{n+1}}{l_{p,n+1}}. \tag{3.70a,b}$$

Consequently, an alternative additive decomposition of the trial elastic strain can be expressed in terms of an increment of permanent strain $\Delta\varepsilon_p$ as

$$\varepsilon_{e,n+1}^{\text{trial}} = \varepsilon_{e,n+1} + \Delta\varepsilon_p; \quad \varepsilon_{e,n+1} = \ln \frac{l_{n+1}}{l_{p,n+1}}. \tag{3.71a,b}$$

Taking logarithms in the above equation provides a useful alternative equation for the incremental plastic strain as

$$\Delta\varepsilon_p = \ln \frac{\lambda_{e,n+1}^{\text{trial}}}{\lambda_{e,n+1}} \tag{3.72a}$$

$$= -\left(\ln \lambda_{e,n+1} - \ln \lambda_{e,n+1}^{\text{trial}}\right) \tag{3.72b}$$

$$= -\left(\varepsilon_{e,n+1} - \varepsilon_{e,n+1}^{\text{trial}}\right). \tag{3.72c}$$

Whereas $\Delta\varepsilon_p = \ln \lambda_{\Delta p}$ is a direct measure of the incremental plastic strain, the alternative expression for $\Delta\varepsilon_p$ given by Equations (3.72b) and (3.72c) is an indirect evaluation which is more suited to the development of the continuum formulation of elasto-plastic behavior to be given in Chapter 7. The equivalence between the two approaches can be clearly see in Figure 3.7. Furthermore note that Equations (3.72b) and (3.72c) are the incremental equivalent of Equation $(3.69\text{a,b})_b$.

How the increment of permanent strain $\Delta\varepsilon_p$ is determined depends upon the flow rule given in Equation (3.54) used in conjunction with the time integration scheme which follows.

### 3.6.4 Time Integration

If plastic strain occurs, then the increment of permanent plastic strain $\Delta\varepsilon_p$ is obtained by integrating in time the rate of plastic strain $\dot{\varepsilon}_p$ over the step as§§

§§ Observe that if $\dot{\varepsilon}_p = \frac{d}{dt}\left(\ln \frac{l_p}{L}\right) = \frac{\dot{l}_p}{l_p}$ is substituted into Equation (3.73a,b), the definition of $\Delta\varepsilon_p$ given in Equation (3.71a,b) is recovered.

$$\Delta\varepsilon_p = \int_{t_n}^{t_{n+1}} \dot{\varepsilon}_p \, dt; \quad \varepsilon_{p,n+1} = \varepsilon_{p,n} + \Delta\varepsilon_p. \tag{3.73a,b}$$

Within the framework of an incremental formulation it is reasonable to approximate the above integral using what is known as the *backward Euler rule*, whereby the integrand $\dot{\varepsilon}_p$ is sampled only at the end of the time step. This, together with the flow rule of Equation (3.54), enables the increment of permanent plastic strain to be calculated as

$$\Delta\varepsilon_p \simeq \dot{\varepsilon}_{p,n+1}\Delta t \tag{3.74a}$$

$$= \dot{\gamma}_{n+1}\,\mathrm{sign}(\tau_{n+1})\Delta t \tag{3.74b}$$

$$= \Delta\gamma\,\mathrm{sign}(\tau_{n+1}); \quad \Delta\gamma = \dot{\gamma}_{n+1}\Delta t, \tag{3.74c}$$

where $\Delta\gamma$ is called the *incremental plastic multiplier*. Note that sampling $\dot{\varepsilon}_p$ at the end of the increment is the simplest way to ensure that at the end of the increment both the flow rule and the yield criterion are satisfied. In a similar manner, the increment in the hardening parameter between steps $n$ and $n+1$ is

$$\bar{\varepsilon}_{p,n+1} = \bar{\varepsilon}_{p,n} + \Delta\gamma, \tag{3.75a}$$

where, from Equations (3.58a,b)$_b$ and (3.74c),

$$\Delta\gamma = \dot{\gamma}_{n+1}\Delta t = \dot{\bar{\varepsilon}}_{p,n+1}\Delta t. \tag{3.75b}$$

With the basic incremental kinematic description and integration scheme in place, consideration is now focused on the determination of the stress.

### 3.6.5 Stress Update and Return Mapping

The stress in the rod at position $n+1$ is given by Equation (3.50) as

$$\tau_{n+1} = E\left(\varepsilon_{n+1} - \varepsilon_{p,n+1}\right). \tag{3.76}$$

This can be rewritten using Equations (3.68a,b,c) and (3.73a,b) to involve the *trial elastic stress* $\tau_{n+1}^{\text{trial}}$ in the following manner:

$$\tau_{n+1} = \tau_{n+1}^{\text{trial}} - E\Delta\varepsilon_p; \quad \tau_{n+1}^{\text{trial}} = E\,\varepsilon_{e,n+1}^{\text{trial}}. \tag{3.77a,b}$$

In the presence of additional permanent deformation occurring during time step $\Delta t$, the term $(-E\Delta\varepsilon_p)$ can be regarded as a correction to the trial stress $\tau_{n+1}^{\text{trial}}$. Whether the correction applies depends upon the loading/unloading condition given by Equation (3.59a,b,c). If the additional permanent deformation has occurred, then $\Delta\varepsilon_p$ will be determined by the incremental form of the flow rule given by Equation (3.74) which requires the evaluation of the plastic multiplier increment $\Delta\gamma$.

Whether or not plastic deformation occurs is determined by substituting the trial elastic stress into the yield criterion of Equation (3.56) to give

$$f\left(\tau_{n+1}^{\text{trial}}, \bar{\varepsilon}_{p,n}\right) = \left|\tau_{n+1}^{\text{trial}}\right| - (\tau_y^0 + H\bar{\varepsilon}_{p,n}), \tag{3.78}$$

where, because of the assumption of no plastic deformation, $\bar{\varepsilon}_{p,n}$ is the hardening parameter at position $n$ at time $t$.

Elastic or elasto-plastic behavior can now be assessed by invoking the loading/unloading condition given by Equation (3.59a,b,c) as follows:

$$\textbf{Elastic}: \quad f\left(\tau_{n+1}^{\text{trial}}, \bar{\varepsilon}_{p,n}\right) \leq 0 \quad \Longrightarrow \begin{cases} \tau_{n+1} = \tau_{n+1}^{\text{trial}} \\ \varepsilon_{p,n+1} = \varepsilon_{p,n} \\ \bar{\varepsilon}_{p,n+1} = \bar{\varepsilon}_{p,n} \end{cases} \tag{3.79}$$

$$\textbf{Plastic}: \quad f\left(\tau_{n+1}^{\text{trial}}, \bar{\varepsilon}_{p,n}\right) > 0 \quad \Longrightarrow \begin{cases} \Delta\gamma > 0 \\ \Delta\varepsilon_p = \Delta\gamma\,\text{sign}(\tau_{n+1}) \\ \varepsilon_{p,n+1} = \varepsilon_{p,n} + \Delta\varepsilon_p \\ \bar{\varepsilon}_{p,n+1} = \bar{\varepsilon}_{p,n} + \Delta\gamma \\ \tau_{n+1} = \tau_{n+1}^{\text{trial}} - E\Delta\varepsilon_p \end{cases} \tag{3.80}$$

In the event of plastic behavior the various updates implied in Equation (3.80) depend upon the determination of the incremental plastic multiplier $\Delta\gamma$ and $\text{sign}(\tau_{n+1})$ (see Equation (3.74)), such that the yield criterion $f(\tau_{n+1}, \bar{\varepsilon}_{p,n+1}) = 0$.

The incremental plastic multiplier is found by following the procedure adopted by Simo and Hughes (1997), with the Kirchhoff stress $\tau$ substituted for the Cauchy stress $\sigma$. From Equations (3.74) and (3.77a,b),

$$\tau_{n+1} = \tau_{n+1}^{\text{trial}} - E\Delta\gamma\,\text{sign}(\tau_{n+1}). \tag{3.81}$$

Rewriting the above equation to reveal the "sign" operator and gathering terms gives

$$(|\tau_{n+1}| + E\Delta\gamma)\,\text{sign}(\tau_{n+1}) = \left|\tau_{n+1}^{\text{trial}}\right|\,\text{sign}\left(\tau_{n+1}^{\text{trial}}\right). \tag{3.82}$$

Enforcing $f(\tau_{n+1}, \bar{\varepsilon}_{p,n+1}) = 0$ implies $\Delta\gamma > 0$; hence the terms multiplying the "signs" are both positive. Consequently,

$$\text{sign}(\tau_{n+1}) = \text{sign}\left(\tau_{n+1}^{\text{trial}}\right) \quad \text{and} \quad |\tau_{n+1}| = \left|\tau_{n+1}^{\text{trial}}\right| - E\Delta\gamma. \tag{3.83a,b}$$

Recall that sign($\tau_{n+1}$) is needed in Equation (3.74) for the evaluation of the plastic strain occurring during the increment. The above equation enables the determination of sign($\tau_{n+1}$) even before consideration of plasticity during the time step.

The yield criterion at position $n+1$ is now written using the hardening parameter update $\Delta\bar{\varepsilon}_p = \Delta\gamma$ from Equation (3.75a) and noting the trial yield criterion from Equation (3.78), to give

$$f(\tau_{n+1}, \bar{\varepsilon}_{p,n+1}) = |\tau_{n+1}| - (\tau_y^0 + H\bar{\varepsilon}_{p,n+1}) \tag{3.84a}$$

$$= \left|\tau_{n+1}^{\text{trial}}\right| - E\Delta\gamma - (\tau_y^0 + H\bar{\varepsilon}_{p,n}) - H\Delta\gamma \tag{3.84b}$$

$$= \left|\tau_{n+1}^{\text{trial}}\right| - (\tau_y^0 + H\bar{\varepsilon}_{p,n}) - (E+H)\Delta\gamma \tag{3.84c}$$

$$= f_{n+1}^{\text{trial}} - (E+H)\Delta\gamma. \tag{3.84d}$$

Since $f(\tau_{n+1}, \bar{\varepsilon}_{p,n+1}) = 0$, the incremental plastic multiplier can be found as

$$\Delta\gamma = \frac{f_{n+1}^{\text{trial}}}{(E+H)}. \tag{3.85}$$

It is now possible to accomplish the various updates given in Equation (3.80), which are frequently collected under the title of the return-mapping algorithm, summarized in Box 3.1. Observe that the incremental plastic multiplier is, somewhat surprisingly, only a function of the trial elastic stress and the hardening parameter at position $n$. The return-mapping algorithm is illustrated in Figure 3.8.

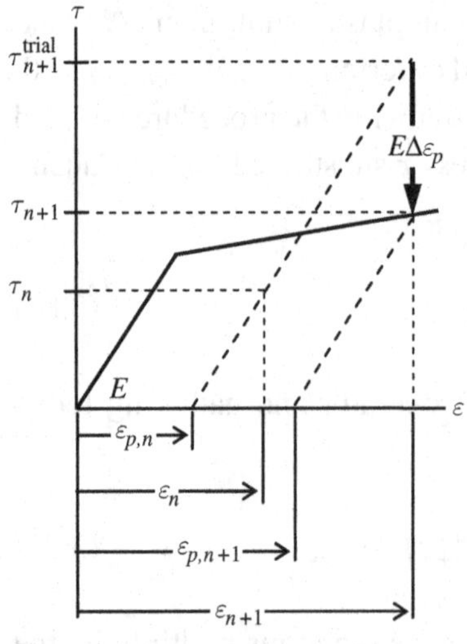


**FIGURE 3.8** Return mapping.

**BOX 3.1: Return-Mapping Algorithm**

**If** $f\left(\tau_{n+1}^{\text{trial}}, \bar{\varepsilon}_{p,n}\right) \leq 0$

$\Delta\gamma = 0$

**Else**

$$\Delta\gamma = \frac{f_{n+1}^{\text{trial}}}{(E+H)}$$

**Endif**

$$\Delta\varepsilon_p = \Delta\gamma\,\text{sign}(\tau_{n+1}^{\text{trial}})$$

$$\tau_{n+1} = \tau_{n+1}^{\text{trial}} - E\Delta\varepsilon_p$$

$$\varepsilon_{p,n+1} = \varepsilon_{p,n} + \Delta\varepsilon_p$$

$$\bar{\varepsilon}_{p,n+1} = \bar{\varepsilon}_{p,n} + \Delta\gamma$$

### 3.6.6 Algorithmic Tangent Modulus

As discussed in Section 3.4, the Newton–Raphson solution procedure required to solve the nonlinear equilibrium equations involves the calculation of the tangent stiffness matrix, which contains (see Equation (3.26)) the tangent modulus term $d\tau/d\varepsilon$. This gradient term could be derived from a continuation of Section 3.6.2; see for example Simo and Hughes (1997). However, as previously explained, the tangent modulus derived from incremental considerations is generally not the same as that obtained from the rate equations. The reason is that, generally, the incremental change in stress imposed by the chosen return-mapping algorithm is different from the continuous change in stress implied by the rate equations. Unfortunately, due to the very simple nature of one-dimensional plasticity, this difference is not apparent in the present situation. Nevertheless, in preparation for the three-dimensional formulations to be given in Chapter 7, the so-called *algorithmic tangent modulus* will be derived.

From the return-mapping algorithm given in Box 3.1 the stress update is given as

$$\tau_{n+1} = \tau_{n+1}^{\text{trial}} - E\Delta\gamma\,\text{sign}(\tau_{n+1}^{\text{trial}}), \tag{3.86}$$

where from Equation (3.76) the stress is $\tau_{n+1} = E\left(\varepsilon_{n+1} - \varepsilon_{p,n+1}\right)$ and from Equation (3.78) the incremental plastic multiplier is $\Delta\gamma = f_{n+1}^{\text{trial}}/(E+H)$. Consequently, since $f_{n+1}^{\text{trial}}$ is a function of $\tau_{n+1}$, both terms are functions of the strain $\varepsilon_{n+1}$. Differentiating $\tau_{n+1}$ with respect to the strain $\varepsilon_{n+1}$ gives

$$\frac{d\,\tau_{n+1}}{d\,\varepsilon_{n+1}} = \frac{d\,\tau^{\rm trial}_{n+1}}{d\,\varepsilon_{n+1}} - E\frac{d\Delta\gamma}{d\,\varepsilon_{n+1}}\,{\rm sign}(\tau^{\rm trial}_{n+1}) - E\Delta\gamma\frac{d\,{\rm sign}(\tau^{\rm trial}_{n+1})}{d\,\varepsilon_{n+1}} \tag{3.87a}$$

$$= E - \frac{E\,{\rm sign}(\tau^{\rm trial}_{n+1})}{E+H}\frac{df^{\rm trial}_{n+1}}{d\varepsilon_{n+1}} - E\Delta\gamma\frac{d\,{\rm sign}(\tau^{\rm trial}_{n+1})}{d\,\varepsilon_{n+1}} \tag{3.87b}$$

$$= E - \frac{E\,{\rm sign}(\tau^{\rm trial}_{n+1})}{E+H}\frac{d\left|\tau^{\rm trial}_{n+1}\right|}{d\varepsilon_{n+1}} - E\Delta\gamma\frac{d\,{\rm sign}(\tau^{\rm trial}_{n+1})}{d\,\varepsilon_{n+1}} \tag{3.87c}$$

$$= E - \frac{E\,{\rm sign}(\tau^{\rm trial}_{n+1})}{E+H}\frac{d\left|\tau^{\rm trial}_{n+1}\right|}{d\tau^{\rm trial}_{n+1}}\frac{d\tau^{\rm trial}_{n+1}}{d\varepsilon_{n+1}} - E\Delta\gamma\frac{d\,{\rm sign}(\tau^{\rm trial}_{n+1})}{d\,\varepsilon_{n+1}} \tag{3.87d}$$

$$= E - \frac{E^2\left({\rm sign}(\tau^{\rm trial}_{n+1})\right)^2}{E+H} - E\Delta\gamma\frac{d\,{\rm sign}(\tau^{\rm trial}_{n+1})}{d\,\varepsilon_{n+1}}. \tag{3.87e}$$

An examination of the sign($\tau$) function given in Equation (3.54) reveals that

$$\left[{\rm sign}(\tau^{\rm trial}_{n+1})\right]^2 = 1 \quad \text{and} \quad \frac{d\,{\rm sign}(\tau^{\rm trial}_{n+1})}{d\,\tau^{\rm trial}_{n+1}} = 0 \quad \left(\tau^{\rm trial}_{n+1} \neq 0\right), \tag{3.88a,b}$$

and therefore the last differential in Equation (3.87) vanishes since

$$\frac{d\,{\rm sign}(\tau^{\rm trial}_{n+1})}{d\,\varepsilon_{n+1}} = \frac{d\,{\rm sign}(\tau^{\rm trial}_{n+1})}{d\,\tau^{\rm trial}_{n+1}}\frac{d\,\tau^{\rm trial}_{n+1}}{d\,\varepsilon_{n+1}} = 0. \tag{3.89}$$

Consequently, the algorithmic tangent modulus emerges as

$$\frac{d\,\tau_{n+1}}{d\,\varepsilon_{n+1}} = \frac{EH}{E+H}. \tag{3.90}$$

When plasticity occurs, this replaces the elastic tangent modulus, $d\tau/d\varepsilon = E$, used in Section 3.4.3.

### 3.6.7 Revised Newton–Raphson Procedure

Satisfaction of either the elastic or the plastic constitutive equation for each axial rod in a loaded truss does not guarantee that the global equilibrium condition given by Equation (3.21a,b,c,d) is satisfied. Consequently, *for given external nodal forces* the nodal positions of the truss need to be adjusted so that both the constitutive equations and the global equilibrium equations are simultaneously satisfied. This is precisely what was achieved in the simple single truss member example discussed in Chapter 1, for which a Newton–Raphson algorithm and a simple computer program were presented in Boxes 1.1 and 1.2 respectively. This Newton–Raphson algorithm is now revised to include elasto-plastic considerations, as shown in Box 3.2.

**BOX 3.2: Elasto-Plastic Newton–Raphson Algorithm**

- INPUT geometry, material properties, and solution parameters
- INITIALIZE $\mathbf{F} = \mathbf{0}$, $\mathbf{x} = \mathbf{X}$ (initial geometry), $\mathbf{R} = \mathbf{0}$
- FIND initial $\mathbf{K}$ (typically assembled from (3.28a,b))
- LOOP over load increments
  - FIND $\Delta\mathbf{F}$ (establish the load increment)
  - SET $\mathbf{F} = \mathbf{F} + \Delta\mathbf{F}$
  - SET $\mathbf{R} = \mathbf{R} - \Delta\mathbf{F}$
  - DO WHILE ($\|\mathbf{R}\|/\|\mathbf{F}\| >$ tolerance)
    - SOLVE $\mathbf{K}\mathbf{u} = -\mathbf{R}$ (typically (3.22a,b,c))
    - UPDATE $\mathbf{x} = \mathbf{x} + \mathbf{u}$ (typically (3.22a,b,c))
      - LOADING/UNLOADING given in Box 3.1
      - UPDATE (if necessary)
        plastic strain & hardening parameter
      - FIND ELEMENT stress and tangent modulus
    - FIND $\mathbf{T}$ and $\mathbf{K}$
      (typically assembled from (3.20) and (3.28a,b))
    - FIND $\mathbf{R} = \mathbf{T} - \mathbf{F}$ (typically (3.21a,b,c,d))
  - ENDDO
- ENDLOOP

## 3.7 EXAMPLES

Two examples illustrate the geometric and material nonlinear structural behavior that can be analyzed using the above formulations. The first is a single inclined axial rod which shows both elastic and elasto-plastic deformation, clearly demonstrating the kinematic discussions of Section 3.6.1. The second example is a two-dimensional trussed frame exhibiting, in the elastic case, highly nonlinear load deflection behavior, and in the elasto-plastic case, the clear emergence of plastic hinges associated with the collapse analysis of frames. In both examples the complete force displacement curve (also known as the *equilibrium path*) requires the algorithm given in Box 3.1 to be augmented by the arc length technique, which will be considered fully in Chapter 9. Although both examples are two-dimensional, the FLagSHyP program can also analyze three-dimensional truss structures.

### 3.7.1 Inclined Axial Rod

The inclined axial rod shown in Figure 3.9(a) is loaded with a downward vertical force $F$, and while the Young's modulus is realistic, $E = 210\,\text{kN/mm}^2$, the yield

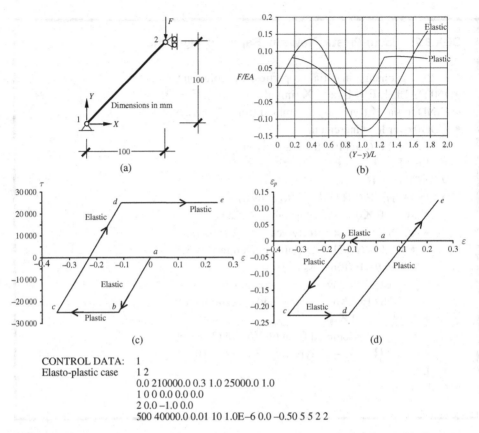

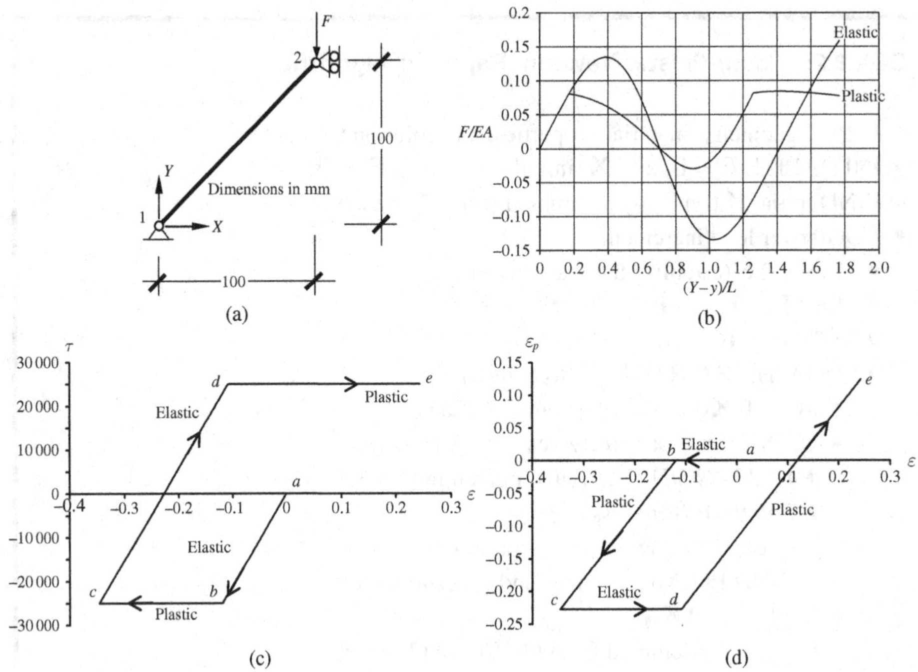


CONTROL DATA: 1
Elasto-plastic case 1 2
0.0 210000.0 0.3 1.0 25000.0 1.0
1 0 0 0.0 0.0 0.0
2 0.0 −1.0 0.0
500 40000.0 0.01 10 1.0E−6 0.0 −0.50 5 5 2 2

**FIGURE 3.9** Large deflection elasto-plastic behavior of a single axial rod: (a) Geometry; (b) Elastic and elasto-plastic force deflection behavior; (c) Kirchhoff stress versus strain; (d) Plastic strain versus strain.

stress is set artificially high at 25 kN/mm$^2$ to permit some degree of geometric nonlinearity to occur prior to the onset of plasticity. The equilibrium path contains positions of stable and unstable equilibrium, the latter for the elastic case being between the limit points. The final deformation is about twice the initial height of the rod, causing the rod to initially undergo compression followed by the development of tension after a vertical deformation of 200 mm. Figure 3.9(b) shows the resulting elastic and elasto-plastic force deflection curves, while Figures 3.9(c,d) refer to the elasto-plastic analysis and show the development of the stress and the plastic strain respectively. Observe that during elastic deformation *a*–*b* and *c*–*d* the corresponding plastic strain is unchanged; however, when the stress is maintained at the effective yield stress *b*–*c* and *d*–*e* the plastic strain changes.

### 3.7.2 Trussed Frame

This example (see Figure 3.10) shows a small strain (except in the region of severe plastic deformation) large deflection of a trussed frame. The frame is loaded

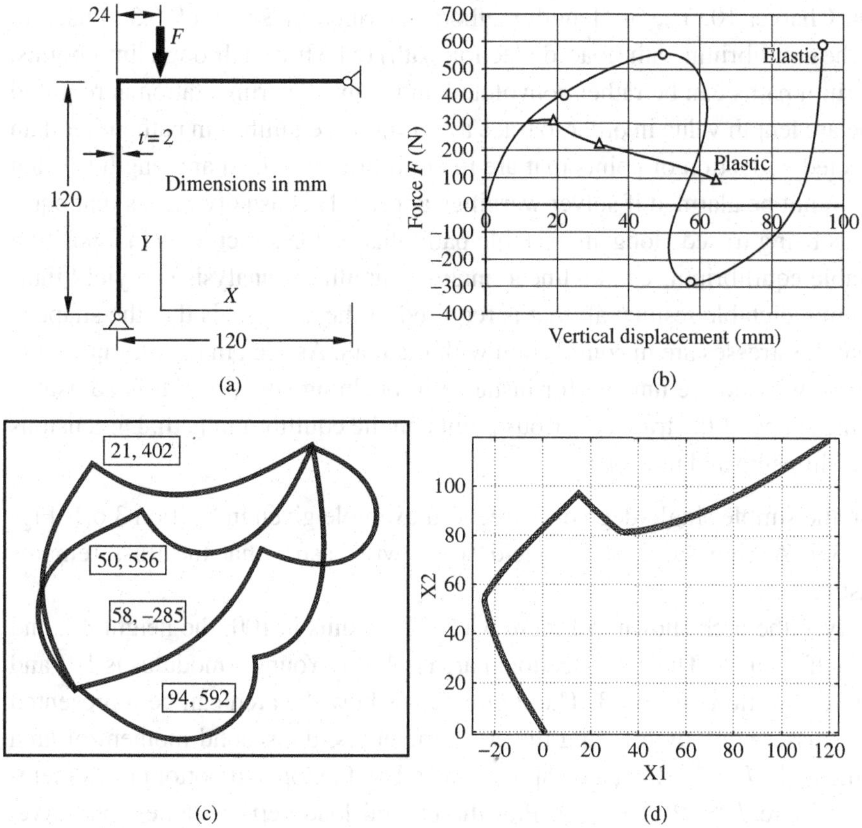


CONTROL DATA: 1
Elasto-plastic case 1 2 0.0 210000.0 0.3 1.0 2500.0 1.0
1 0 0 0.0 0.0 0.0
144 0.0 −100.0 0.0
500 25000.0 0.1 100 1.0E−6 0.0 −10.0 5 5 144 2

**FIGURE 3.10** Large deflection elastic behavior of a trussed frame: (a) Geometry; (b) Force $F$ vertical deflection (under force) behavior; (c) Elastic deformed shape at [u2,F]; (d) Elasto-plastic deformed shape at [32, 222]. The paired quantities in brackets or boxes refer to the vertical displacement u2 and corresponding vertical force F.

with a nominal downward vertical force of 100 N at node 144 having coordinates (24.0, 120.0, 0.0); see Figure 3.10(a). The elastic equilibrium path shown in Figure 3.10(b) is an example of *snap-back behavior*, where both load and deflection reduce to maintain an equilibrium configuration. Figure 3.10(b) clearly reveals the reduction in maximum load caused by the inclusion of plasticity. Figures 3.10(c,d) show various elastic and elasto-plastic equilibrium configurations.

## Exercises

These structures can be analyzed using the truss facility in the program FLagSHyP, which can be downloaded from `www.flagshyp.com`. The user instructions are given in Chapter 10. The arc length method described in Section 9.6.3 is used to enable the equilibrium path (load-deflection path) to be traced through limit points. Equilibrium paths can be rather convoluted and some experimentation is required with the arc length value in order to trace a continuous equilibrium path rather than a connected collection of points that are in equilibrium; a fixed arc length is often better. Do not be alarmed if solver warnings appear. This usually means that equilibrium is being traced along an unstable path; that is, the structure is in a position of unstable equilibrium. Unlike linear analysis, nonlinear analysis can yield intuitively unpredictable results; all that is required of the analysis is that the shape is such that the stresses are in equilibrium with the load. As seen in the examples, the shape may well not be unique, for instance in buckling situations. It is advisable to plot the shape of the truss at various points on the equilibrium path. Dimensions are in millimeters and newtons.

1. Run the simple single-degree-of-freedom example given in Section 3.6.1, Figure 3.9. A high value of the yield stress will ensure that the truss remains elastic.
2. Analyze the arch shown in Figure 3.11. The radius is 100, the height 40, and the half span 80. The cross-sectional area is $1 \times 1$. Young's modulus is $10^7$ and Poisson's ratio is $\nu = 0.3$. The figure shows how the arch can be represented as a truss where, by ignoring the cross members, the second moment of area of the arch, $I = 1/12$, can be approximated by the top and bottom truss members, where $I = 2(A(t/2)^2)$. Plot the central load vertical deflection curve. Slight imperfections in the symmetry of the geometry may cause unsymmetric deformations; otherwise these can be initiated by a very small horizontal load being placed with the vertical load.

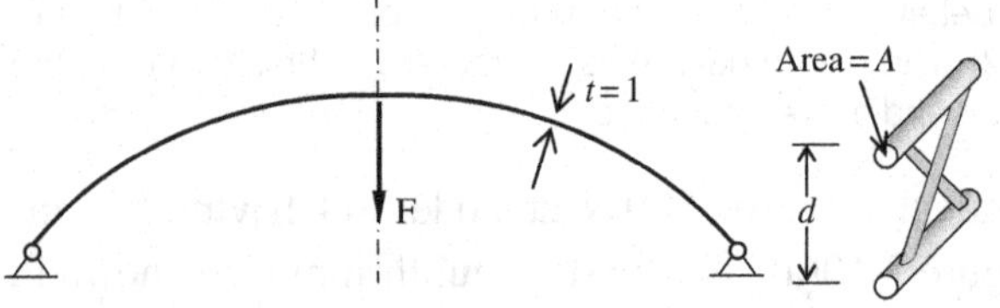


**FIGURE 3.11** Exercise 2 – arch.

3. Analyze the shallow trussed dome shown in Figure 3.12. The outer radius is 50 and height 0, the inner radius is 25 and height 6.216, and the apex height is 8.216. The cross-sectional area of each truss member is unity. The figure is approximate in that the apparent major triangles spanning the outer circle do not have straight sides as shown. Young's modulus is $8 \times 10^7$ and Poisson's ratio is 0.5, indicating incompressible behavior. Plot the vertical downward

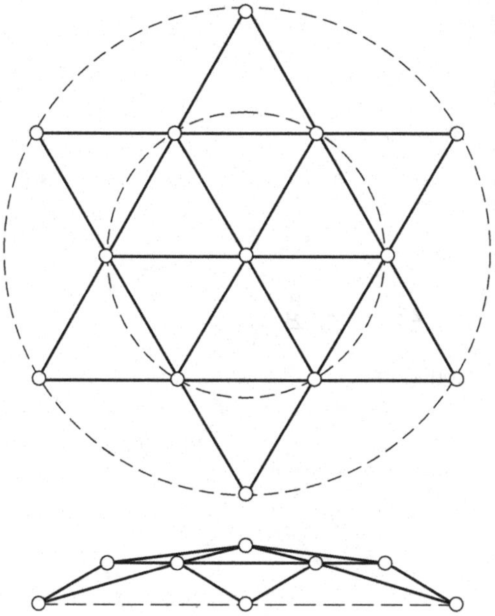

**FIGURE 3.12** Exercise 3 – shallow dome.

load deflection behavior at the apex. This is a good example of snap-through behavior. The equilibrium path is very convoluted but upon examination the corresponding dome shapes are perfectly reasonable.

4. Run the trussed frame example given in Figure 3.10, initially as shown and then with clamped supports. The cross-sectional area is 6, giving a truss member area of 1.
5. This example, devised by Crisfield,¶¶ demonstrates snap-through behavior. The layout is shown in Figure 3.13, where elements $1, 2$, and 4 support an effectively rigid element 3. In the original example, elements $1, 2$, and 4 were linear springs; consequently the lengths and material properties of these elements are chosen to model linear springs in which the initial stiffness term is negligible.

The lengths of elements $1, 2$, and 4 are each $10^7$ and that of the sloping element 3 is approximately 2500. Other material properties are such that

$$\left(\frac{EA}{L}\right)_1 = 1.00; \quad \left(\frac{EA}{L}\right)_2 = 0.25; \quad \left(\frac{EA}{L}\right)_4 = 1.50. \tag{3.91}$$

A high value, $10^{10}$, is chosen for the yield stress to ensure elasticity. The nodal coordinate and boundary code data are given below.

¶¶ Crisfield (1991), p. 100.

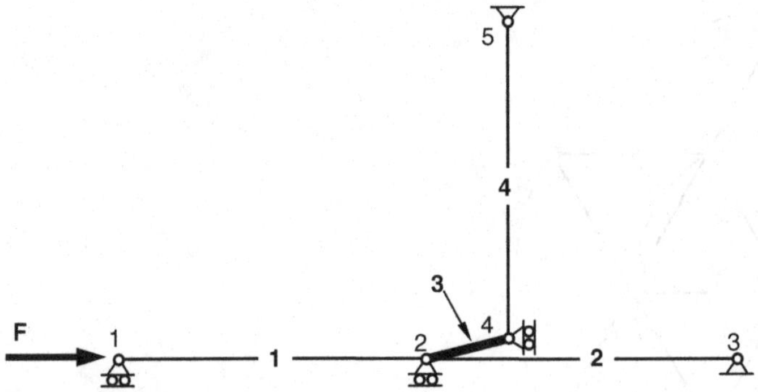

**FIGURE 3.13** Crisfield's snap-back problem – configuration.

```
1 6 -10000000.0 0.0    0.0
2 6          0.0 0.0    0.0
3 7 10000000.0  0.0    0.0
4 3    2500.0     0.0  25.0
5 7 2500.0 0.0 10000025.0
```

6. This is an interesting example of the use of the directional derivative to find an equilibrium equation and the corresponding tangent stiffness term. Figure 3.14 represents two configurations of a one-dimensional structural system comprised of a truss member joining nodes $a$ and $b$ of initial length $L$ and cross-sectional area $A$ connected to a piston chamber of constant section $A_p$ filled with a gas. An external force $F^{ext}$ is applied at node $b$. The constant mass of gas enclosed in the piston is considered to satisfy Boyle's law, which states that

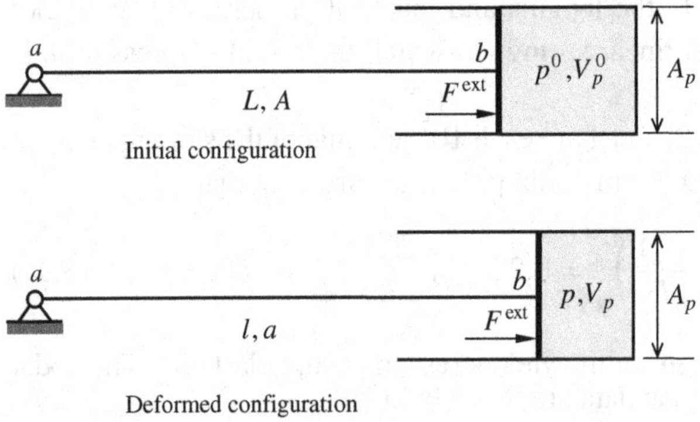

**FIGURE 3.14** Truss–piston system.

$$p\,V_p = p^0\,V_p^0 = K; \quad K \text{ is a constant;}$$

where $p$ and $V_p$ denote the pressure and the volume of the gas within the chamber in the deformed configuration, respectively. This example employs the strain energy per unit initial volume given by Equation (3.17) as $\psi = E(\ln\lambda)^2/2$, where $E$ is a Young's modulus-type constitutive term and $\lambda$ is the stretch of the axial rod. Consequently the total energy functional for the overall system can be written as a function of the spatial position of point $b$ denoted by the deformed length $l$ and the initial volume $AL$ as follows:

$$\begin{aligned}\Pi(l) &= \Pi^{\text{truss}}(l) + \Pi^{\text{piston}}(l) - \Pi^{\text{ext}}\\ &= \frac{1}{2}EAL\,(\ln\lambda)^2 - \int_{V_p} p(V)dV - F^{\text{ext}}l,\end{aligned} \tag{3.92}$$

in which $\lambda = \frac{l}{L}$ is the stretch ratio of the truss member.

(a) Obtain the stationary point of the above energy functional and derive the principle of virtual work.

(b) Obtain the equilibrium equation at node $b$ and show that it is indeed a nonlinear equation.

(c) Obtain the tangent stiffness matrix required for a Newton–Raphson algorithm after suitable linearization.

(d) Explain whether the inclusion of the piston chamber increases or decreases the stiffness of the truss member.

# CHAPTER FOUR

# KINEMATICS

## 4.1 INTRODUCTION

It is almost a tautology to say that a proper description of motion is fundamental to finite deformation analysis, but such an emphasis is necessary because infinitesimal deformation analysis implies a host of assumptions that we take for granted and seldom articulate. For example, we have seen in Chapter 1, in the simple truss example, that care needs to be exercised when large deformations are anticipated and that a linear definition of strain is totally inadequate in the context of a finite rotation. A study of finite deformation will require that cherished assumptions be abandoned and a fresh start made with an open (but not empty!) mind.

Kinematics is the study of motion and deformation without reference to the cause. We shall see immediately that consideration of finite deformation enables alternative coordinate systems to be employed, namely, material and spatial descriptions associated with the names of Lagrange and Euler respectively.

Although we are not directly concerned with inertial effects, nevertheless time derivatives of various kinematic quantities enrich our understanding and also provide the basis for the formulation of the virtual work expression of equilibrium, which uses the notion of virtual velocity and associated kinematic quantities.

Wherever appropriate, nonlinear kinematic quantities are linearized in preparation for inclusion in the linearized equilibrium equations that form the basis of the Newton–Raphson solution to the finite element equilibrium equations.

## 4.2 THE MOTION

Figure 4.1 shows the general motion of a deformable body. The body is imagined as being an assemblage of material particles that are labeled by the coordinates $\boldsymbol{X}$, with respect to Cartesian basis $\boldsymbol{E}_I$, at their initial positions at time $t = 0$. Generally, the current positions of these particles are located, at time $= t$, by the coordinates

| Time | Length | Area | Volume | Density |
|---|---|---|---|---|
| 0 | $L$ | $A$ | $V$ | $\rho_0$ |
| $t$ | $l$ | $a$ | $v$ | $\rho$ |

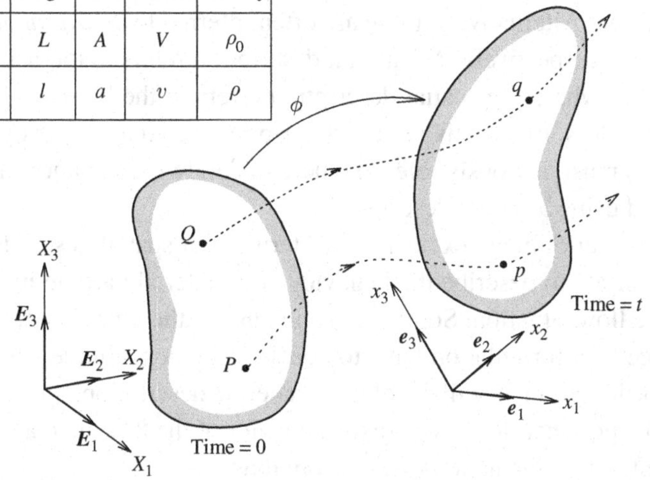


**FIGURE 4.1** General motion of a deformable body.

$\boldsymbol{x}$ with respect to an alternative Cartesian basis $\boldsymbol{e}_i$. In the remainder of this chapter the bases $\boldsymbol{E}_I$ and $\boldsymbol{e}_i$ will be taken to be coincident. However, the notational distinction between $\boldsymbol{E}_I$ and $\boldsymbol{e}_i$ will be retained in order to identify the association of quantities with initial or current configurations. The motion can be mathematically described by a mapping $\boldsymbol{\phi}$ between initial and current particle positions as

$$\boldsymbol{x} = \boldsymbol{\phi}(\boldsymbol{X}, t). \tag{4.1}$$

For a fixed value of $t$ the above equations represent a mapping between the undeformed and deformed bodies. Additionally, for a fixed particle $\boldsymbol{X}$, Equation (4.1) describes the motion or trajectory of this particle as a function of time. In finite deformation analysis no assumptions are made regarding the magnitude of the displacement $\boldsymbol{x} - \boldsymbol{X}$; indeed the displacement may well be of the order of or even exceed the initial dimensions of the body as is the case, for example, in metal forming. In infinitesimal deformation analysis the displacement $\boldsymbol{x} - \boldsymbol{X}$ is assumed to be small in comparison with the dimensions of the body, and geometrical changes are ignored.

## 4.3 MATERIAL AND SPATIAL DESCRIPTIONS

In finite deformation analysis a careful distinction has to be made between the coordinate systems that can be chosen to describe the behavior of the body whose motion is under consideration. Roughly speaking, relevant quantities, such as density, can be described in terms of where the body was before deformation or where it is during deformation; the former is called a *material* description, and the latter is called a *spatial* description. Alternatively, these are often referred to as *Lagrangian* and *Eulerian* descriptions respectively. A material description refers to the behavior of a material particle, whereas a spatial description refers to the behavior at a spatial position. Nevertheless, irrespective of the description eventually employed, the governing equations must obviously refer to where the body is and hence must primarily be formulated using a spatial description.

Fluid mechanicians work almost exclusively in terms of a spatial description because it is not appropriate to describe the behavior of a material particle in, for example, a steady-state flow situation. Solid mechanicians, on the other hand, will generally at some stage of a formulation have to consider the constitutive behavior of the material particle, which will involve a material description. In many instances – for example, polymer flow – where the behavior of the flowing material may be time-dependent, these distinctions are less obvious.

In order to understand the difference between a material and a spatial description, consider a simple scalar quantity such as the current density $\rho$ of the material:

(a) *Material description*: the variation of $\rho$ over the body is described with respect to the original (or initial) coordinate $\boldsymbol{X}$ used to label a material particle in the continuum at time $t = 0$ as

$$\rho = \rho(\boldsymbol{X}, t). \tag{4.2a}$$

(b) *Spatial description*: $\rho$ is described with respect to the position in space, $\boldsymbol{x}$, currently occupied by a material particle in the continuum at time $t$ as

$$\rho = \rho(\boldsymbol{x}, t). \tag{4.2b}$$

In Equation (4.2a) a change in time $t$ implies that the same material particle $\boldsymbol{X}$ has a different density $\rho$. Consequently, interest is focused on the material particle $\boldsymbol{X}$. In Equation (4.2b), however, a change in the time $t$ implies that a different density is observed at the same spatial position $\boldsymbol{x}$, now probably occupied by a different particle. Consequently, interest is focused on a spatial position $\boldsymbol{x}$.

**EXAMPLE 4.1: Uniaxial motion**

This example illustrates the difference between a material and a spatial description of motion. Consider the mapping $x = (1+t)X$ defining the motion of a rod of initial length 2 units. The rod experiences a temperature distribution given by the material description $T = Xt^2$ or by the spatial description $T = xt^2/(1+t)$; see the diagram below.

*(continued)*

**EXAMPLE 4.1:** *(cont.)*

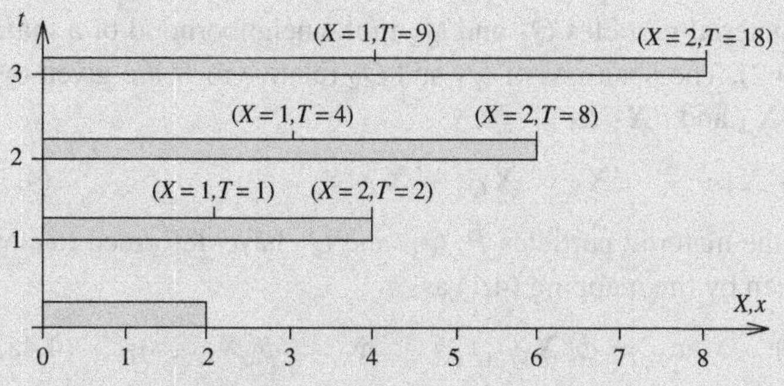


The diagram makes it clear that the particle material coordinates (label) $X$ remains associated with the particle while its spatial position $x$ changes. The temperature at a given time can be found in two ways. For example, at time $t = 3$ the temperature of the particle labeled $X = 2$ is $T = 2 \times 3^2 = 18$. Alternatively, the temperature of the same particle which at $t = 3$ is at the spatial position $x = 8$ is $T = 8 \times 3^2/(1+3) = 18$. Note that, whatever the time, it makes no sense to inquire about particles for which $X > 2$, nor, for example, at time $t = 3$ does it make sense to inquire about the temperature at $x > 8$.

Often it is necessary to transform between the material and spatial descriptions for relevant quantities. For instance, given a scalar quantity such as the density, a material description can be easily obtained from a spatial description by using motion Equation (4.1) as

$$\rho(\boldsymbol{X}, t) = \rho(\boldsymbol{\phi}(\boldsymbol{X}, t), t). \tag{4.2c}$$

Certain magnitudes, irrespective of whether they are materially or spatially described, are naturally associated with the current or initial configurations of the body. For instance, the initial density of the body is a material magnitude, whereas the current density is intrinsically a spatial quantity. Nevertheless, Equations (4.2a–c) clearly show that spatial quantities can, if desired, be expressed in terms of the initial coordinates.

## 4.4 DEFORMATION GRADIENT

A key quantity in finite deformation analysis is the deformation gradient $\boldsymbol{F}$, which is involved in all equations relating quantities before deformation to corresponding quantities after (or during) deformation. The deformation gradient tensor enables the relative spatial position of two neighboring particles after deformation to be described in terms of their relative material position before deformation; consequently, it is central to the description of deformation and hence strain.

Consider two material particles $Q_1$ and $Q_2$ in the neighborhood of a material particle $P$ (Figure 4.2). The positions of $Q_1$ and $Q_2$ relative to $P$ are given by the elemental vectors $d\boldsymbol{X}_1$ and $d\boldsymbol{X}_2$ as

$$d\boldsymbol{X}_1 = \boldsymbol{X}_{Q_1} - \boldsymbol{X}_P; \qquad d\boldsymbol{X}_2 = \boldsymbol{X}_{Q_2} - \boldsymbol{X}_P. \tag{4.3a,b}$$

After deformation, the material particles $P$, $Q_1$, and $Q_2$ have deformed to current spatial positions given by the mapping (4.1) as

$$\boldsymbol{x}_p = \boldsymbol{\phi}(\boldsymbol{X}_P, t); \qquad \boldsymbol{x}_{q_1} = \boldsymbol{\phi}(\boldsymbol{X}_{Q_1}, t); \qquad \boldsymbol{x}_{q_2} = \boldsymbol{\phi}(\boldsymbol{X}_{Q_2}, t); \tag{4.4a,b,c}$$

and the corresponding elemental vectors become

$$d\boldsymbol{x}_1 = \boldsymbol{x}_{q_1} - \boldsymbol{x}_p = \boldsymbol{\phi}(\boldsymbol{X}_P + d\boldsymbol{X}_1, t) - \boldsymbol{\phi}(\boldsymbol{X}_P, t); \tag{4.5a}$$

$$d\boldsymbol{x}_2 = \boldsymbol{x}_{q_2} - \boldsymbol{x}_p = \boldsymbol{\phi}(\boldsymbol{X}_P + d\boldsymbol{X}_2, t) - \boldsymbol{\phi}(\boldsymbol{X}_P, t). \tag{4.5b}$$

Defining the *deformation gradient tensor* $\boldsymbol{F}$ as

$$\boldsymbol{F} = \frac{\partial \boldsymbol{\phi}}{\partial \boldsymbol{X}} = \boldsymbol{\nabla}_0 \boldsymbol{\phi}, \tag{4.6}$$

where $\boldsymbol{\nabla}_0$ denotes the gradient with respect to the material configuration, the elemental vectors $d\boldsymbol{x}_1$ and $d\boldsymbol{x}_2$ can be obtained in terms of $d\boldsymbol{X}_1$ and $d\boldsymbol{X}_2$ as

$$d\boldsymbol{x}_1 = \boldsymbol{F} d\boldsymbol{X}_1; \qquad d\boldsymbol{x}_2 = \boldsymbol{F} d\boldsymbol{X}_2. \tag{4.7a,b}$$

Note that $\boldsymbol{F}$ transforms vectors in the initial or reference configuration into vectors in the current configuration and is therefore said to be a *two-point* tensor.

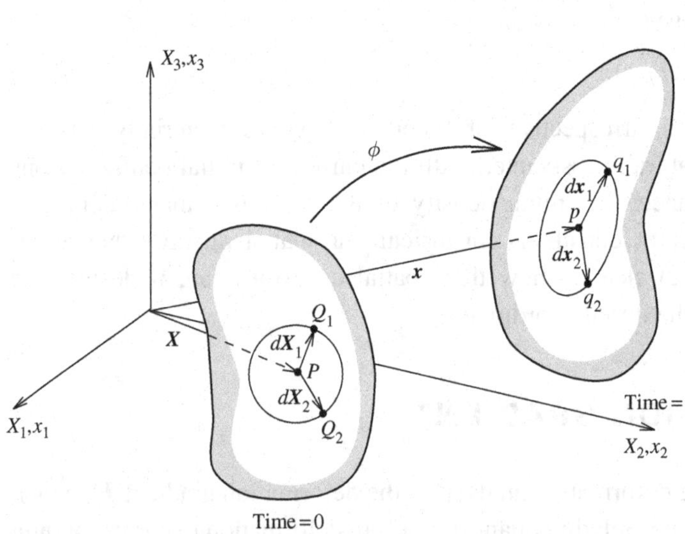


**FIGURE 4.2** General motion in the neighborhood of a particle.

**Remark 4.1**: In many textbooks the motion is expressed as

$$\boldsymbol{x} = \boldsymbol{x}(\boldsymbol{X}, t), \tag{4.8}$$

which allows the deformation gradient tensor to be written, perhaps in a clearer manner, as

$$\boldsymbol{F} = \frac{\partial \boldsymbol{x}}{\partial \boldsymbol{X}}. \tag{4.9a}$$

In indicial notation the deformation gradient tensor is expressed as

$$\boldsymbol{F} = \sum_{i,I=1}^{3} F_{iI} \boldsymbol{e}_i \otimes \boldsymbol{E}_I; \qquad F_{iI} = \frac{\partial x_i}{\partial X_I}; \qquad i, I = 1, 2, 3; \tag{4.9b}$$

where lower-case indices refer to current (spatial) Cartesian coordinates, whereas upper-case indices refer to initial (material) Cartesian coordinates.

Confining attention to a single elemental material vector $d\boldsymbol{X}$, the corresponding vector $d\boldsymbol{x}$ in the spatial configuration is conveniently written as

$$d\boldsymbol{x} = \boldsymbol{F} d\boldsymbol{X}. \tag{4.10}$$

The inverse of $\boldsymbol{F}$ is

$$\boldsymbol{F}^{-1} = \frac{\partial \boldsymbol{X}}{\partial \boldsymbol{x}} = \boldsymbol{\nabla} \phi^{-1}, \tag{4.11a}$$

which in indicial notation is

$$\boldsymbol{F}^{-1} = \sum_{I,i=1}^{3} \frac{\partial X_I}{\partial x_i} \boldsymbol{E}_I \otimes \boldsymbol{e}_i. \tag{4.11b}$$

**Remark 4.2**: Much research literature expresses the relationship between quantities in the material and spatial configurations in terms of the general concepts of *push forward* and *pull back*. For example, the elemental spatial vector $d\boldsymbol{x}$ can be considered as the push-forward equivalent of the material vector $d\boldsymbol{X}$. This can be expressed in terms of the operation

$$d\boldsymbol{x} = \phi_*[d\boldsymbol{X}] = \boldsymbol{F} d\boldsymbol{X}. \tag{4.12}$$

Inversely, the material vector $d\boldsymbol{X}$ is the pull-back equivalent of the spatial vector $d\boldsymbol{x}$, which is expressed as*

$$d\boldsymbol{X} = \phi_*^{-1}[d\boldsymbol{x}] = \boldsymbol{F}^{-1} d\boldsymbol{x}. \tag{4.13}$$

Observe that in (4.12) the nomenclature $\phi_*[\,]$ implies an *operation* that will be evaluated in different ways for different operands $[\,]$.

* In the literature $\phi_*[\,]$ and $\phi_*^{-1}[\,]$ are often written as $\phi_*$ and $\phi^*$ respectively.

**EXAMPLE 4.2: Uniform deformation**

This example illustrates the role of the deformation gradient tensor $\boldsymbol{F}$. Consider the uniform deformation given by the mapping

$$x_1 = \frac{1}{4}(18 + 4X_1 + 6X_2);$$

$$x_2 = \frac{1}{4}(14 + 6X_2);$$

which, for a square of side 2 units initially centered at $\boldsymbol{X} = (0,0)$, produces the deformation shown below.

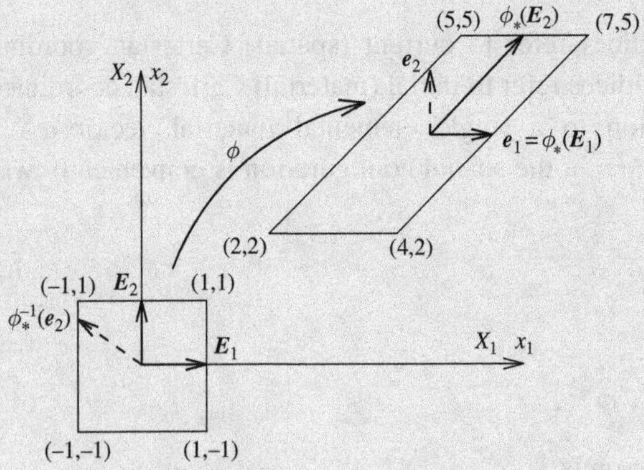


$$\boldsymbol{F} = \begin{bmatrix} \frac{\partial x_1}{\partial X_1} & \frac{\partial x_1}{\partial X_2} \\ \frac{\partial x_2}{\partial X_1} & \frac{\partial x_2}{\partial X_2} \end{bmatrix} = \frac{1}{2}\begin{bmatrix} 2 & 3 \\ 0 & 3 \end{bmatrix}; \quad \boldsymbol{F}^{-1} = \frac{1}{3}\begin{bmatrix} 3 & -3 \\ 0 & 2 \end{bmatrix}.$$

Unit vectors $\boldsymbol{E}_1$ and $\boldsymbol{E}_2$ in the initial configuration deform to

$$\boldsymbol{\phi}_*[\boldsymbol{E}_1] = \boldsymbol{F}\begin{bmatrix} 1 \\ 0 \end{bmatrix} = \begin{bmatrix} 1 \\ 0 \end{bmatrix}; \quad \boldsymbol{\phi}_*[\boldsymbol{E}_2] = \boldsymbol{F}\begin{bmatrix} 0 \\ 1 \end{bmatrix} = \begin{bmatrix} 1.5 \\ 1.5 \end{bmatrix};$$

and unit vectors in the current (deformed) configuration deform from

$$\boldsymbol{\phi}_*^{-1}[\mathrm{e}_1] = \boldsymbol{F}^{-1}\begin{bmatrix} 1 \\ 0 \end{bmatrix} = \begin{bmatrix} 1 \\ 0 \end{bmatrix}; \quad \boldsymbol{\phi}_*^{-1}[\mathrm{e}_2] = \boldsymbol{F}^{-1}\begin{bmatrix} 0 \\ 1 \end{bmatrix} = \begin{bmatrix} -1 \\ \frac{2}{3} \end{bmatrix}.$$

## 4.5 STRAIN

As a general measure of deformation, consider the change in the scalar product of the two elemental vectors $d\boldsymbol{X}_1$ and $d\boldsymbol{X}_2$ shown in Figure 4.2 as they deform to $d\boldsymbol{x}_1$ and $d\boldsymbol{x}_2$. This change will involve both the stretching (that is, change in length) and changes in the enclosed angle between the two vectors. Recalling Equation (4.7a,b), the spatial scalar product $d\boldsymbol{x}_1 \cdot d\boldsymbol{x}_2$ can be found in terms of the material vectors $d\boldsymbol{X}_1$ and $d\boldsymbol{X}_2$ as

$$d\boldsymbol{x}_1 \cdot d\boldsymbol{x}_2 = d\boldsymbol{X}_1 \cdot \boldsymbol{C}\, d\boldsymbol{X}_2, \tag{4.14}$$

where $\boldsymbol{C}$ is the *right Cauchy–Green deformation tensor*, which is given in terms of the deformation gradient as $\boldsymbol{F}$ as

$$\boldsymbol{C} = \boldsymbol{F}^T \boldsymbol{F}. \tag{4.15}$$

Note that in Equation (4.15) the tensor $\boldsymbol{C}$ operates on the material vectors $d\boldsymbol{X}_1$ and $d\boldsymbol{X}_2$, and consequently $\boldsymbol{C}$ is called a material tensor quantity.

Alternatively, the initial material scalar product $d\boldsymbol{X}_1 \cdot d\boldsymbol{X}_2$ can be obtained in terms of the spatial vectors $d\boldsymbol{x}_1$ and $d\boldsymbol{x}_2$ via the *left Cauchy–Green* or *Finger tensor* $\boldsymbol{b}$ as†

$$d\boldsymbol{X}_1 \cdot d\boldsymbol{X}_2 = d\boldsymbol{x}_1 \cdot \boldsymbol{b}^{-1} d\boldsymbol{x}_2, \tag{4.16}$$

where $\boldsymbol{b}$ is

$$\boldsymbol{b} = \boldsymbol{F}\boldsymbol{F}^T. \tag{4.17}$$

Observe that in Equation (4.16) $\boldsymbol{b}^{-1}$ operates on the spatial vectors $d\boldsymbol{x}_1$ and $d\boldsymbol{x}_2$, and consequently $\boldsymbol{b}^{-1}$, or indeed $\boldsymbol{b}$ itself, is a spatial tensor quantity.

The change in scalar product can now be found in terms of the material vectors $d\boldsymbol{X}_1$ and $d\boldsymbol{X}_2$ and the *Lagrangian* or *Green strain tensor* $\boldsymbol{E}$ as

$$\frac{1}{2}(d\boldsymbol{x}_1 \cdot d\boldsymbol{x}_2 - d\boldsymbol{X}_1 \cdot d\boldsymbol{X}_2) = d\boldsymbol{X}_1 \cdot \boldsymbol{E}\, d\boldsymbol{X}_2, \tag{4.18a}$$

where the material tensor $\boldsymbol{E}$ is

$$\boldsymbol{E} = \frac{1}{2}(\boldsymbol{C} - \boldsymbol{I}). \tag{4.18b}$$

Alternatively, the same change in scalar product can be expressed with reference to the spatial elemental vectors $d\boldsymbol{x}_1$ and $d\boldsymbol{x}_2$ and the *Eulerian* or *Almansi strain tensor* $\boldsymbol{e}$ as

$$\frac{1}{2}(d\boldsymbol{x}_1 \cdot d\boldsymbol{x}_2 - d\boldsymbol{X}_1 \cdot d\boldsymbol{X}_2) = d\boldsymbol{x}_1 \cdot \boldsymbol{e}\, d\boldsymbol{x}_2, \tag{4.19a}$$

where the spatial tensor $\boldsymbol{e}$ is

$$\boldsymbol{e} = \frac{1}{2}(\boldsymbol{I} - \boldsymbol{b}^{-1}). \tag{4.19b}$$

† In $\boldsymbol{C} = \boldsymbol{F}^T \boldsymbol{F}$, $\boldsymbol{F}$ is on the right and in $\boldsymbol{b} = \boldsymbol{F}\boldsymbol{F}^T$, $\boldsymbol{F}$ is on the left.

**EXAMPLE 4.3: Green and Almansi strain tensors**

For the deformation given in Example 4.2 the right and left Cauchy–Green deformation tensors are respectively

$$\boldsymbol{C} = \boldsymbol{F}^T\boldsymbol{F} = \frac{1}{2}\begin{bmatrix} 2 & 3 \\ 3 & 9 \end{bmatrix} \quad \text{and} \quad \boldsymbol{b} = \boldsymbol{F}\boldsymbol{F}^T = \frac{1}{4}\begin{bmatrix} 13 & 9 \\ 9 & 9 \end{bmatrix},$$

from which the Green's strain tensor is simply

$$\boldsymbol{E} = \frac{1}{4}\begin{bmatrix} 0 & 3 \\ 3 & 7 \end{bmatrix}$$

and the Almansi strain tensor is

$$\boldsymbol{e} = \frac{1}{18}\begin{bmatrix} 0 & 9 \\ 9 & -4 \end{bmatrix}.$$

The physical interpretation of these strain measures will be demonstrated in the next example.

**Remark 4.3**: The general nature of the scalar product as a measure of deformation can be clarified by taking $d\boldsymbol{X}_2$ and $d\boldsymbol{X}_1$ equal to $d\boldsymbol{X}$ and consequently $d\boldsymbol{x}_1 = d\boldsymbol{x}_2 = d\boldsymbol{x}$. This enables initial (material) and current (spatial) elemental lengths squared to be determined as (Figure 4.3)

$$dL^2 = d\boldsymbol{X} \cdot d\boldsymbol{X}; \qquad dl^2 = d\boldsymbol{x} \cdot d\boldsymbol{x}. \tag{4.20a,b}$$

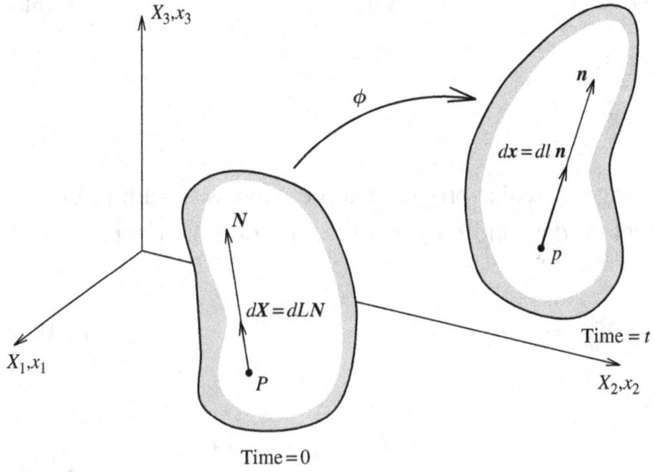


**FIGURE 4.3** Change in length.

The change in the squared lengths that occurs as the body deforms from the initial to the current configuration can now be written in terms of the elemental material vector $d\boldsymbol{X}$ as

$$\frac{1}{2}(dl^2 - dL^2) = d\boldsymbol{X} \cdot \boldsymbol{E}\, d\boldsymbol{X}, \tag{4.21}$$

which, upon division by $dL^2$, gives the scalar Green's strain as

$$\frac{dl^2 - dL^2}{2\, dL^2} = \frac{d\boldsymbol{X}}{dL} \cdot \boldsymbol{E} \frac{d\boldsymbol{X}}{dL}, \tag{4.22}$$

where $d\boldsymbol{X}/dL$ is a unit material vector $\boldsymbol{N}$ in the direction of $d\boldsymbol{X}$. Hence, finally,

$$\frac{1}{2}\left(\frac{dl^2 - dL^2}{dL^2}\right) = \boldsymbol{N} \cdot \boldsymbol{E}\boldsymbol{N}. \tag{4.23}$$

Using Equation (4.19a), a similar expression involving the Almansi strain tensor can be derived as

$$\frac{1}{2}\left(\frac{dl^2 - dL^2}{dl^2}\right) = \boldsymbol{n} \cdot \boldsymbol{e}\boldsymbol{n}, \tag{4.24}$$

where $\boldsymbol{n}$ is a unit vector in the direction of $d\boldsymbol{x}$.

**EXAMPLE 4.4: Physical interpretation of strain tensors**

Referring to Example 4.2, the magnitude of the elemental vector $d\boldsymbol{x}_2$ is $dl_2 = 4.5^{1/2}$. Using (4.23), the scalar value of Green's strain associated with the elemental material vector $d\boldsymbol{X}_2$ is

$$\varepsilon_G = \frac{1}{2}\left(\frac{dl^2 - dL^2}{dL^2}\right) = \frac{7}{4}.$$

Again using Equation (4.23) and Example 4.3, the same strain can be determined from the Green's strain tensor $\boldsymbol{E}$ as

$$\varepsilon_G = \boldsymbol{N}^T \boldsymbol{E} \boldsymbol{N} = [0,1]\frac{1}{4}\begin{bmatrix} 0 & 3 \\ 3 & 7 \end{bmatrix}\begin{bmatrix} 0 \\ 1 \end{bmatrix} = \frac{7}{4}.$$

Using Equation (4.24) the scalar value of the Almansi strain associated with the elemental spatial vector $d\boldsymbol{x}_2$ is

$$\varepsilon_A = \frac{1}{2}\left(\frac{dl^2 - dL^2}{dl^2}\right) = \frac{7}{18}.$$

*(continued)*

**EXAMPLE 4.4:** *(cont.)*

Alternatively, again using Equation (4.24) and Example 4.3 the same strain is determined from the Almansi strain tensor $\boldsymbol{e}$ as

$$\varepsilon_A = \boldsymbol{n}^T \boldsymbol{e} \boldsymbol{n} = \left[\frac{1}{\sqrt{2}}, \frac{1}{\sqrt{2}}\right] \frac{1}{18} \begin{bmatrix} 0 & 9 \\ 9 & -4 \end{bmatrix} \begin{bmatrix} \frac{1}{\sqrt{2}} \\ \frac{1}{\sqrt{2}} \end{bmatrix} = \frac{7}{18}.$$

**Remark 4.4**: In terms of the language of pull back and push forward, the material and spatial strain measures can be related through the operator $\phi_*$. Precisely how this operator works in this case can be discovered by recognizing, because of their definitions, the equality

$$d\boldsymbol{x}_1 \cdot \boldsymbol{e}\, d\boldsymbol{x}_2 = d\boldsymbol{X}_1 \cdot \boldsymbol{E}\, d\boldsymbol{X}_2 \tag{4.25}$$

for any corresponding pairs of elemental vectors. Recalling Equations (4.12) and (4.13) enables the push-forward and pull-back operations to be written as

*Push forward*

$$\boldsymbol{e} = \phi_* [\boldsymbol{E}] = \boldsymbol{F}^{-T} \boldsymbol{E} \boldsymbol{F}^{-1}. \tag{4.26a}$$

*Pull back*

$$\boldsymbol{E} = \phi_*^{-1}[\boldsymbol{e}] = \boldsymbol{F}^T \boldsymbol{e} \boldsymbol{F}. \tag{4.26b}$$

## 4.6 POLAR DECOMPOSITION

The deformation gradient tensor $\boldsymbol{F}$ discussed in the previous sections transforms a material vector $d\boldsymbol{X}$ into the corresponding spatial vector $d\boldsymbol{x}$. The crucial role of $\boldsymbol{F}$ is further disclosed in terms of its decomposition into stretch and rotation components. The use of the physical terminology "stretch" and "rotation" will become clearer later. For the moment, from a purely mathematical point of view, the tensor $\boldsymbol{F}$ is expressed as the product of a *rotation tensor* $\boldsymbol{R}$ times a *stretch tensor* $\boldsymbol{U}$ to define the polar decomposition as

$$\boldsymbol{F} = \boldsymbol{R}\boldsymbol{U}. \tag{4.27}$$

For the purpose of evaluating these tensors, recall the definition of the right Cauchy–Green tensor $\boldsymbol{C}$ as

$$\boldsymbol{C} = \boldsymbol{F}^T \boldsymbol{F} = \boldsymbol{U}^T \boldsymbol{R}^T \boldsymbol{R} \boldsymbol{U}. \tag{4.28}$$

Given that $\boldsymbol{R}$ is an orthogonal rotation tensor as defined in Equation (2.26), that is, $\boldsymbol{R}^T\boldsymbol{R} = \boldsymbol{I}$, choosing $\boldsymbol{U}$ to be a symmetric tensor gives a unique definition of the *material stretch tensor* $\boldsymbol{U}$ in terms of $\boldsymbol{C}$ as

$$\boldsymbol{U}^2 = \boldsymbol{U}\boldsymbol{U} = \boldsymbol{C}. \tag{4.29}$$

In order to actually obtain $\boldsymbol{U}$ from this equation, it is first necessary to evaluate the principal directions of $\boldsymbol{C}$, denoted here by the eigenvector triad $\{\boldsymbol{N}_1, \boldsymbol{N}_2, \boldsymbol{N}_3\}$ and their corresponding eigenvalues $\lambda_1^2$, $\lambda_2^2$, and $\lambda_3^2$, which enable $\boldsymbol{C}$ to be expressed as

$$\boldsymbol{C} = \sum_{\alpha=1}^{3} \lambda_\alpha^2 \boldsymbol{N}_\alpha \otimes \boldsymbol{N}_\alpha, \tag{4.30}$$

where, because of the symmetry of $\boldsymbol{C}$, the triad $\{\boldsymbol{N}_1, \boldsymbol{N}_2, \boldsymbol{N}_3\}$ are orthogonal unit vectors. Combining Equations (4.29) and (4.30), the material stretch tensor $\boldsymbol{U}$ can be easily obtained as

$$\boldsymbol{U} = \sum_{\alpha=1}^{3} \lambda_\alpha \boldsymbol{N}_\alpha \otimes \boldsymbol{N}_\alpha. \tag{4.31}$$

Once the stretch tensor $\boldsymbol{U}$ is known, the rotation tensor $\boldsymbol{R}$ can be easily evaluated from Equation (4.27) as $\boldsymbol{R} = \boldsymbol{F}\boldsymbol{U}^{-1}$.

In terms of this polar decomposition, typical material and spatial elemental vectors are related as

$$d\boldsymbol{x} = \boldsymbol{F}d\boldsymbol{X} = \boldsymbol{R}(\boldsymbol{U}d\boldsymbol{X}). \tag{4.32}$$

In the above equation, the material vector $d\boldsymbol{X}$ is first stretched to give $\boldsymbol{U}d\boldsymbol{X}$ and then rotated to the spatial configuration by $\boldsymbol{R}$. Note that $\boldsymbol{U}$ is a material tensor whereas $\boldsymbol{R}$ transforms material vectors into spatial vectors and is therefore, like $\boldsymbol{F}$, a two-point tensor.

**EXAMPLE 4.5: Polar decomposition (i)**

This example illustrates the decomposition of the deformation gradient tensor $\boldsymbol{F} = \boldsymbol{R}\boldsymbol{U}$ using the deformation shown below as

$$x_1 = \frac{1}{4}(4X_1 + (9 - 3X_1 - 5X_2 - X_1X_2)t);$$

$$x_2 = \frac{1}{4}(4X_2 + (16 + 8X_1)t).$$

*(continued)*

**EXAMPLE 4.5:** *(cont.)*

For $\boldsymbol{X} = (0, 0)$ and time $t = 1$ the deformation gradient $\boldsymbol{F}$ and right Cauchy–Green tensor $\boldsymbol{C}$ are

$$\boldsymbol{F} = \frac{1}{4}\begin{bmatrix} 1 & -5 \\ 8 & 4 \end{bmatrix}; \quad \boldsymbol{C} = \frac{1}{16}\begin{bmatrix} 65 & 27 \\ 27 & 41 \end{bmatrix},$$

from which the stretches $\lambda_1$ and $\lambda_2$ and principal material vectors $\boldsymbol{N}_1$ and $\boldsymbol{N}_2$ are found as

$$\lambda_1 = 2.2714; \quad \lambda_2 = 1.2107; \quad \boldsymbol{N}_1 = \begin{bmatrix} 0.8385 \\ 0.5449 \end{bmatrix}; \quad \boldsymbol{N}_2 = \begin{bmatrix} -0.5449 \\ 0.8385 \end{bmatrix}.$$

Hence, using (4.31) and $\boldsymbol{R} = \boldsymbol{F}\boldsymbol{U}^{-1}$, the stretch and rotation tensors can be found as

$$\boldsymbol{U} = \begin{bmatrix} 1.9564 & 0.4846 \\ 0.4846 & 1.5257 \end{bmatrix}; \quad \boldsymbol{R} = \begin{bmatrix} 0.3590 & -0.9333 \\ 0.9333 & 0.3590 \end{bmatrix}.$$

It is also possible to decompose $\boldsymbol{F}$ in terms of the same rotation tensor $\boldsymbol{R}$ followed by a stretch in the spatial configuration as

$$\boldsymbol{F} = \boldsymbol{V}\boldsymbol{R}, \tag{4.33}$$

which can now be interpreted as first rotating the material vector $d\boldsymbol{X}$ to the spatial configuration, where it is then stretched to give $d\boldsymbol{x}$ as

$$d\boldsymbol{x} = \boldsymbol{F}d\boldsymbol{X} = \boldsymbol{V}(\boldsymbol{R}d\boldsymbol{X}), \tag{4.34}$$

where the *spatial stretch tensor* $\boldsymbol{V}$ can be obtained in terms of $\boldsymbol{U}$ by combining Equations (4.27) and (4.33) to give

$$\boldsymbol{V} = \boldsymbol{R}\boldsymbol{U}\boldsymbol{R}^T. \tag{4.35}$$

Additionally, recalling Equation (4.17) for the left Cauchy–Green or Finger tensor $\boldsymbol{b}$ gives

$$\boldsymbol{b} = \boldsymbol{F}\boldsymbol{F}^T = (\boldsymbol{V}\boldsymbol{R})(\boldsymbol{R}^T\boldsymbol{V}) = \boldsymbol{V}^2. \tag{4.36}$$

Consequently, if the principal directions of $\boldsymbol{b}$ are given by the orthogonal spatial vectors $\{\boldsymbol{n}_1, \boldsymbol{n}_2, \boldsymbol{n}_3\}$ with associated eigenvalues $\bar{\lambda}_1^2$, $\bar{\lambda}_2^2$, and $\bar{\lambda}_3^2$, then the spatial stretch tensor can be expressed as

$$\boldsymbol{V} = \sum_{\alpha=1}^{3} \bar{\lambda}_\alpha \, \boldsymbol{n}_\alpha \otimes \boldsymbol{n}_\alpha. \tag{4.37}$$

Substituting Equation (4.31) for $U$ into expression (4.35) for $V$ gives $V$ in terms of the vector triad in the undeformed configuration as

$$V = \sum_{\alpha=1}^{3} \lambda_\alpha (RN_\alpha) \otimes (RN_\alpha). \tag{4.38}$$

Comparing this expression with Equation (4.37) and noting that $(RN_\alpha)$ remain unit vectors, it must follow that

$$\lambda_\alpha = \bar{\lambda}_\alpha; \qquad n_\alpha = RN_\alpha; \qquad \alpha = 1, 2, 3. \tag{4.39a,b}$$

This equation implies that the two-point tensor $R$ rotates the *material vector triad* $\{N_1, N_2, N_3\}$ into the *spatial vector triad* $\{n_1, n_2, n_3\}$ as shown in Figure 4.4. Furthermore, the unique eigenvalues $\lambda_1^2$, $\lambda_2^2$, and $\lambda_3^2$ are the squares of the stretches in the principal directions in the sense that taking a material vector $dX_1$ of length $dL_1$ in the direction of $N_1$, its corresponding push-forward spatial vector $dx_1$ of length $dl_1$ is given as

$$dx_1 = F dX_1 = RU(dL_1 N_1). \tag{4.40}$$

Given that $UN_1 = \lambda_1 N_1$ and recalling Equation (4.39a,b) gives the spatial vector $dx_1$ as

$$dx_1 = (\lambda_1 dL_1) n_1 = dl_1 n_1. \tag{4.41}$$

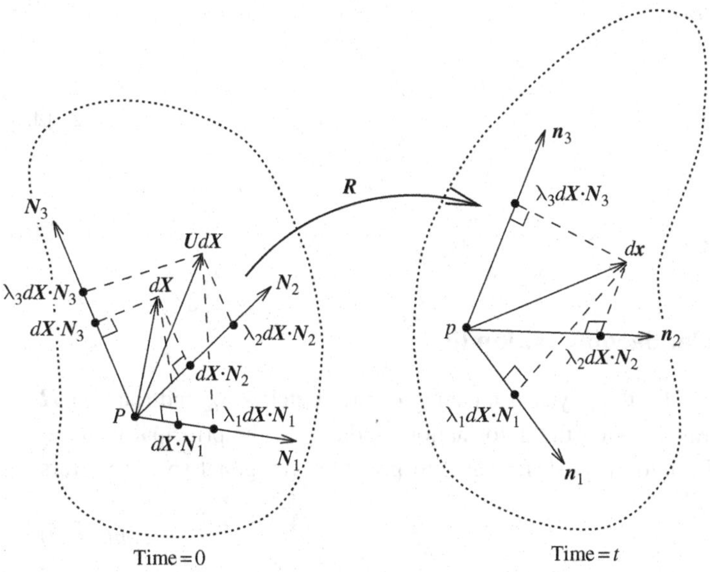


**FIGURE 4.4** Material and spatial vector triads.

Hence, the stretch $\lambda_1$ gives the ratio between current and initial lengths as

$$\lambda_1 = \frac{dl_1}{dL_1}. \tag{4.42}$$

It is instructive to express the deformation gradient tensor in terms of the principal stretches and principal directions. To this end, substitute Equation (4.31) for $\boldsymbol{U}$ into Equation (4.27) for $\boldsymbol{F}$ and use (4.39a,b) to give

$$\boldsymbol{F} = \sum_{\alpha=1}^{3} \lambda_\alpha \, \boldsymbol{n}_\alpha \otimes \boldsymbol{N}_\alpha. \tag{4.43}$$

This expression clearly reveals the two-point nature of the deformation gradient tensor in that it involves the eigenvectors in both the initial and final configurations.

It will be seen later that it is often convenient, and indeed more natural, to describe the material behavior in terms of principal directions. Consequently, it is pertinent to develop the relationships inherent in Equation (4.43) a little further. For this purpose, consider the mapping of the unit vector $\boldsymbol{N}_\alpha$ given by the tensor $\boldsymbol{F}$, which on substituting the polar decomposition $\boldsymbol{F} = \boldsymbol{RU}$ gives

$$\begin{aligned} \boldsymbol{F}\boldsymbol{N}_\alpha &= \boldsymbol{R}\boldsymbol{U}\boldsymbol{N}_\alpha \\ &= \lambda_\alpha \boldsymbol{R}\boldsymbol{N}_\alpha \\ &= \lambda_\alpha \boldsymbol{n}_\alpha. \end{aligned} \tag{4.44a}$$

Alternative expressions relating $\boldsymbol{N}_\alpha$ and $\boldsymbol{n}_\alpha$ can be similarly obtained as

$$\boldsymbol{F}^{-T}\boldsymbol{N}_\alpha = \frac{1}{\lambda_\alpha}\boldsymbol{n}_\alpha; \tag{4.44b}$$

$$\boldsymbol{F}^{-1}\boldsymbol{n}_\alpha = \frac{1}{\lambda_\alpha}\boldsymbol{N}_\alpha; \tag{4.44c}$$

$$\boldsymbol{F}^{T}\boldsymbol{n}_\alpha = \lambda_\alpha \boldsymbol{N}_\alpha. \tag{4.44d}$$

**EXAMPLE 4.6: Polar decomposition (ii)**

Building on Example 4.5 the physical meaning of the stretches $\lambda_\alpha$ and rotation $\boldsymbol{R}$ can easily be illustrated. Using the deformation gradient $\boldsymbol{F}$, the principal material vectors $\boldsymbol{N}_1$ and $\boldsymbol{N}_2$ deform (push forward) to give the orthogonal spatial vectors $\phi_*[\boldsymbol{N}_1]$ and $\phi_*[\boldsymbol{N}_2]$ as

*(continued)*

**EXAMPLE 4.6:** *(cont.)*

$$\phi_*[\boldsymbol{N}_1] = \begin{bmatrix} -0.4715 \\ 2.2219 \end{bmatrix}; \qquad \phi_*[\boldsymbol{N}_2] = \begin{bmatrix} -1.1843 \\ -0.2513 \end{bmatrix};$$

$$\phi_*[\boldsymbol{N}_1] \cdot \phi_*[\boldsymbol{N}_2] = 0.$$

However, these two vectors may alternatively emerge by, first, stretching the material vectors $\boldsymbol{N}_1$ and $\boldsymbol{N}_2$ to give

$$\lambda_1 \boldsymbol{N}_1 = \begin{bmatrix} 1.9046 \\ 1.2377 \end{bmatrix}; \qquad \lambda_2 \boldsymbol{N}_2 = \begin{bmatrix} -0.6597 \\ 1.0152 \end{bmatrix};$$

and, second, rotating these stretched vectors using the rotation tensor $\boldsymbol{R}$ (see Equation (4.44)) to give

$$\phi_*[\boldsymbol{N}_1] = \boldsymbol{R}\lambda_1 \boldsymbol{N}_1 = \begin{bmatrix} 0.3590 & -0.9333 \\ 0.9333 & 0.3590 \end{bmatrix} \begin{bmatrix} 1.9046 \\ 1.2377 \end{bmatrix} = \begin{bmatrix} -0.4715 \\ 2.2219 \end{bmatrix};$$

similarly for $\phi_*[\boldsymbol{N}_2]$. Hence the deformation of the eigenvectors $\boldsymbol{N}_1$ and $\boldsymbol{N}_2$ associated with $\boldsymbol{F}$, at a particular material position, can be interpreted as a stretch followed by a rotation of (about) 69°. Finally, it is easy to confirm Equation (4.39a,b) that the spatial unit vectors $\boldsymbol{n}_\alpha = \boldsymbol{R}\boldsymbol{N}_\alpha$.

Equations (4.44a–b) can be interpreted in terms of the push-forward of vectors in the initial configuration to vectors in the current configuration. Likewise, (4.44c–d) can be interpreted as alternative pull-back operations.

**Remark 4.5**: The Lagrangian and Eulerian strain tensors, defined in Section 4.5, can now be expressed in terms of $\boldsymbol{U}$ and $\boldsymbol{V}$ as

$$\boldsymbol{E} = \frac{1}{2}(\boldsymbol{U}^2 - \boldsymbol{I}) = \sum_{\alpha=1}^{3} \frac{1}{2}(\lambda_\alpha^2 - 1)\, \boldsymbol{N}_\alpha \otimes \boldsymbol{N}_\alpha; \tag{4.45}$$

$$\boldsymbol{e} = \frac{1}{2}(\boldsymbol{I} - \boldsymbol{V}^{-2}) = \sum_{\alpha=1}^{3} \frac{1}{2}(1 - \lambda_\alpha^{-2})\, \boldsymbol{n}_\alpha \otimes \boldsymbol{n}_\alpha. \tag{4.46}$$

These expressions motivate the definition of generalized material and spatial strain measures of order $n$ as

$$\boldsymbol{E}^{(n)} = \frac{1}{n}(\boldsymbol{U}^n - \boldsymbol{I}) = \sum_{\alpha=1}^{3} \frac{1}{n}(\lambda_\alpha^n - 1)\, \boldsymbol{N}_\alpha \otimes \boldsymbol{N}_\alpha; \tag{4.47}$$

$$e^{(n)} = \frac{1}{n}(\boldsymbol{I} - \boldsymbol{V}^{-n}) = \sum_{\alpha=1}^{3} \frac{1}{n}\left(1 - \lambda_\alpha^{-n}\right) \boldsymbol{n}_\alpha \otimes \boldsymbol{n}_\alpha; \tag{4.48}$$

$$e^{(-n)} = \boldsymbol{R}\boldsymbol{E}^{(n)}\boldsymbol{R}^T. \tag{4.49}$$

In particular, the case $n \to 0$ gives the *material* and *spatial logarithmic strain tensors*

$$\boldsymbol{E}^{(0)} = \sum_{\alpha=1}^{3} \ln \lambda_\alpha \boldsymbol{N}_\alpha \otimes \boldsymbol{N}_\alpha = \ln \boldsymbol{U}; \tag{4.50}$$

$$e^{(0)} = \sum_{\alpha=1}^{3} \ln \lambda_\alpha \boldsymbol{n}_\alpha \otimes \boldsymbol{n}_\alpha = \ln \boldsymbol{V}. \tag{4.51}$$

## 4.7 VOLUME CHANGE

Consider an infinitesimal volume element in the material configuration with edges parallel to the Cartesian axes given by $d\boldsymbol{X}_1 = dX_1\boldsymbol{E}_1$, $d\boldsymbol{X}_2 = dX_2\boldsymbol{E}_2$, and $d\boldsymbol{X}_3 = dX_3\boldsymbol{E}_3$, where $\boldsymbol{E}_1$, $\boldsymbol{E}_2$, and $\boldsymbol{E}_3$ are the orthogonal unit vectors (Figure 4.5). The elemental material volume $dV$ defined by these three vectors is clearly given as

$$dV = dX_1 dX_2 dX_3. \tag{4.52}$$

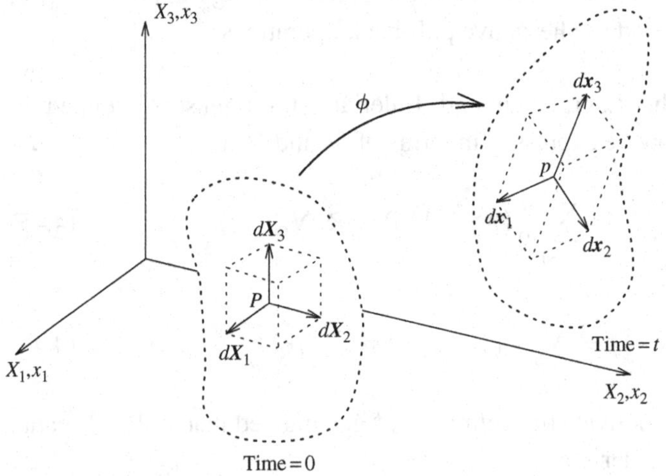


**FIGURE 4.5** Volume change.

In order to obtain the corresponding deformed volume $dv$ in the spatial configuration, note first that the spatial vectors obtained by pushing forward the previous material vectors are

$$d\boldsymbol{x}_1 = \boldsymbol{F} d\boldsymbol{X}_1 = \frac{\partial \phi}{\partial X_1} dX_1; \tag{4.53}$$

$$d\boldsymbol{x}_2 = \boldsymbol{F} d\boldsymbol{X}_2 = \frac{\partial \phi}{\partial X_2} dX_2; \tag{4.54}$$

$$d\boldsymbol{x}_3 = \boldsymbol{F} d\boldsymbol{X}_3 = \frac{\partial \phi}{\partial X_3} dX_3. \tag{4.55}$$

The triple product of these elemental vectors gives the deformed volume as

$$dv = d\boldsymbol{x}_1 \cdot (d\boldsymbol{x}_2 \times d\boldsymbol{x}_3) = \frac{\partial \phi}{\partial X_1} \cdot \left( \frac{\partial \phi}{\partial X_2} \times \frac{\partial \phi}{\partial X_3} \right) dX_1 \, dX_2 \, dX_3. \tag{4.56}$$

Noting that the above triple product is the determinant of $\boldsymbol{F}$ gives the volume change in terms of the Jacobian $J$ as

$$dv = J\, dV; \quad J = \det \boldsymbol{F}. \tag{4.57}$$

Finally, the element of mass can be related to the volume element in terms of the initial and current densities as

$$dm = \rho_0 \, dV = \rho \, dv. \tag{4.58}$$

Hence, the conservation of mass or continuity equation can be expressed as

$$\rho_0 = \rho J. \tag{4.59}$$

## 4.8 DISTORTIONAL COMPONENT OF THE DEFORMATION GRADIENT

When dealing with incompressible and nearly incompressible materials it is necessary to separate the volumetric from the distortional (or isochoric) components of the deformation. Such a separation must ensure that the distortional component, namely $\hat{\boldsymbol{F}}$, does not imply any change in volume. Noting that the determinant of the deformation gradient gives the volume ratio, the determinant of $\hat{\boldsymbol{F}}$ must therefore satisfy

$$\det \hat{\boldsymbol{F}} = 1. \tag{4.60}$$

This condition can be achieved by choosing $\hat{\boldsymbol{F}}$ as

$$\hat{\boldsymbol{F}} = J^{-1/3} \boldsymbol{F}. \tag{4.61}$$

The fact that the condition (4.60) is satisfied is demonstrated as

$$\begin{aligned}\det \hat{\boldsymbol{F}} &= \det(J^{-1/3}\boldsymbol{F}) \\ &= (J^{-1/3})^3 \det \boldsymbol{F} \\ &= J^{-1}J \\ &= 1.\end{aligned} \tag{4.62}$$

The deformation gradient $\boldsymbol{F}$ can now be expressed in terms of the volumetric and distortional components, $J$ and $\hat{\boldsymbol{F}}$ respectively, as

$$\boldsymbol{F} = J^{1/3}\hat{\boldsymbol{F}}. \tag{4.63}$$

This decomposition is illustrated for a two-dimensional case in Figure 4.6.

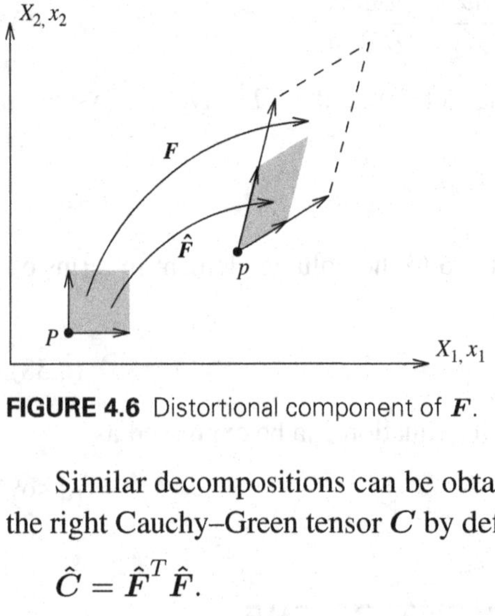


**FIGURE 4.6** Distortional component of $\boldsymbol{F}$.

Similar decompositions can be obtained for other strain-based tensors such as the right Cauchy–Green tensor $\boldsymbol{C}$ by defining its distortional component $\hat{\boldsymbol{C}}$ as

$$\hat{\boldsymbol{C}} = \hat{\boldsymbol{F}}^T\hat{\boldsymbol{F}}. \tag{4.64}$$

Substituting for $\hat{\boldsymbol{F}}$ from Equation (4.61) gives an alternative expression for $\hat{\boldsymbol{C}}$ as

$$\hat{\boldsymbol{C}} = (\det \boldsymbol{C})^{-1/3}\boldsymbol{C}; \quad \det \boldsymbol{C} = J^2. \tag{4.65}$$

**EXAMPLE 4.7: Distortional component of $F$**

Again using Example 4.5, the function of the isochoric component $\hat{\boldsymbol{F}}$ of $\boldsymbol{F}$ can be demonstrated. However, to proceed correctly it is necessary to introduce the third $X_3$ dimension into the formulation, giving

$$\boldsymbol{F} = \frac{1}{4}\begin{bmatrix} 1 & -5 & 0 \\ 8 & 4 & 0 \\ 0 & 0 & 4 \end{bmatrix}; \qquad J = \det \boldsymbol{F} = 2.75;$$

*(continued)*

**EXAMPLE 4.7:** *(cont.)*

from which $\hat{\boldsymbol{F}}$ is found as

$$\hat{\boldsymbol{F}} = J^{-\frac{1}{3}}\boldsymbol{F} = \begin{bmatrix} 0.1784 & -0.8922 & 0 \\ 1.4276 & 0.7138 & 0 \\ 0 & 0 & 0.7138 \end{bmatrix}.$$

Without loss of generality, consider the isochoric deformation at $\boldsymbol{X}=\boldsymbol{0}$ of the orthogonal unit material vectors $\boldsymbol{N}_1=(0.8385, 0.5449, 0)^T$, $\boldsymbol{N}_2=(-0.5449, 0.8385, 0)^T$, and $\boldsymbol{N}_3 = (0, 0, 1)^T$, for which the associated elemental material volume is $dV = 1$. After deformation the material unit vectors push forward to give

$$\hat{\boldsymbol{n}}_1 = \hat{\boldsymbol{F}}\boldsymbol{N}_1 = \begin{bmatrix} -0.3366 \\ 1.5856 \\ 0 \end{bmatrix}; \qquad \hat{\boldsymbol{n}}_2 = \hat{\boldsymbol{F}}\boldsymbol{N}_2 = \begin{bmatrix} -0.8453 \\ -0.1794 \\ 0 \end{bmatrix};$$

$$\hat{\boldsymbol{n}}_3 = \hat{\boldsymbol{F}}\boldsymbol{N}_3 = \begin{bmatrix} 0 \\ 0 \\ 0.7138 \end{bmatrix}.$$

Since $\boldsymbol{N}_\alpha$ are principal directions, $\hat{\boldsymbol{n}}_\alpha$ are orthogonal vectors and the corresponding elemental spatial volume is, conveniently,

$$dv = \|\hat{\boldsymbol{n}}_1\|\|\hat{\boldsymbol{n}}_2\|\|\hat{\boldsymbol{n}}_3\| = 1,$$

thus demonstrating the isochoric nature of $\hat{\boldsymbol{F}}$.

**EXAMPLE 4.8: Simple shear**

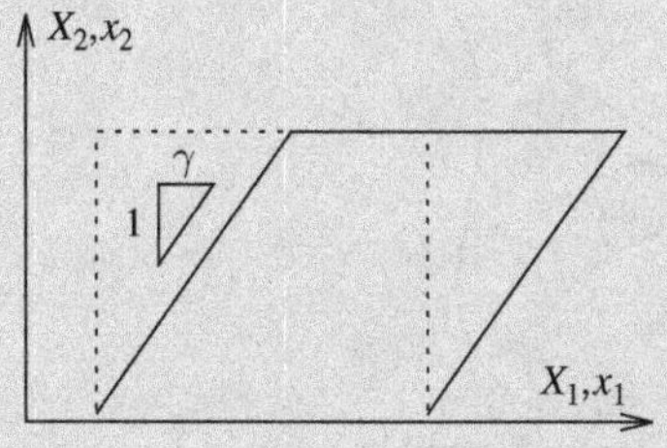


Sometimes the motion of a body is isochoric and the distortional component of $\boldsymbol{F}$ coincides with $\boldsymbol{F}$. A well-known example is the simple shear of a two-dimensional block as defined by the motion

*(continued)*

**EXAMPLE 4.8:** *(cont.)*

$$x_1 = X_1 + \gamma X_2,$$
$$x_2 = X_2,$$

for any arbitrary value of $\gamma$. A simple derivation gives the deformation gradient and its Jacobean $J$ as

$$\boldsymbol{F} = \begin{bmatrix} 1 & \gamma \\ 0 & 1 \end{bmatrix}; \quad J = \det \boldsymbol{F} = 1;$$

and the Lagrangian and Eulerian deformation tensors are

$$\boldsymbol{E} = \frac{1}{2}\begin{bmatrix} 0 & \gamma \\ \gamma & \gamma^2 \end{bmatrix}; \quad \boldsymbol{e} = \frac{1}{2}\begin{bmatrix} 0 & \gamma \\ \gamma & -\gamma^2 \end{bmatrix}.$$

## 4.9 AREA CHANGE

Consider an element of area in the initial configuration $d\boldsymbol{A} = dA\,\boldsymbol{N}$ which after deformation becomes $d\boldsymbol{a} = da\,\boldsymbol{n}$, as shown in Figure 4.7. For the purpose of obtaining a relationship between these two vectors, consider an arbitrary material vector $d\boldsymbol{L}$, which after deformation pushes forward to $d\boldsymbol{l}$. The corresponding initial and current volume elements are

$$dV = d\boldsymbol{L} \cdot d\boldsymbol{A}; \tag{4.66a}$$

$$dv = d\boldsymbol{l} \cdot d\boldsymbol{a}. \tag{4.66b}$$

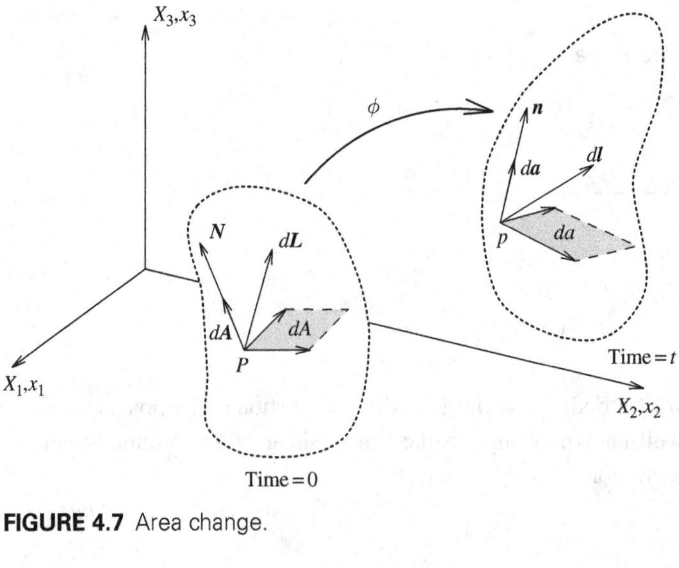


**FIGURE 4.7** Area change.

Relating the current and initial volumes in terms of the Jacobian $J$ and recalling that $dl = \boldsymbol{F}\, d\boldsymbol{L}$ gives

$$J d\boldsymbol{L} \cdot d\boldsymbol{A} = (\boldsymbol{F} d\boldsymbol{L}) \cdot d\boldsymbol{a}. \tag{4.67}$$

The fact that the above expression is valid for any vector $d\boldsymbol{L}$ enables the elements of area to be related as

$$d\boldsymbol{a} = J\boldsymbol{F}^{-T} d\boldsymbol{A}. \tag{4.68}$$

## 4.10 LINEARIZED KINEMATICS

The strain quantities defined in the previous section are nonlinear expressions in terms of the motion $\phi$ and will lead to nonlinear governing equations. These governing equations will need to be linearized in order to enable a Newton–Raphson solution process. It is therefore essential to derive equations for the linearization of the above strain quantities with respect to small changes in the motion.

### 4.10.1 Linearized Deformation Gradient

Consider a small displacement $\boldsymbol{u}(\boldsymbol{x})$ from the current configuration $\boldsymbol{x} = \phi_t(\boldsymbol{X}) = \phi(\boldsymbol{X}, t)$ as shown in Figure 4.8. The deformation gradient $\boldsymbol{F}$ can be linearized in the direction of $\boldsymbol{u}$ at this position as

$$\begin{aligned}
D\boldsymbol{F}(\phi_t)[\boldsymbol{u}] &= \frac{d}{d\epsilon}\bigg|_{\epsilon=0} \boldsymbol{F}(\phi_t + \epsilon \boldsymbol{u}) \\
&= \frac{d}{d\epsilon}\bigg|_{\epsilon=0} \frac{\partial(\phi_t + \epsilon \boldsymbol{u})}{\partial \boldsymbol{X}} \\
&= \frac{d}{d\epsilon}\bigg|_{\epsilon=0} \left(\frac{\partial \phi_t}{\partial \boldsymbol{X}} + \epsilon \frac{\partial \boldsymbol{u}}{\partial \boldsymbol{X}}\right) \\
&= \frac{\partial \boldsymbol{u}}{\partial \boldsymbol{X}} \\
&= (\boldsymbol{\nabla} \boldsymbol{u})\boldsymbol{F}.
\end{aligned} \tag{4.69}$$

Note that if $\boldsymbol{u}$ is given as a function of the initial position $\boldsymbol{X}$ of the body particles (the material description), then

$$D\boldsymbol{F}[\boldsymbol{u}] = \frac{\partial \boldsymbol{u}(\boldsymbol{X})}{\partial \boldsymbol{X}} = \boldsymbol{\nabla}_0 \boldsymbol{u}, \tag{4.70}$$

where $\boldsymbol{\nabla}_0$ indicates the gradient with respect to the coordinates at the initial configuration.

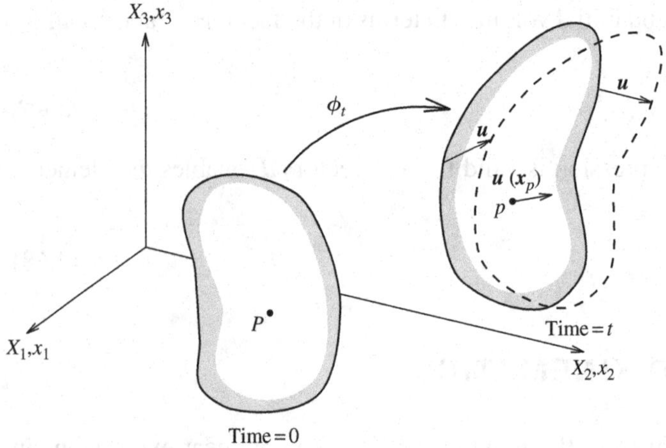


**FIGURE 4.8** Linearized kinematics.

### 4.10.2 Linearized Strain

Using Equation (4.69) and the product rule seen in Section 2.3.3, the Lagrangian (or Green's) strain can be linearized at the current configuration in the direction $\boldsymbol{u}$ as

$$\begin{aligned} D\boldsymbol{E}[\boldsymbol{u}] &= \frac{1}{2}(\boldsymbol{F}^T D\boldsymbol{F}[\boldsymbol{u}] + D\boldsymbol{F}^T[\boldsymbol{u}]\boldsymbol{F}) \\ &= \frac{1}{2}[\boldsymbol{F}^T \boldsymbol{\nabla u F} + \boldsymbol{F}^T(\boldsymbol{\nabla u})^T \boldsymbol{F}] \\ &= \frac{1}{2}\boldsymbol{F}^T[\boldsymbol{\nabla u} + (\boldsymbol{\nabla u})^T]\boldsymbol{F}. \end{aligned} \tag{4.71}$$

Note that half the tensor inside [ ] is the small strain tensor $\boldsymbol{\varepsilon}$, and therefore $D\boldsymbol{E}[\boldsymbol{u}]$ can be interpreted as the pull-back of the small strain tensor $\boldsymbol{\varepsilon}$ as

$$D\boldsymbol{E}[\boldsymbol{u}] = \phi_*^{-1}[\boldsymbol{\varepsilon}] = \boldsymbol{F}^T \boldsymbol{\varepsilon} \boldsymbol{F}. \tag{4.72}$$

In particular, if the linearization of $\boldsymbol{E}$ is performed at the initial material configuration, that is, when $\boldsymbol{x} = \boldsymbol{X}$ and therefore $\boldsymbol{F} = \boldsymbol{I}$, then

$$D\boldsymbol{E}_0[\boldsymbol{u}] = \boldsymbol{\varepsilon}. \tag{4.73}$$

Similarly, the right and left Cauchy–Green deformation tensors defined in Equations (4.15) and (4.17) can be linearized to give

$$D\boldsymbol{C}[\boldsymbol{u}] = 2\boldsymbol{F}^T \boldsymbol{\varepsilon} \boldsymbol{F}; \tag{4.74a}$$

$$D\boldsymbol{b}[\boldsymbol{u}] = (\boldsymbol{\nabla u})\boldsymbol{b} + \boldsymbol{b}(\boldsymbol{\nabla u})^T. \tag{4.74b}$$

### 4.10.3 Linearized Volume Change

The volume change has been shown earlier to be given by the Jacobian $J = \det \boldsymbol{F}$. Using the chain rule given in Section 2.3.3, the directional derivative of $J$ with respect to an increment $\boldsymbol{u}$ in the spatial configuration is

$$DJ[\boldsymbol{u}] = D\det(\boldsymbol{F})[D\boldsymbol{F}[\boldsymbol{u}]]. \tag{4.75}$$

Recalling the directional derivative of the determinant from (2.119) and the linearization of $\boldsymbol{F}$ from Equation (4.69) gives

$$\begin{aligned} DJ[\boldsymbol{u}] &= J\,\mathrm{tr}\left(\boldsymbol{F}^{-1}\frac{\partial \boldsymbol{u}}{\partial \boldsymbol{X}}\right) \\ &= J\,\mathrm{tr}\boldsymbol{\nabla}\boldsymbol{u} \\ &= J\mathrm{div}\,\boldsymbol{u}. \end{aligned} \tag{4.76}$$

Alternatively, the above equation can be expressed in terms of the linear strain tensor $\boldsymbol{\varepsilon}$ as

$$DJ[\boldsymbol{u}] = J\,\mathrm{tr}\boldsymbol{\varepsilon}. \tag{4.77}$$

Finally, the directional derivative of the volume element in the direction of $\boldsymbol{u}$ emerges from Equation (4.57) as

$$D(dv)[\boldsymbol{u}] = (\mathrm{tr}\,\boldsymbol{\varepsilon})\;dv. \tag{4.78}$$

## 4.11 VELOCITY AND MATERIAL TIME DERIVATIVES

### 4.11.1 Velocity

Obviously, many nonlinear processes are time-dependent; consequently, it is necessary to consider velocity and material time derivatives of various quantities. However, even if the process is not rate-dependent it is nevertheless convenient to establish the equilibrium equations in terms of virtual velocities and associated virtual time-dependent quantities. For this purpose consider the usual motion of the body given by Equation (4.1) as

$$\boldsymbol{x} = \boldsymbol{\phi}(\boldsymbol{X}, t), \tag{4.79}$$

from which the velocity of a particle is defined as the time derivative of $\boldsymbol{\phi}$ (Figure 4.9) as

$$\boldsymbol{v}(\boldsymbol{X}, t) = \frac{\partial \boldsymbol{\phi}(\boldsymbol{X}, t)}{\partial t}. \tag{4.80}$$

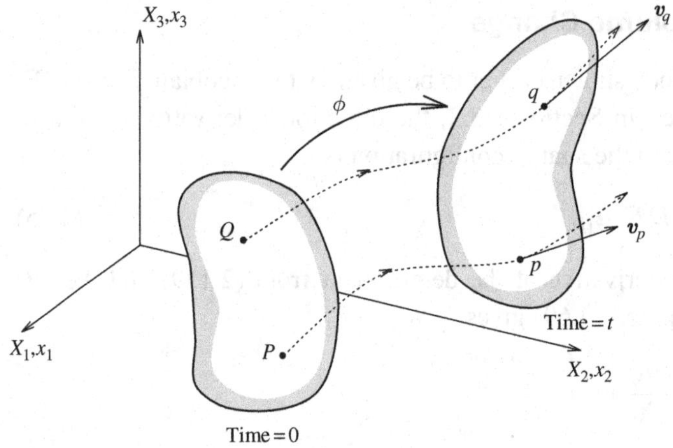


**FIGURE 4.9** Particle velocity.

Observe that the velocity is a spatial vector despite the fact that the equation has been expressed in terms of the material coordinates of the particle $\boldsymbol{X}$. In fact, by inverting Equation (4.79) the velocity can be more consistently expressed as a function of the spatial position $\boldsymbol{x}$ and time as

$$\boldsymbol{v}(\boldsymbol{x}, t) = \boldsymbol{v}(\boldsymbol{\phi}^{-1}(\boldsymbol{x}, t), t). \tag{4.81}$$

### 4.11.2 Material Time Derivative

Given a general scalar or tensor quantity $\boldsymbol{g}$ expressed in terms of the material coordinates $\boldsymbol{X}$, the time derivative of $\boldsymbol{g}(\boldsymbol{X}, t)$, denoted henceforth by $\dot{\boldsymbol{g}}(\boldsymbol{X}, t)$ or $d\boldsymbol{g}(\boldsymbol{X}, t)/dt$, is defined as

$$\dot{\boldsymbol{g}} = \frac{d\boldsymbol{g}}{dt} = \frac{\partial \boldsymbol{g}(\boldsymbol{X}, t)}{\partial t}. \tag{4.82}$$

This expression measures the change in $\boldsymbol{g}$ associated with a specific particle initially located at $\boldsymbol{X}$, and it is known as the *material time derivative* of $\boldsymbol{g}$. Invariably, however, spatial quantities are expressed as functions of the spatial position $\boldsymbol{x}$, in which case the material derivative is more complicated to establish. The complication arises because as time progresses the specific particle being considered changes spatial position. Consequently, the material time derivative in this case is obtained from a careful consideration of the motion of the particle as

$$\dot{\boldsymbol{g}}(\boldsymbol{x}, t) = \lim_{\Delta t \to 0} \frac{\boldsymbol{g}(\boldsymbol{\phi}(\boldsymbol{X}, t + \Delta t), t + \Delta t) - \boldsymbol{g}(\boldsymbol{\phi}(\boldsymbol{X}, t), t)}{\Delta t}. \tag{4.83}$$

This equation clearly illustrates that $\boldsymbol{g}$ changes in time (i) as a result of a change in time but with the particle remaining in the same spatial position, and (ii) because of the change in spatial position of the specific particle. Using the chain rule, Equation (4.83) gives the material derivative of $\boldsymbol{g}(\boldsymbol{x},t)$ as

$$\dot{\boldsymbol{g}}(\boldsymbol{x},t)=\frac{\partial \boldsymbol{g}(\boldsymbol{x},t)}{\partial t}+\frac{\partial \boldsymbol{g}(\boldsymbol{x},t)}{\partial \boldsymbol{x}}\frac{\partial \boldsymbol{\phi}(\boldsymbol{X},t)}{\partial t}=\frac{\partial \boldsymbol{g}(\boldsymbol{x},t)}{\partial t}+(\boldsymbol{\nabla}\boldsymbol{g})\boldsymbol{v}. \tag{4.84}$$

The second term, involving the particle velocity in Equation (4.84), is often referred to as the *convective derivative*.

**EXAMPLE 4.9: Material time derivative**

Here Example 4.1 is revisited to illustrate the calculation of a material time derivative based on either a material or spatial description. The material description of the temperature distribution along the rod is $T = Xt^2$, yielding $\dot{T}$ directly as $\dot{T} = 2Xt$. From the description of motion, $x = (1+t)X$, the velocity is expressed as $v = X$ or $v = x/(1+t)$ in the material and spatial descriptions respectively. Using the spatial description, $T = xt^2/(1+t)$ gives

$$\frac{\partial T(x,t)}{\partial t}=\frac{(2t+t^2)x}{(1+t)^2};\quad \nabla T=\frac{\partial T(x,t)}{\partial x}=\frac{t^2}{(1+t)}.$$

Hence from Equation (4.84), $\dot{T} = 2xt/(1+t) = 2Xt$.

### 4.11.3 Directional Derivative and Time Rates

Traditionally, linearization has been implemented in terms of an artificial time and associated rates. This procedure, however, leads to confusion when real rates are involved in the problem. It transpires that linearization as defined in Equation (2.101) avoids this confusion and leads to a much clearer finite element formulation. Nevertheless, it is valuable to appreciate the relationship between linearization and the material time derivative. For this purpose consider a general operator $\mathcal{F}$ that applies to the motion $\boldsymbol{\phi}(\boldsymbol{X},t)$. The directional derivative of $\mathcal{F}$ in the direction of $\boldsymbol{v}$ coincides with the time derivative of $\mathcal{F}$, that is,

$$D\mathcal{F}[\boldsymbol{v}]=\frac{d}{dt}\mathcal{F}(\boldsymbol{\phi}(\boldsymbol{X},t)). \tag{4.85}$$

To prove this, let $\boldsymbol{\phi}_{\boldsymbol{X}}(t)$ denote the motion of a given particle and $F(t)$ the function of time obtained by applying the operator $\mathcal{F}$ on this motion as

$$F(t)=\mathcal{F}(\boldsymbol{\phi}_{\boldsymbol{X}}(t));\quad \boldsymbol{\phi}_{\boldsymbol{X}}(t)=\boldsymbol{\phi}(\boldsymbol{X},t). \tag{4.86}$$

Note first that the derivative with respect to time of a function $f(t)$ is related to the directional derivative of this function in the direction of an increment in time $\Delta t$ as

$$Df[\Delta t] = \frac{d}{d\epsilon}\bigg|_{\epsilon=0} f(t + \epsilon\Delta t) = \frac{df}{dt}\Delta t. \tag{4.87}$$

Using this equation with $\Delta t = 1$ for the functions $F(t)$ and $\phi_{\boldsymbol{X}}(t)$ and recalling the chain rule for directional derivatives given by Equation (2.105c) gives

$$\begin{aligned}\frac{d}{dt}\mathcal{F}(\boldsymbol{\phi}(\boldsymbol{X}, t)) &= \frac{df}{dt} \\ &= DF[1] \\ &= D\mathcal{F}(\boldsymbol{\phi}_{\boldsymbol{X}}(t))[1] \\ &= D\mathcal{F}[D\boldsymbol{\phi}_{\boldsymbol{X}}[1]] \\ &= D\mathcal{F}[\boldsymbol{v}].\end{aligned} \tag{4.88}$$

A simple illustration of Equation (4.85) emerges from the time derivative of the deformation gradient tensor $\boldsymbol{F}$, which can be easily obtained from Equations (4.6) and (4.80) as

$$\dot{\boldsymbol{F}} = \frac{d}{dt}\left(\frac{\partial \boldsymbol{\phi}}{\partial \boldsymbol{X}}\right) = \frac{\partial}{\partial \boldsymbol{X}}\left(\frac{\partial \boldsymbol{\phi}}{\partial t}\right) = \boldsymbol{\nabla}_0 \boldsymbol{v}. \tag{4.89}$$

Alternatively, recalling Equation (4.70) for the linearized deformation gradient $D\boldsymbol{F}$ gives

$$D\boldsymbol{F}[\boldsymbol{v}] = \boldsymbol{\nabla}_0 \boldsymbol{v} = \dot{\boldsymbol{F}}. \tag{4.90}$$

### 4.11.4 Velocity Gradient

We have defined velocity as a spatial vector. Consequently, velocity was expressed in Equation (4.81) as a function of the spatial coordinates as $\boldsymbol{v}(\boldsymbol{x}, t)$. The derivative of this expression with respect to the spatial coordinates defines the *velocity gradient tensor* $\boldsymbol{l}$ as

$$\boldsymbol{l} = \frac{\partial \boldsymbol{v}(\boldsymbol{x}, t)}{\partial \boldsymbol{x}} = \boldsymbol{\nabla}\boldsymbol{v}. \tag{4.91}$$

This is clearly a spatial tensor, which, as Figure 4.10 shows, gives the relative velocity of a particle currently at point $q$ with respect to a particle currently at $p$ as $d\boldsymbol{v} = \boldsymbol{l}d\boldsymbol{x}$. The tensor $\boldsymbol{l}$ enables the time derivative of the deformation gradient given by Equation (4.89) to be more usefully written as

$$\dot{\boldsymbol{F}} = \frac{\partial \boldsymbol{v}}{\partial \boldsymbol{X}} = \frac{\partial \boldsymbol{v}}{\partial \boldsymbol{x}}\frac{\partial \boldsymbol{\phi}}{\partial \boldsymbol{X}} = \boldsymbol{l}\boldsymbol{F}, \tag{4.92}$$

from which an alternative expression for $\boldsymbol{l}$ emerges as

$$\boldsymbol{l} = \dot{\boldsymbol{F}}\boldsymbol{F}^{-1}. \tag{4.93}$$

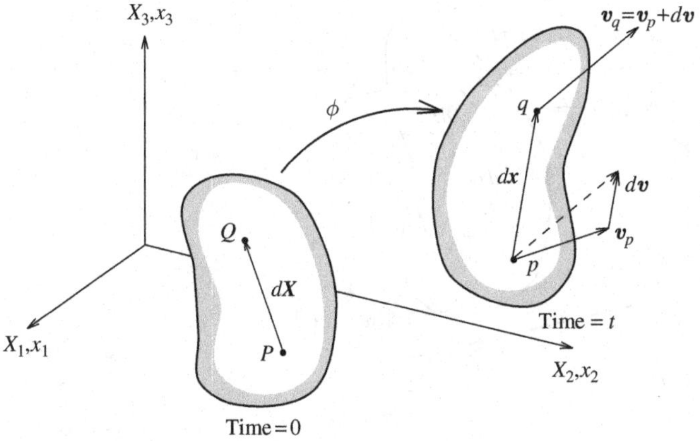


**FIGURE 4.10** Velocity gradient.

## 4.12 RATE OF DEFORMATION

Consider again the initial elemental vectors $d\boldsymbol{X}_1$ and $d\boldsymbol{X}_2$ introduced in Section 4.4 and their corresponding pushed-forward spatial counterparts $d\boldsymbol{x}_1$ and $d\boldsymbol{x}_2$ given as (Figure 4.11)

$$d\boldsymbol{x}_1 = \boldsymbol{F}d\boldsymbol{X}_1; \quad d\boldsymbol{x}_2 = \boldsymbol{F}d\boldsymbol{X}_2. \tag{4.94a,b}$$

In Section 4.5 strain was defined and measured as the change in the scalar product of two arbitrary vectors. Similarly, strain rate can now be defined as the rate of change of the scalar product of any pair of vectors. For the purpose of measuring this rate of change, recall from Section 4.5 that the current scalar product could be expressed in terms of the material vectors $d\boldsymbol{X}_1$ and $d\boldsymbol{X}_2$ (which are not functions of time) and the time-dependent right Cauchy–Green tensor $\boldsymbol{C}$ as

$$d\boldsymbol{x}_1 \cdot d\boldsymbol{x}_2 = d\boldsymbol{X}_1 \cdot \boldsymbol{C}\, d\boldsymbol{X}_2. \tag{4.95}$$

Differentiating this expression with respect to time and recalling the relationship between the Lagrangian strain tensor $\boldsymbol{E}$ and the right Cauchy–Green tensor as $2\boldsymbol{E} = (\boldsymbol{C} - \boldsymbol{I})$ gives the current rate of change of the scalar product in terms of the initial elemental vectors as

$$\frac{d}{dt}(d\boldsymbol{x}_1 \cdot d\boldsymbol{x}_2) = d\boldsymbol{X}_1 \cdot \dot{\boldsymbol{C}}\, d\boldsymbol{X}_2 = 2\, d\boldsymbol{X}_1 \cdot \dot{\boldsymbol{E}}\, d\boldsymbol{X}_2, \tag{4.96}$$

where $\dot{\boldsymbol{E}}$, the derivative with respect to time of the Lagrangian strain tensor, is known as the *material strain rate tensor* and can be easily obtained in terms of $\dot{\boldsymbol{F}}$ as

$$\dot{\boldsymbol{E}} = \frac{1}{2}\dot{\boldsymbol{C}} = \frac{1}{2}(\dot{\boldsymbol{F}}^T\boldsymbol{F} + \boldsymbol{F}^T\dot{\boldsymbol{F}}). \tag{4.97}$$

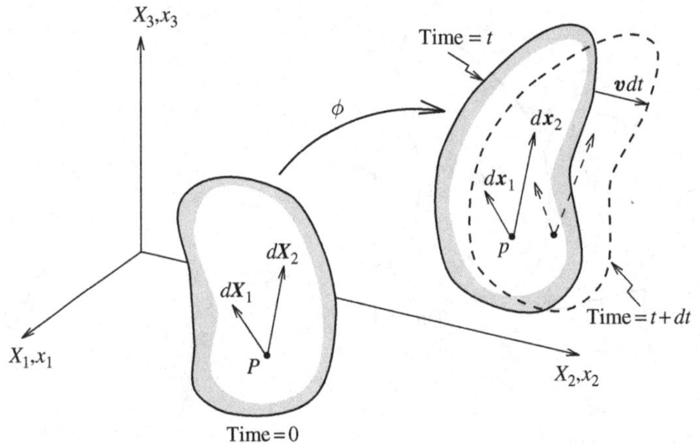


**FIGURE 4.11** Rate of deformation.

The material strain rate tensor $\dot{\boldsymbol{E}}$ gives the current rate of change of the scalar product in terms of the initial elemental vectors. Alternatively, it is often convenient to express the same rate of change in terms of the current spatial vectors. For this purpose, recall first from Section 4.4 that Equations (4.94a,b) can be inverted as

$$d\boldsymbol{X}_1 = \boldsymbol{F}^{-1} d\boldsymbol{x}_1; \quad d\boldsymbol{X}_2 = \boldsymbol{F}^{-1} d\boldsymbol{x}_2. \tag{4.98a,b}$$

Introducing these expressions into Equation (4.96) gives the rate of change of the scalar product in terms of $d\boldsymbol{x}_1$ and $d\boldsymbol{x}_2$ as

$$\frac{1}{2}\frac{d}{dt}(d\boldsymbol{x}_1 \cdot d\boldsymbol{x}_2) = d\boldsymbol{x}_1 \cdot (\boldsymbol{F}^{-T}\dot{\boldsymbol{E}}\boldsymbol{F}^{-1}) d\boldsymbol{x}_2. \tag{4.99}$$

The tensor in the expression on the right-hand side is simply the pushed-forward spatial counterpart of $\dot{\boldsymbol{E}}$ and is known as the *rate of deformation tensor* **d**, given as

$$\boldsymbol{d} = \phi_*[\dot{\boldsymbol{E}}] = \boldsymbol{F}^{-T}\dot{\boldsymbol{E}}\boldsymbol{F}^{-1}; \quad \dot{\boldsymbol{E}} = \phi_*^{-1}[\boldsymbol{d}] = \boldsymbol{F}^T\boldsymbol{d}\boldsymbol{F}. \tag{4.100a,b}$$

A more conventional expression for $\boldsymbol{d}$ emerges from the combination of Equations (4.92) for $\dot{\boldsymbol{F}}$ and (4.97) for $\dot{\boldsymbol{E}}$ to give, after simple algebra, the tensor $\boldsymbol{d}$ as the symmetric part of $\boldsymbol{l}$ as

$$\boldsymbol{d} = \frac{1}{2}(\boldsymbol{l} + \boldsymbol{l}^T). \tag{4.101}$$

**Remark 4.6**: A simple physical interpretation of the tensor $\boldsymbol{d}$ can be obtained by taking $d\boldsymbol{x}_1 = d\boldsymbol{x}_2 = d\boldsymbol{x}$, as shown in Figure 4.12, to give

$$\frac{1}{2}\frac{d}{dt}(d\boldsymbol{x} \cdot d\boldsymbol{x}) = d\boldsymbol{x} \cdot \boldsymbol{d}\, d\boldsymbol{x}. \tag{4.102}$$

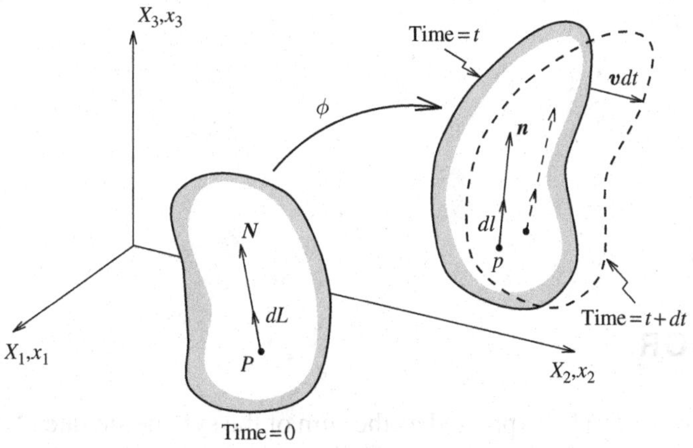


**FIGURE 4.12** Rate of change of length.

Expressing $d\boldsymbol{x}$ as $dl\,\boldsymbol{n}$, where $\boldsymbol{n}$ is a unit vector in the direction of $d\boldsymbol{x}$, as shown in Figure 4.12, gives

$$\frac{1}{2}\frac{d}{dt}(dl^2) = dl^2\,\boldsymbol{n}\cdot\boldsymbol{d}\boldsymbol{n}, \tag{4.103}$$

which, noting that $d(dl^2)/dt = 2dl\,d(dl)/dt$, finally yields

$$\boldsymbol{n}\cdot\boldsymbol{d}\boldsymbol{n} = \frac{1}{dl}\frac{d}{dt}(dl) = \frac{d\ln(dl)}{dt}. \tag{4.104}$$

Hence the rate of deformation tensor $\boldsymbol{d}$ gives the rate of extension per unit current length of a line element having a current direction defined by $\boldsymbol{n}$. In particular, for a rigid body motion $\boldsymbol{d} = \boldsymbol{0}$.

**Remark 4.7**: Note that the spatial rate of deformation tensor $\boldsymbol{d}$ is not the material derivative of the Almansi or spatial strain tensor $\boldsymbol{e}$ introduced in Section 4.5. Instead, $\boldsymbol{d}$ is the push-forward of $\dot{\boldsymbol{E}}$, which is the derivative with respect to time of the pull-back of $\boldsymbol{e}$, that is,

$$\boldsymbol{d} = \phi_*\left[\frac{d}{dt}(\phi_*^{-1}[\boldsymbol{e}])\right]. \tag{4.105}$$

This operation is illustrated in Figure 4.13. It is known as the *Lie derivative* of a tensor quantity over the mapping $\phi$ and is generally expressed as

$$\mathcal{L}_\phi[\boldsymbol{g}] = \phi_*\left[\frac{d}{dt}(\phi_*^{-1}[\boldsymbol{g}])\right]. \tag{4.106}$$

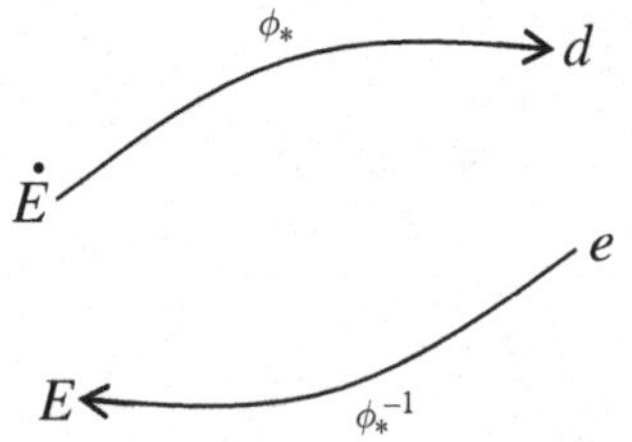


**FIGURE 4.13** Lie derivative.

## 4.13 SPIN TENSOR

The velocity gradient tensor $\boldsymbol{l}$ can be expressed as the sum of the symmetric rate of deformation tensor $\boldsymbol{d}$ plus an additional antisymmetric component $\boldsymbol{w}$ as

$$\boldsymbol{l} = \boldsymbol{d} + \boldsymbol{w}, \quad \boldsymbol{d}^T = \boldsymbol{d}, \quad \boldsymbol{w}^T = -\boldsymbol{w}, \tag{4.107}$$

where the antisymmetric tensor $\boldsymbol{w}$ is known as the *spin tensor* and can be obtained as

$$\boldsymbol{w} = \frac{1}{2}(\boldsymbol{l} - \boldsymbol{l}^T). \tag{4.108}$$

The terminology employed for $\boldsymbol{w}$ can be justified by obtaining a relationship between the spin tensor and the rate of change of the rotation tensor $\boldsymbol{R}$ introduced in Section 4.6. For this purpose, note that $\boldsymbol{l}$ can be obtained from Equation (4.93), thereby enabling Equation (4.108) to be rewritten as

$$\boldsymbol{w} = \frac{1}{2}(\dot{\boldsymbol{F}}\boldsymbol{F}^{-1} - \boldsymbol{F}^{-T}\dot{\boldsymbol{F}}^T). \tag{4.109}$$

Combining this equation with the polar decomposition of $\boldsymbol{F}$ and its time derivative as

$$\boldsymbol{F} = \boldsymbol{R}\boldsymbol{U}, \tag{4.110a}$$

$$\dot{\boldsymbol{F}} = \dot{\boldsymbol{R}}\boldsymbol{U} + \boldsymbol{R}\dot{\boldsymbol{U}} \tag{4.110b}$$

yields, after some simple algebra, $\boldsymbol{w}$ as

$$\boldsymbol{w} = \frac{1}{2}(\dot{\boldsymbol{R}}\boldsymbol{R}^T - \boldsymbol{R}\dot{\boldsymbol{R}}^T) + \frac{1}{2}\boldsymbol{R}(\dot{\boldsymbol{U}}\boldsymbol{U}^{-1} - \boldsymbol{U}^{-1}\dot{\boldsymbol{U}})\boldsymbol{R}^T. \tag{4.111}$$

Finally, differentiation with respect to time of the expression $\boldsymbol{R}\boldsymbol{R}^T = \boldsymbol{I}$ easily shows that the tensor $\dot{\boldsymbol{R}}\boldsymbol{R}^T$ is antisymmetric, that is,

$$\boldsymbol{R}\dot{\boldsymbol{R}}^T = -\dot{\boldsymbol{R}}\boldsymbol{R}^T, \tag{4.112}$$

thereby allowing Equation (4.111) to be rewritten as

$$\boldsymbol{w} = \dot{\boldsymbol{R}}\boldsymbol{R}^T + \frac{1}{2}\boldsymbol{R}(\dot{\boldsymbol{U}}\boldsymbol{U}^{-1} - \boldsymbol{U}^{-1}\dot{\boldsymbol{U}})\boldsymbol{R}^T. \tag{4.113}$$

The second term in the above equation vanishes in several cases for instance, rigid body motion. A more realistic example arises when the principal directions of strain given by the Lagrangian triad remain constant; such a case is the deformation of a cylindrical rod. Under such circumstances $\dot{U}$ can be derived from Equation (4.31) as

$$\dot{U} = \sum_{\alpha=1}^{3} \dot{\lambda}_\alpha \boldsymbol{N}_\alpha \otimes \boldsymbol{N}_\alpha. \tag{4.114}$$

Note that this implies that $\dot{U}$ has the same principal directions as $\boldsymbol{U}$. Expressing the inverse stretch tensor as $\boldsymbol{U}^{-1} = \sum_{\alpha=1}^{3} \lambda_\alpha^{-1} \boldsymbol{N}_\alpha \otimes \boldsymbol{N}_\alpha$ gives

$$\dot{U}\boldsymbol{U}^{-1} = \sum_{\alpha=1}^{3} \lambda_\alpha^{-1}\dot{\lambda}_\alpha \boldsymbol{N}_\alpha \otimes \boldsymbol{N}_\alpha = \boldsymbol{U}^{-1}\dot{U}. \tag{4.115}$$

Consequently, the spin tensor $\boldsymbol{w}$ becomes

$$\boldsymbol{w} = \dot{\boldsymbol{R}}\boldsymbol{R}^T. \tag{4.116}$$

Often the spin tensor $\boldsymbol{w}$ is physically interpreted in terms of its associated *angular velocity vector* $\boldsymbol{\omega}$ (see Section 2.2.2), defined as

$$\omega_1 = w_{32} = -w_{23}, \tag{4.117a}$$

$$\omega_2 = w_{13} = -w_{31}, \tag{4.117b}$$

$$\omega_3 = w_{21} = -w_{12}, \tag{4.117c}$$

so that, in the case of a rigid body motion where $\boldsymbol{l} = \boldsymbol{w}$, the incremental or relative velocity of a particle $q$ in the neighborhood of particle $p$ shown in Figure 4.14 can be expressed as

$$d\boldsymbol{v} = \boldsymbol{w}\, d\boldsymbol{x} = \boldsymbol{\omega} \times d\boldsymbol{x}. \tag{4.118}$$

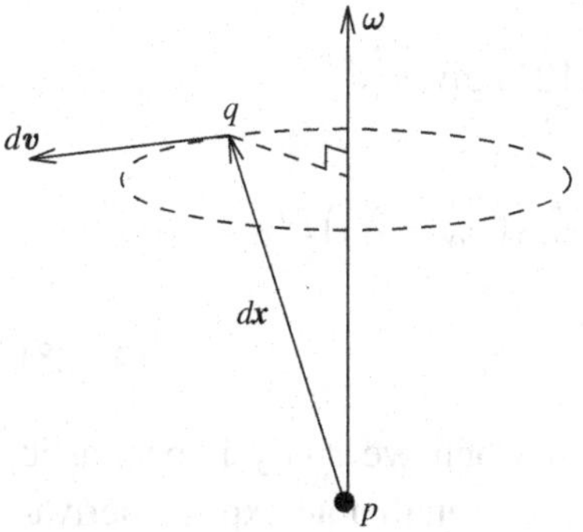


**FIGURE 4.14** Angular velocity vector.

**Remark 4.8**: In the case of a constant Lagrangian triad, useful equations similar to (4.114) can be obtained for the material strain rate tensor $\dot{\boldsymbol{E}}$ by differentiating Equation (4.45) with respect to time to give

$$\dot{\boldsymbol{E}} = \sum_{\alpha=1}^{3} \frac{1}{2}\frac{d\lambda_\alpha^2}{dt}\, \boldsymbol{N}_\alpha \otimes \boldsymbol{N}_\alpha. \tag{4.119}$$

Furthermore, pushing this expression forward to the spatial configuration with the aid of Equations (4.100a,b)$_a$ and (4.44b) enables the rate of deformation tensor to be expressed in terms of the time rate of the logarithmic stretches as

$$\boldsymbol{d} = \sum_{\alpha=1}^{3} \frac{d \ln \lambda_\alpha}{dt}\, \boldsymbol{n}_\alpha \otimes \boldsymbol{n}_\alpha. \tag{4.120}$$

In general, however, the Lagrangian triad changes with time, and both the material strain rate and rate of deformation tensors exhibit off-diagonal terms (that is, shear terms) when expressed in the corresponding material and spatial principal axes. The general equation for $\dot{\boldsymbol{E}}$ is easily obtained from Equation (4.45) as

$$\dot{\boldsymbol{E}} = \sum_{\alpha=1}^{3} \frac{1}{2}\frac{d\lambda_\alpha^2}{dt}\, \boldsymbol{N}_\alpha \otimes \boldsymbol{N}_\alpha + \sum_{\alpha=1}^{3} \frac{1}{2}\lambda_\alpha^2 (\dot{\boldsymbol{N}}_\alpha \otimes \boldsymbol{N}_\alpha + \boldsymbol{N}_\alpha \otimes \dot{\boldsymbol{N}}_\alpha), \tag{4.121}$$

where time differentiation of the expression $\boldsymbol{N}_\alpha \cdot \boldsymbol{N}_\beta = \delta_{\alpha\beta}$ to give $\dot{\boldsymbol{N}}_\alpha \cdot \boldsymbol{N}_\beta = -\dot{\boldsymbol{N}}_\beta \cdot \boldsymbol{N}_\alpha$ reveals that the rate of change of the Lagrangian triad can be expressed in terms of the components of a skew symmetric tensor $\boldsymbol{W}$ as

$$\dot{\boldsymbol{N}}_\alpha = \sum_{\beta=1}^{3} W_{\alpha\beta} \boldsymbol{N}_\beta; \quad W_{\alpha\beta} = -W_{\beta\alpha}. \tag{4.122}$$

Substituting this expression into Equation (4.121) gives

$$\dot{\boldsymbol{E}} = \sum_{\alpha=1}^{3} \frac{1}{2}\frac{d\lambda_\alpha^2}{dt}\, \boldsymbol{N}_\alpha \otimes \boldsymbol{N}_\alpha + \sum_{\substack{\alpha,\beta=1\\ \alpha\neq\beta}}^{3} \frac{1}{2} W_{\alpha\beta}(\lambda_\alpha^2 - \lambda_\beta^2)\boldsymbol{N}_\alpha \otimes \boldsymbol{N}_\beta. \tag{4.123}$$

This equation will prove useful in Chapter 6 when we study hyperelastic materials in principal directions, where it will be seen that an explicit derivation of $W_{\alpha\beta}$ is unnecessary.

## 4.14 RATE OF CHANGE OF VOLUME

The volume change between the initial and current configuration was given in Section 4.7 in terms of the Jacobian $J$ as

$$dv = J\,dV; \quad J = \det \boldsymbol{F}. \tag{4.124}$$

Differentiating this expression with respect to time gives the material rate of change of the volume element as‡ (Figure 4.15)

$$\frac{d}{dt}(dv) = \dot{J}\,dV = \frac{\dot{J}}{J}\,dv. \tag{4.125}$$

The relationship between time and directional derivatives discussed in Section 4.11.3 can now be used to enable the material rate of change of the Jacobian to be evaluated as

$$\dot{J} = DJ[\boldsymbol{v}]. \tag{4.126}$$

Recalling Equations (4.76) and (4.77) for the linearized volume change $DJ[\boldsymbol{u}]$ gives a similar expression for $\dot{J}$, where now the linear strain tensor $\boldsymbol{\varepsilon}$ has been replaced by the rate of deformation tensor $\boldsymbol{d}$ to give

$$\dot{J} = J\,\mathrm{tr}\boldsymbol{d}. \tag{4.127}$$

Alternatively, noting that the trace of $\boldsymbol{d}$ is the divergence of $\boldsymbol{v}$ gives

$$\dot{J} = J\,\mathrm{div}\,\boldsymbol{v}. \tag{4.128}$$

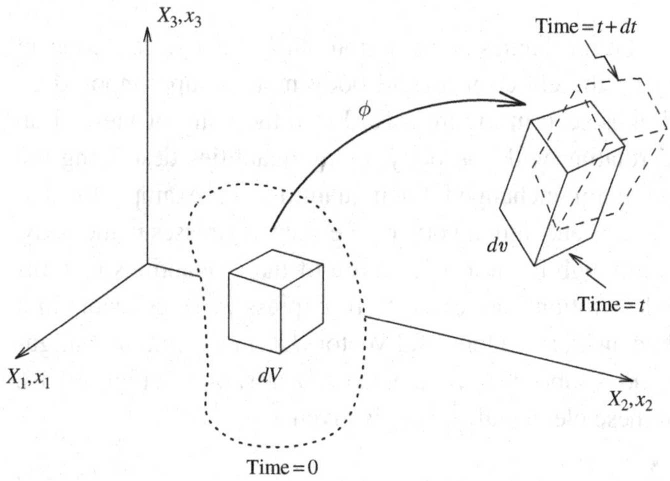


**FIGURE 4.15** Material rate of change of volume.

‡ Note that the spatial rate of change of the volume element is zero, that is, $\partial(dv)/\partial t = 0$.

An alternative equation for $\dot{J}$ can be derived in terms of the material rate tensors $\dot{\boldsymbol{C}}$ or $\dot{\boldsymbol{E}}$ from Equations (4.127), (4.100a,b), and (4.97) to give

$$\begin{aligned}\dot{J} &= J\text{tr}\boldsymbol{d}\\ &= J\text{tr}(\boldsymbol{F}^{-T}\dot{\boldsymbol{E}}\boldsymbol{F}^{-1})\\ &= J\text{tr}(\boldsymbol{C}^{-1}\dot{\boldsymbol{E}})\\ &= J\boldsymbol{C}^{-1} : \dot{\boldsymbol{E}}\\ &= \frac{1}{2}J\boldsymbol{C}^{-1} : \dot{\boldsymbol{C}}. \end{aligned} \tag{4.129}$$

This alternative expression for $\dot{J}$ is used later, in Chapter 6, when we consider the important topic of incompressible elasticity.

Finally, taking the material derivative of Equation (4.59) for the current density enables the conservation of mass equation to be written in a rate form as

$$\frac{d\rho}{dt} + \rho\,\text{div}\,\boldsymbol{v} = 0. \tag{4.130}$$

Alternatively, expressing the material rate of $\rho$ in terms of the spatial rate $\partial\rho/\partial t$ using Equation (4.84) gives the continuity equation in a form often found in the fluid dynamics literature as

$$\frac{\partial\rho}{\partial t} + \text{div}\,(\rho\boldsymbol{v}) = 0. \tag{4.131}$$

## 4.15 SUPERIMPOSED RIGID BODY MOTIONS AND OBJECTIVITY

An important concept in solid mechanics is the notion of *objectivity*. This concept can be explored by studying the effect of a rigid body motion superimposed on the deformed configuration as seen in Figure 4.16. From the point of view of an observer attached to and rotating with the body, many quantities describing the behavior of the solid will remain unchanged. Such quantities, for example the distance between any two particles and, among others, the state of stresses in the body, are said to be *objective*. Although the intrinsic nature of these quantities remains unchanged, their spatial description may change. To express these concepts in a mathematical framework, consider an elemental vector $d\boldsymbol{X}$ in the initial configuration that deforms to $d\boldsymbol{x}$ and is subsequently rotated to $d\tilde{\boldsymbol{x}}$ as shown in Figure 4.16. The relationship between these elemental vectors is given as

$$d\tilde{\boldsymbol{x}} = \boldsymbol{Q}d\boldsymbol{x} = \boldsymbol{Q}\boldsymbol{F}d\boldsymbol{X}, \tag{4.132}$$

where $\boldsymbol{Q}$ is an orthogonal tensor describing the superimposed rigid body rotation. Although the vector $d\tilde{\boldsymbol{x}}$ is different from $d\boldsymbol{x}$, their magnitudes are obviously equal.

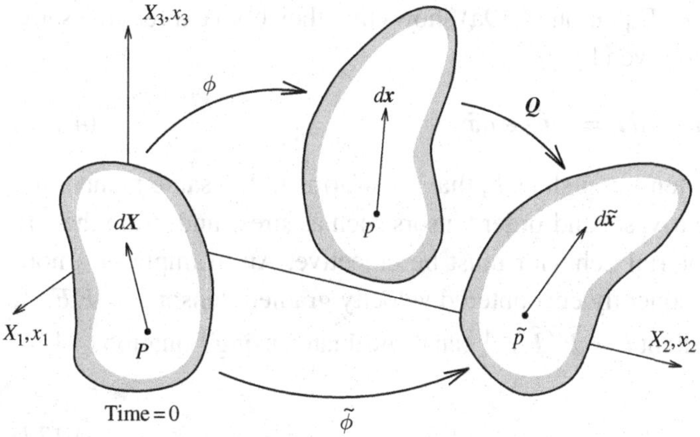


**FIGURE 4.16** Superimposed rigid body motion.

In this sense it can be said the $d\boldsymbol{x}$ is objective under rigid body motions. This definition is extended to any vector $\boldsymbol{a}$ that transforms according to $\tilde{\boldsymbol{a}} = \boldsymbol{Q}\boldsymbol{a}$. Velocity is an example of a non-objective vector because differentiating the rotated mapping $\tilde{\boldsymbol{\phi}} = \boldsymbol{Q}\boldsymbol{\phi}$ with respect to time gives

$$\begin{aligned}\tilde{\boldsymbol{v}} &= \frac{\partial \tilde{\boldsymbol{\phi}}}{\partial t} \\ &= \boldsymbol{Q}\frac{\partial \boldsymbol{\phi}}{\partial t} + \dot{\boldsymbol{Q}}\boldsymbol{\phi} \\ &= \boldsymbol{Q}\boldsymbol{v} + \dot{\boldsymbol{Q}}\boldsymbol{\phi}. \end{aligned} \tag{4.133}$$

Obviously, the magnitudes of $\boldsymbol{v}$ and $\tilde{\boldsymbol{v}}$ are not equal as a result of the presence of the term $\dot{\boldsymbol{Q}}\boldsymbol{\phi}$, which violates the objectivity criteria.

For the purpose of extending the definition of objectivity to second-order tensors, note first from Equation (4.132) that the deformation gradients with respect to the current and rotated configurations are related as

$$\tilde{\boldsymbol{F}} = \boldsymbol{Q}\boldsymbol{F}. \tag{4.134}$$

Using this expression together with Equations (4.15) and (4.18b) shows that material strain tensors such as $\boldsymbol{C}$ and $\boldsymbol{E}$ remain unaltered by the rigid body motion. In contrast, introducing Equation (4.132) into Equations (4.17) and (4.19b) for $\boldsymbol{b}$ and $\boldsymbol{e}$ gives

$$\tilde{\boldsymbol{b}} = \boldsymbol{Q}\boldsymbol{b}\boldsymbol{Q}^T; \tag{4.135a}$$

$$\tilde{\boldsymbol{e}} = \boldsymbol{Q}\boldsymbol{e}\boldsymbol{Q}^T. \tag{4.135b}$$

Note that, although $\tilde{e} \neq e$, Equation (4.19a) shows that they both express the same intrinsic change in length given by

$$\frac{1}{2}(dl^2 - dL^2) = d\boldsymbol{x} \cdot \boldsymbol{e}\, d\boldsymbol{x} = d\tilde{\boldsymbol{x}} \cdot \tilde{\boldsymbol{e}}\, d\tilde{\boldsymbol{x}}. \tag{4.136}$$

In this sense, $\boldsymbol{e}$ and any tensor, such as $\boldsymbol{b}$, that transforms in the same manner are said to be objective. Clearly, second-order tensors such as stress and strain that are used to describe the material behavior must be objective. An example of a non-objective tensor is the frequently encountered velocity gradient tensor $\boldsymbol{l} = \dot{\boldsymbol{F}}\boldsymbol{F}^{-1}$. The rotated velocity gradient $\tilde{\boldsymbol{l}} = \dot{\tilde{\boldsymbol{F}}}(\tilde{\boldsymbol{F}})^{-1}$ can be evaluated using Equation (4.134) to give

$$\tilde{\boldsymbol{l}} = \boldsymbol{Q}\boldsymbol{l}\boldsymbol{Q}^T + \dot{\boldsymbol{Q}}\boldsymbol{Q}^T. \tag{4.137}$$

Again, it is the presence of the second term in the above equation that renders the spatial velocity gradient non-objective. Fortunately, it transpires that the rate of deformation tensor $\boldsymbol{d}$ is objective. This is easily demonstrated by writing the rotated rate of deformation $\tilde{\boldsymbol{d}}$ in terms of $\tilde{\boldsymbol{l}}$ as

$$\tilde{\boldsymbol{d}} = \frac{1}{2}(\tilde{\boldsymbol{l}} + \tilde{\boldsymbol{l}}^T) = \boldsymbol{Q}\boldsymbol{d}\boldsymbol{Q}^T + \frac{1}{2}(\dot{\boldsymbol{Q}}\boldsymbol{Q}^T + \boldsymbol{Q}\dot{\boldsymbol{Q}}^T). \tag{4.138}$$

Observing that the term in brackets is the time derivative of $\boldsymbol{Q}\boldsymbol{Q}^T = \boldsymbol{I}$ and is consequently zero shows that the rate of deformation satisfies Equation (4.135) and is therefore objective.

## Exercises

1. (a) For the uniaxial strain case find the engineering, Green, and Almansi strains in terms of the stretch $\lambda$.
   (b) Using these expressions show that when the engineering strain is small, all three strain measures converge to the same value (see Chapter 1, Equations (1.6) and (1.8)).
2. A continuum body (see Figure (4.17)) undergoes a rigid body rotation $\theta$ about the origin defined by

$$\boldsymbol{x} = \boldsymbol{R}\boldsymbol{X}\ ; \quad \boldsymbol{R} = \begin{bmatrix} \cos\theta & -\sin\theta \\ \sin\theta & \cos\theta \end{bmatrix}, \tag{4.139}$$

   where $\boldsymbol{R}$ is a rotation matrix. In other words, material point $P(X_1, X_2)$ rotates to spatial point $p(x_1, x_2)$.
   (a) Demonstrate why the engineering or small strain tensor $\boldsymbol{\varepsilon}$ is not a valid measure of strain when the rotation $\theta$ is large.
   (b) Demonstrate that Green's strain $\boldsymbol{E}$ is a valid measure of strain for the above motion regardless of the magnitude of $\theta$.

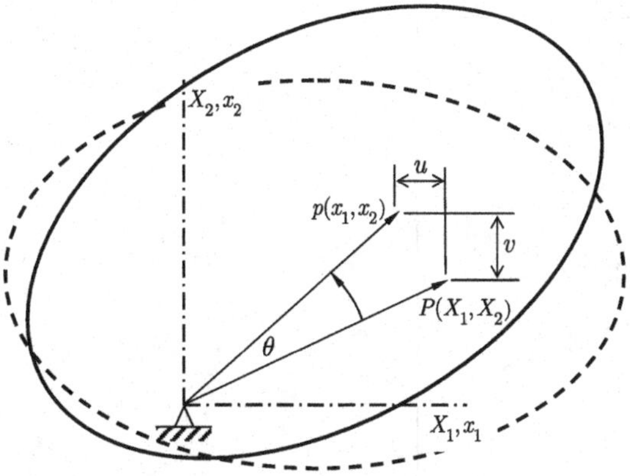


**FIGURE 4.17** Rigid body rotation.

3. A single four-node isoparametric element can be used to illustrate the material and spatial coordinates used in finite deformation analysis.[§] The non-dimensional coordinates $\xi$ and $\eta$ can be replaced by the material coordinates $X_1$ and $X_2$ giving the resulting single square element with initial dimensions $2 \times 2$ and centered at $X_1 = X_2 = 0$; see Figure 4.18. The shape functions are employed to define the spatial coordinates $\boldsymbol{x} = (x_1, x_2)^T$ in terms of the material coordinates $\boldsymbol{X} = (X_1, X_2)^T$ and $\boldsymbol{x}_a = (x_1, x_2)_a^T$, where $\boldsymbol{x}_a$ contains the spatial coordinates of nodes $a = 1, 4$ in the deformed configuration. Consequently the mapping $\boldsymbol{x} = \phi(\boldsymbol{X})$ is

$$\boldsymbol{x} = \phi(\boldsymbol{X}) = N_1\boldsymbol{x}_1 + N_2\boldsymbol{x}_2 + N_3\boldsymbol{x}_3 + N_4\boldsymbol{x}_4, \tag{4.140}$$

where

$$N_1 = \frac{1}{4}(1 - X_1)(1 - X_2); \quad N_2 = \frac{1}{4}(1 + X_1)(1 - X_2); \tag{4.141a}$$

$$N_3 = \frac{1}{4}(1 + X_1)(1 + X_2)\,; \quad N_4 = \frac{1}{4}(1 - X_1)(1 + X_2). \tag{4.141b}$$

(a) For $\boldsymbol{X} = (0, 0)^T$ and $\boldsymbol{x}_1 = (4, 2)^T$, $\boldsymbol{x}_2 = (8, 4)^T$, $\boldsymbol{x}_3 = (6, 8)^T$, $\boldsymbol{x}_4 = (4, 6)^T$, find the deformation gradient $\boldsymbol{F}$ and Green's strain $\boldsymbol{E}$.
(b) Find the principal stretches $\lambda_1$ and $\lambda_2$ and the corresponding principal material and spatial unit vectors $\boldsymbol{N}_1$, $\boldsymbol{N}_2$ and $\boldsymbol{n}_1$, $\boldsymbol{n}_2$ respectively.
(c) Using the above calculations show that Green's strain tensor can be expressed in terms of the principal material directions as

$$\boldsymbol{E} = \frac{1}{2}\left[(\lambda_1^2 - 1)\boldsymbol{N}_1 \otimes \boldsymbol{N}_1 + (\lambda_2^2 - 1)\boldsymbol{N}_2 \otimes \boldsymbol{N}_2\right]. \tag{4.142}$$

[§] This example can be illustrated and calculated using the MATLAB program `polar_decomposition.m`, which can be downloaded from the website `www.flagshyp.com`. This program can be used to explore the deformation using user chosen spatial coordinates.

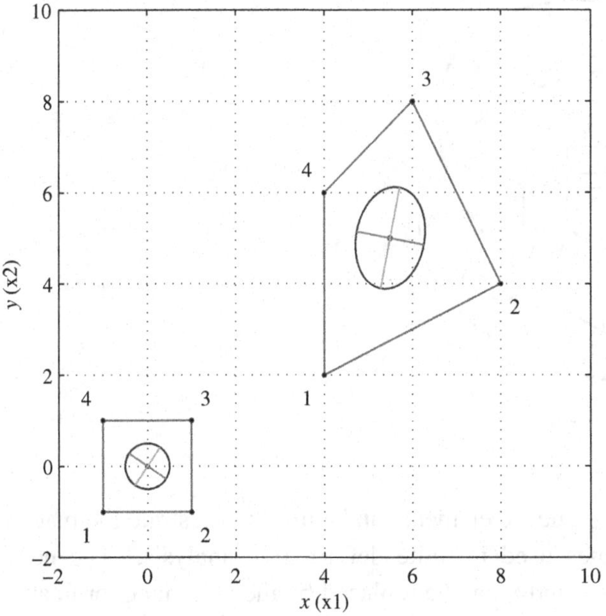


**FIGURE 4.18** Polar decomposition.

4. The finite deformation of a two-dimensional continuum from initial position $\boldsymbol{X} = (X_1, X_2)^T$ to a final configuration $\boldsymbol{x} = (x_1, x_2)^T$ is given as

$$x_1 = 4 - 2X_1 - X_2\,; \quad x_2 = 2 + \frac{3}{2}X_1 - \frac{1}{2}X_2. \tag{4.143}$$

(a) Calculate the deformation gradient tensor $\boldsymbol{F}$, the right Cauchy–Green strain tensor $\boldsymbol{C}$, and the left Cauchy–Green strain tensor $\boldsymbol{b}$.
(b) Calculate the stretch undergone by a material unit material vector $\boldsymbol{a}_0 = (\frac{3}{5}, \frac{4}{5})^T$ as a result of the deformation.
(c) A pair of orthonormal material vectors $\boldsymbol{b}_0 = (1, 0)^T$ and $\boldsymbol{c}_0 = (0, 1)^T$ are subjected to the deformation process. Calculate the new angle formed between these two vectors after deformation.
(d) Demonstrate whether the deformation is isochoric.
(e) Calculate the principal stretches $\lambda_\alpha$ and the principal material directions $\boldsymbol{N}_\alpha$, with $\alpha = 1, 2$.
(f) Calculate the principal spatial directions $\boldsymbol{n}_\alpha$, with $\alpha = 1, 2$.

5. (a) If the deformation gradients at times $t$ and $t + \Delta t$ are $\boldsymbol{F}_t$ and $\boldsymbol{F}_{t+\Delta t}$ respectively, show that the deformation gradient $\Delta\boldsymbol{F}$ relating the incremental motion from configuration at $t$ to $t + \Delta t$ is $\Delta\boldsymbol{F} = \boldsymbol{F}_{t+\Delta t}\boldsymbol{F}_t^{-1}$.
(b) Using the deformation given in Example 4.5 with $\boldsymbol{X} = (0, 0)$, $t = 1$, $\Delta t = 1$, show that $\Delta\boldsymbol{F} = \boldsymbol{F}_{t+\Delta t}\boldsymbol{F}_t^{-1}$ is correct by pushing forward the

initial vector $\boldsymbol{G} = [1,1]^T$ to vectors $\boldsymbol{g}_t$ and $\boldsymbol{g}_{t+\Delta t}$ at times $t$ and $t+\Delta t$ respectively, and checking that $\boldsymbol{g}_{t+\Delta t} = \Delta\boldsymbol{F}\boldsymbol{g}_t$.
(c) Using the deformation given in Example 4.5 with $\boldsymbol{X} = (0,0)$ and $t = 1$, calculate the velocity gradient $\boldsymbol{l}$ and the rate of deformation $\boldsymbol{d}$.

6. Using Equation (4.68) prove that the area ratio can be alternatively expressed as
$$\frac{da}{dA} = J\sqrt{\boldsymbol{N}\cdot\boldsymbol{C}^{-1}\boldsymbol{N}}.$$

7. Consider the planar (1-2) deformation for which the deformation gradient is
$$\boldsymbol{F} = \begin{bmatrix} F_{11} & F_{12} & 0 \\ F_{21} & F_{22} & 0 \\ 0 & 0 & \lambda_3 \end{bmatrix},$$
where $\lambda_3$ is the stretch in the thickness direction normal to the plane (1-2). If $dA$ and $da$ are the elemental areas in the (1-2) plane and $H$ and $h$ the thicknesses before and after deformation, show that
$$\frac{da}{dA} = j \quad \text{and} \quad h = H\frac{J}{j},$$
where $j = \det(F_{kl})$, $k, l = 1, 2$.

8. Using Figure 4.4 as a guide, draw a similar diagram that interprets the polar decomposition Equation (4.34), $d\boldsymbol{x} = \boldsymbol{V}(\boldsymbol{R}d\boldsymbol{X})$.

9. Show that the condition for an elemental material vector $d\boldsymbol{X} = \boldsymbol{N}dL$ to exhibit zero extension is $\boldsymbol{N}\cdot\boldsymbol{C}\boldsymbol{N} = 1$, where $\boldsymbol{C} = \boldsymbol{F}^T\boldsymbol{F}$.

10. Prove Equation (4.44b), that is,
$$\boldsymbol{F}^{-T}\boldsymbol{N}_\alpha = \frac{1}{\lambda_\alpha}\boldsymbol{n}_\alpha.$$

11. The motion of a body at time $t$ is given by
$$\boldsymbol{x} = \boldsymbol{F}(t)\boldsymbol{X}; \qquad \boldsymbol{F}(t) = \begin{bmatrix} 1 & t & t^2 \\ t^2 & 1 & t \\ t & t^2 & 1 \end{bmatrix};$$
$$\boldsymbol{F}^{-1}(t) = \frac{1}{(t^3-1)}\begin{bmatrix} -1 & t & 0 \\ 0 & -1 & t \\ t & 0 & -1 \end{bmatrix}.$$
Find the velocity of the particle (a) initially at $\boldsymbol{X} = (1,1,1)$ at time $t = 0$ and (b) currently at $\boldsymbol{x} = (1,1,1)$ at time $t = 2$. Using $J = dv/dV$, show that at time $t = 1$ the motion is not realistic.

12. For a pure expansion the deformation gradient is $\boldsymbol{F} = \alpha \boldsymbol{I}$, where $\alpha$ is a scalar. Show that the rate of deformation is

$$\boldsymbol{d} = \frac{\dot{\alpha}}{\alpha}\boldsymbol{I}.$$

13. Show that at the initial configuration ($\boldsymbol{F} = \boldsymbol{I}$) the linearization of $\hat{\boldsymbol{C}}$ in the direction of a displacement $\boldsymbol{u}$ is

$$D\hat{\boldsymbol{C}}[\boldsymbol{u}] = 2\boldsymbol{\varepsilon}' = 2\left[\boldsymbol{\varepsilon} - \frac{1}{3}(\mathrm{tr}\boldsymbol{\varepsilon})\boldsymbol{I}\right].$$

CHAPTER FIVE

# STRESS AND EQUILIBRIUM

## 5.1 INTRODUCTION

This chapter will introduce the stress and equilibrium concepts for a deformable body undergoing a finite motion. Stress is first defined in the current configuration in the standard way as force per unit area. This leads to the well-known Cauchy stress tensor as used in linear analysis. We will then derive the differential equations enforcing translational and rotational equilibrium and the equivalent principle of virtual work.

In contrast to linear small displacement analysis, stress quantities that refer back to the initial body configuration can also be defined. This will be achieved using work conjugacy concepts that will lead to the Piola–Kirchhoff stress tensors and alternative equilibrium equations. Finally, the objectivity of several stress rate tensors is considered.

## 5.2 CAUCHY STRESS TENSOR

### 5.2.1 Definition

Consider a general deformable body at its current position as shown in Figure 5.1. In order to develop the concept of stress it is necessary to study the action of the forces applied by one region $R_1$ of the body on the remaining part $R_2$ of the body with which it is in contact. For this purpose consider the element of area $\Delta a$ normal to $\boldsymbol{n}$ in the neighborhood of spatial point $p$. If the resultant force on this area is $\Delta\boldsymbol{p}$, the traction vector $\boldsymbol{t}$ corresponding to the normal $\boldsymbol{n}$ at $p$ is defined as

$$\boldsymbol{t}(\boldsymbol{n}) = \lim_{\Delta a \to 0} \frac{\Delta \boldsymbol{p}}{\Delta a}, \tag{5.1}$$

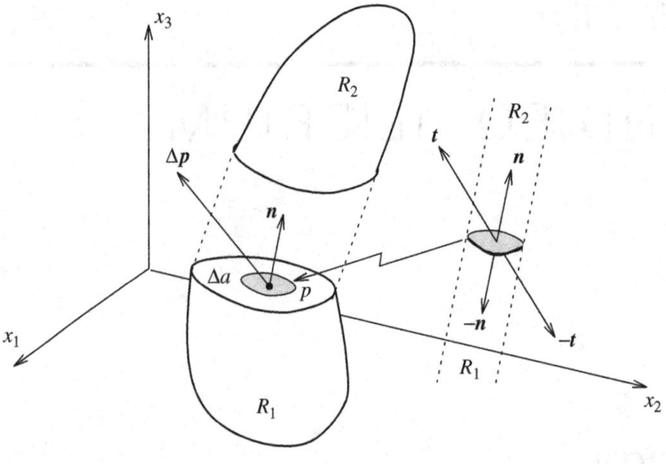


**FIGURE 5.1** Traction vector.

where the relationship between $\boldsymbol{t}$ and $\boldsymbol{n}$ must be such that it satisfies Newton's third law of action and reaction, which is expressed as (see Figure 5.1)

$$\boldsymbol{t}(-\boldsymbol{n}) = -\boldsymbol{t}(\boldsymbol{n}). \tag{5.2}$$

To develop the idea of a stress tensor, let the three traction vectors associated with the three Cartesian directions $\boldsymbol{e}_1$, $\boldsymbol{e}_2$, and $\boldsymbol{e}_3$ be expressed in a component form as (Figure 5.2)

$$\boldsymbol{t}(\boldsymbol{e}_1) = \sigma_{11}\boldsymbol{e}_1 + \sigma_{21}\boldsymbol{e}_2 + \sigma_{31}\boldsymbol{e}_3; \tag{5.3a}$$

$$\boldsymbol{t}(\boldsymbol{e}_2) = \sigma_{12}\boldsymbol{e}_1 + \sigma_{22}\boldsymbol{e}_2 + \sigma_{32}\boldsymbol{e}_3; \tag{5.3b}$$

$$\boldsymbol{t}(\boldsymbol{e}_3) = \sigma_{13}\boldsymbol{e}_1 + \sigma_{23}\boldsymbol{e}_2 + \sigma_{33}\boldsymbol{e}_3. \tag{5.3c}$$

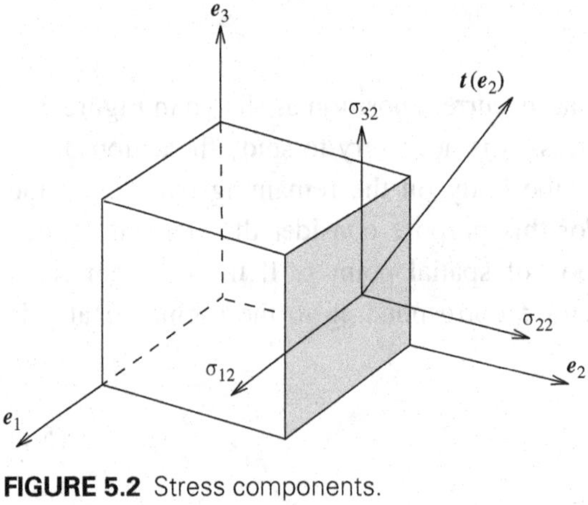


**FIGURE 5.2** Stress components.

Although the general equilibrium of a deformable body will be discussed in detail in the next section, a relationship between the traction vector $\boldsymbol{t}$ corresponding to a general direction $\boldsymbol{n}$ and the components $\sigma_{ij}$ can be obtained only by studying the translational equilibrium of the elemental tetrahedron shown in Figure 5.3. Letting $\boldsymbol{f}$ be the force per unit volume acting on the body at point $p$ (which in general could also include inertia terms), the equilibrium of the tetrahedron is given as

$$\boldsymbol{t}(\boldsymbol{n})\, da + \sum_{i=1}^{3} \boldsymbol{t}(-\boldsymbol{e}_i)\, da_i + \boldsymbol{f}\, dv = \boldsymbol{0}, \tag{5.4}$$

where $da_i = (\boldsymbol{n} \cdot \boldsymbol{e}_i)\, da$ is the projection of the area $da$ onto the plane orthogonal to the Cartesian direction $i$ (see Figure 5.3) and $dv$ is the volume of the tetrahedron. Dividing Equation (5.4) by $da$, recalling Newton's third law, using Equations (5.3a–c), and noting that $dv/da \to 0$ gives

$$\begin{aligned} \boldsymbol{t}(\boldsymbol{n}) &= -\sum_{j=1}^{3} \boldsymbol{t}(-\boldsymbol{e}_j)\, \frac{da_j}{da} - \boldsymbol{f}\, \frac{dv}{da} \\ &= \sum_{j=1}^{3} \boldsymbol{t}(\boldsymbol{e}_j)\, (\boldsymbol{n} \cdot \boldsymbol{e}_j) \\ &= \sum_{i,j=1}^{3} \sigma_{ij} (\boldsymbol{e}_j \cdot \boldsymbol{n})\, \boldsymbol{e}_i. \end{aligned} \tag{5.5}$$

Observing that $(\boldsymbol{e}_j \cdot \boldsymbol{n})\boldsymbol{e}_i$ can be rewritten in terms of the tensor product as $(\boldsymbol{e}_i \otimes \boldsymbol{e}_j)\boldsymbol{n}$ gives

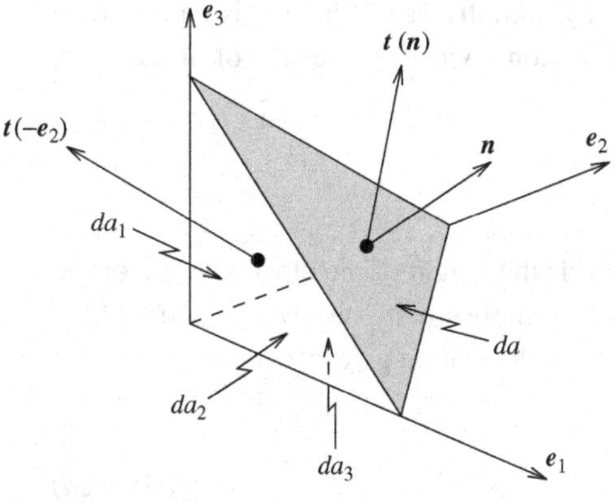


**FIGURE 5.3** Elemental tetrahedron.

$$\begin{aligned} \boldsymbol{t}(\boldsymbol{n}) &= \sum_{i,j=1}^{3} \sigma_{ij}(\boldsymbol{e}_j \cdot \boldsymbol{n})\, \boldsymbol{e}_i \\ &= \sum_{i,j=1}^{3} \sigma_{ij}(\boldsymbol{e}_i \otimes \boldsymbol{e}_j)\boldsymbol{n} \\ &= \left[ \sum_{i,j=1}^{3} \sigma_{ij}(\boldsymbol{e}_i \otimes \boldsymbol{e}_j) \right] \boldsymbol{n}, \end{aligned} \tag{5.6}$$

which clearly identifies a tensor $\boldsymbol{\sigma}$, known as the *Cauchy stress tensor*, that relates the normal vector $\boldsymbol{n}$ to the traction vector $\boldsymbol{t}$ as

$$\boldsymbol{t}(\boldsymbol{n}) = \boldsymbol{\sigma}\boldsymbol{n}; \quad \boldsymbol{\sigma} = \sum_{i,j=1}^{3} \sigma_{ij}\, \boldsymbol{e}_i \otimes \boldsymbol{e}_j\, . \tag{5.7a,b}$$

**EXAMPLE 5.1: Rectangular block under self-weight (i)**

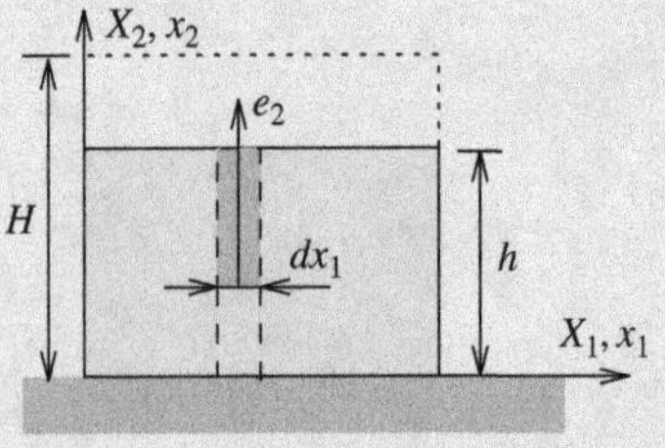


A simple example of a two-dimensional stress tensor results from the self-weight of a block of uniform initial density $\rho_0$ resting on a frictionless surface as shown in the figure above. For simplicity we will assume that there is no lateral deformation (in linear elasticity this would imply that Poisson's ratio $\nu = 0$).

Using the definition (5.1), the traction vector $\boldsymbol{t}$ associated with the unit vertical vector $\boldsymbol{e}_2$ at an arbitrary point at height $x_2$, initially at height $X_2$, is equal to the weight of material above an infinitesimal section divided by the area of this section. This gives

$$\boldsymbol{t}(\boldsymbol{e}_2) = \frac{(-\int_y^h \rho g\, dx_2)\, \boldsymbol{e}_2 dx_1}{dx_1},$$

where $g$ is the acceleration of gravity and $h$ is the height of the block after deformation. The mass conservation Equation (4.57) implies that $\rho dx_1 dx_2 = \rho_0 dX_1 dX_2$, which in conjunction with the lack of lateral deformation gives

$$\boldsymbol{t}(\boldsymbol{e}_2) = \rho_0 g(H - X_2)\, \boldsymbol{e}_2.$$

*(continued)*

**EXAMPLE 5.1:** *(cont.)*

Combining this equation with the fact that the stress components $\sigma_{12}$ and $\sigma_{22}$ are defined in Equation (5.3) by the expression $\boldsymbol{t}(\boldsymbol{e}_2) = \sigma_{12}\boldsymbol{e}_1 + \sigma_{22}\boldsymbol{e}_2$ gives $\sigma_{12} = 0$ and $\sigma_{22} = -\rho_0 g(H - X_2)$. Using a similar process and given the absence of horizontal forces, it is easy to show that the traction vector associated with the horizontal unit vector is zero and consequently $\sigma_{11} = \sigma_{21} = 0$. The complete stress tensor in Cartesian components is therefore

$$[\sigma] = \begin{bmatrix} 0 & 0 \\ 0 & \rho_0 g(X_2 - H) \end{bmatrix}.$$

The Cauchy stress tensor can alternatively be expressed in terms of its principal directions $\boldsymbol{m}_1, \boldsymbol{m}_2, \boldsymbol{m}_3$ and principal stresses $\sigma_{\alpha\alpha}$ for $\alpha = 1, 2, 3$ as

$$\boldsymbol{\sigma} = \sum_{\alpha=1}^{3} \sigma_{\alpha\alpha}\, \boldsymbol{m}_\alpha \otimes \boldsymbol{m}_\alpha, \tag{5.8}$$

where from Equations (2.57a–b) the eigenvectors $\boldsymbol{m}_\alpha$ and eigenvalues $\sigma_{\alpha\alpha}$ satisfy

$$\boldsymbol{\sigma}\boldsymbol{m}_\alpha = \sigma_{\alpha\alpha}\boldsymbol{m}_\alpha. \tag{5.9}$$

In Chapter 6 we shall show that for isotropic materials the principal directions $\boldsymbol{m}_\alpha$ of the Cauchy stress coincide with the principal Eulerian triad $\boldsymbol{n}_\alpha$ introduced in Chapter 4.

Note that $\boldsymbol{\sigma}$ is a spatial tensor; equivalent material stress measures associated with the initial configuration of the body will be discussed later. Note also that the well-known symmetry of $\boldsymbol{\sigma}$ has not yet been established. In fact this results from the rotational equilibrium equation, which is discussed in the following section.

### 5.2.2 Stress Objectivity

Because the Cauchy stress tensor is a key feature of any equilibrium or material equation, it is important to inquire whether $\boldsymbol{\sigma}$ is objective as defined in Section 4.15. For this purpose consider the transformations of the normal and traction vectors implied by the superimposed rigid body motion $\boldsymbol{Q}$ shown in Figure 5.4 as

$$\tilde{\boldsymbol{t}}(\tilde{\boldsymbol{n}}) = \boldsymbol{Q}\boldsymbol{t}(\boldsymbol{n}); \tag{5.10a}$$

$$\tilde{\boldsymbol{n}} = \boldsymbol{Q}\boldsymbol{n}. \tag{5.10b}$$

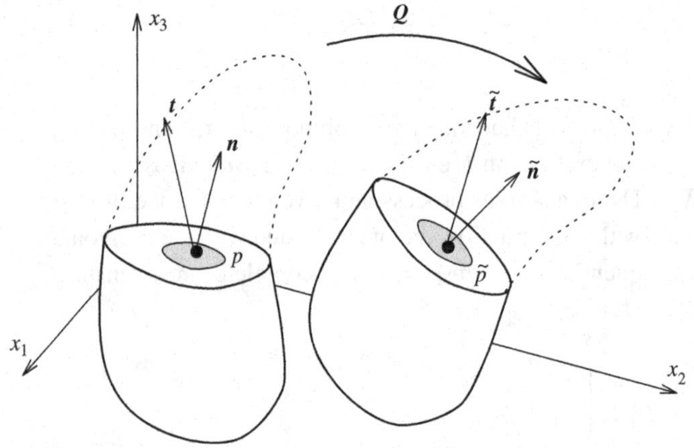


**FIGURE 5.4** Superimposed rigid body motion.

Using the relationship between the traction vector and stress tensor given by Equation $(5.7\text{a,b})_a$ in conjunction with the above equation gives

$$\tilde{\boldsymbol{\sigma}} = \boldsymbol{Q}\boldsymbol{\sigma}\boldsymbol{Q}^T. \tag{5.11}$$

The rotation of $\boldsymbol{\sigma}$ given by the above equation conforms with the definition of objectivity given by Equation (4.135), and hence $\boldsymbol{\sigma}$ is objective and a valid candidate for inclusion in a material description. It will be shown later that the material rate of change of stress is not an objective tensor.

## 5.3 EQUILIBRIUM

### 5.3.1 Translational Equilibrium

In order to derive the differential static equilibrium equations, consider the spatial configuration of a general deformable body defined by a volume $v$ with boundary area $\partial v$ as shown in Figure 5.5. We can assume that the body is under the action of body forces $\boldsymbol{f}$ per unit volume and traction forces $\boldsymbol{t}$ per unit area acting on the boundary. For simplicity, however, inertia forces will be ignored, and therefore translational equilibrium implies that the sum of all forces acting on the body vanishes. This gives

$$\int_{\partial v} \boldsymbol{t}\, da + \int_v \boldsymbol{f}\, dv = \boldsymbol{0}. \tag{5.12}$$

Using Equation $(5.7\text{a,b})_a$ for the traction vector enables Equation (5.12) to be expressed in terms of the Cauchy stresses as

$$\int_{\partial v} \boldsymbol{\sigma}\boldsymbol{n}\, da + \int_v \boldsymbol{f}\, dv = \boldsymbol{0}. \tag{5.13}$$

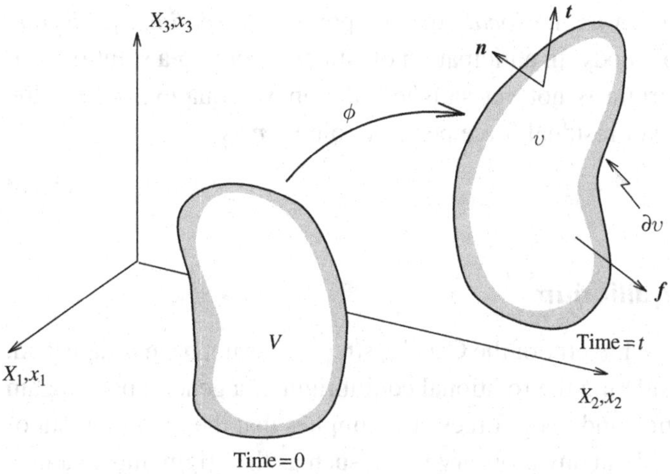


**FIGURE 5.5** Equilibrium.

The first term in this equation can be transformed into a volume integral by using the Gauss theorem in Equation (2.139) to give

$$\int_v (\operatorname{div} \boldsymbol{\sigma} + \boldsymbol{f})\, dv = 0, \tag{5.14}$$

where the vector div $\boldsymbol{\sigma}$ is defined in Section 2.4.1. The fact that the above equation can be equally applied to any enclosed region of the body implies that the integrand function must vanish, that is,

$$\operatorname{div} \boldsymbol{\sigma} + \boldsymbol{f} = \boldsymbol{0}. \tag{5.15}$$

**EXAMPLE 5.2: Rectangular block under self-weight (ii)**

It is easy to show that the stress tensor given in Example 5.1 satisfies the equilibrium equation. For this purpose, note first that in this particular case the forces $\boldsymbol{f}$ per unit volume are $\boldsymbol{f} = -\rho g \boldsymbol{e}_2$ or, in component form,

$$[\boldsymbol{f}] = \begin{bmatrix} 0 \\ -\rho g \end{bmatrix}.$$

Additionally, using the definition (2.134), the two-dimensional components of the divergence of $\boldsymbol{\sigma}$ are

$$[\operatorname{div} \boldsymbol{\sigma}] = \begin{bmatrix} \frac{\partial \sigma_{11}}{\partial x_1} + \frac{\partial \sigma_{12}}{\partial x_2} \\ \frac{\partial \sigma_{21}}{\partial x_1} + \frac{\partial \sigma_{22}}{\partial x_2} \end{bmatrix} = \begin{bmatrix} 0 \\ \rho_0 g \frac{dX_2}{dx_2} \end{bmatrix},$$

which combined with the mass conservation equation $\rho dx_1 dx_2 = \rho_0 dX_1 dX_2$ and the lack of lateral deformation implies that Equation (5.14) is satisfied.

This equation is known as the *local* (that is, pointwise) *spatial equilibrium equation* for a deformable body. In anticipation of situations during a solution procedure in which equilibrium is not yet satisfied, the above equation defines the pointwise out-of-balance or residual force per unit volume $\boldsymbol{r}$ as

$$\boldsymbol{r} = \operatorname{div}\boldsymbol{\sigma} + \boldsymbol{f}. \tag{5.16}$$

### 5.3.2 Rotational Equilibrium

Thus far the well-known symmetry of the Cauchy stresses has not been established. This is achieved by considering the rotational equilibrium of a general body, again under the action of traction and body forces. This implies that the total moment of body and traction forces about any arbitrary point, such as the origin, must vanish, that is,

$$\int_{\partial v} \boldsymbol{x} \times \boldsymbol{t}\, da + \int_v \boldsymbol{x} \times \boldsymbol{f}\, dv = \boldsymbol{0}, \tag{5.17}$$

where it should be recalled that the cross product of a force with a position vector $\boldsymbol{x}$ yields the moment of that force about the origin. Equation (5.7a,b)$_a$ for the traction vector in terms of the Cauchy stress tensor enables the above equation to be rewritten as

$$\int_{\partial v} \boldsymbol{x} \times (\boldsymbol{\sigma}\boldsymbol{n})\, da + \int_v \boldsymbol{x} \times \boldsymbol{f}\, dv = \boldsymbol{0}. \tag{5.18}$$

Using the Gauss theorem and after some algebra, the equation becomes*

$$\int_v \boldsymbol{x} \times (\operatorname{div}\boldsymbol{\sigma})\, dv + \int_v \boldsymbol{\mathcal{E}} : \boldsymbol{\sigma}^T\, dv + \int_v \boldsymbol{x} \times \boldsymbol{f}\, dv = \boldsymbol{0}, \tag{5.19}$$

where $\boldsymbol{\mathcal{E}}$ is the third-order alternating tensor, defined in Section 2.2.4 ($\mathcal{E}_{ijk} = 1$ if the permutation $\{i, j, k\}$ is even, –1 if it is odd, and 0 if any indices are repeated), so that the vector $\boldsymbol{\mathcal{E}} : \boldsymbol{\sigma}^T$ is

$$\boldsymbol{\mathcal{E}} : \boldsymbol{\sigma}^T = \begin{bmatrix} \sigma_{32} - \sigma_{23} \\ \sigma_{13} - \sigma_{31} \\ \sigma_{21} - \sigma_{12} \end{bmatrix}. \tag{5.20}$$

* To show this it is convenient to use indicial notation and the summation convention whereby repeated indices imply addition. Equation (2.136) then gives

$$\begin{aligned} \int_{\partial v} \mathcal{E}_{ijk} x_j \sigma_{kl} n_l\, da &= \int_v \frac{\partial}{\partial x_l}(\mathcal{E}_{ijk} x_j \sigma_{kl})\, dv \\ &= \int_v \mathcal{E}_{ijk} x_j \frac{\partial \sigma_{kl}}{\partial x_l}\, dv + \int_v \mathcal{E}_{ijk} \sigma_{kj}\, dv \\ &= \int_v (\boldsymbol{x} \times \operatorname{div}\boldsymbol{\sigma})_i\, dv + \int_v (\boldsymbol{\mathcal{E}} : \boldsymbol{\sigma}^T)_i\, dv. \end{aligned}$$

Rearranging terms in Equation (5.19) to take into account the translational equilibrium Equation (5.15) and noting that the resulting equation is valid for any enclosed region of the body gives

$$\boldsymbol{\mathcal{E}} : \boldsymbol{\sigma}^T = 0, \tag{5.21}$$

which, in view of Equation (5.20), clearly implies the symmetry of the Cauchy stress tensor $\boldsymbol{\sigma}$.

## 5.4 PRINCIPLE OF VIRTUAL WORK

Generally, the finite element formulation is established in terms of a weak form of the differential equations under consideration. In the context of solid mechanics this implies the use of the virtual work equation. For this purpose, let $\delta\boldsymbol{v}$ denote an arbitrary virtual velocity from the current position of the body as shown in Figure 5.6. The virtual work $\delta w$ per unit volume and time done by the residual force $\boldsymbol{r}$ during this virtual motion is $\boldsymbol{r} \cdot \delta\boldsymbol{v}$, and equilibrium implies

$$\delta w = \boldsymbol{r} \cdot \delta\boldsymbol{v} = 0. \tag{5.22}$$

Note that the above scalar equation is fully equivalent to the vector equation $\boldsymbol{r} = \boldsymbol{0}$. This is because $\delta\boldsymbol{v}$ is arbitrary, and hence by choosing $\delta\boldsymbol{v} = [1, 0, 0]^T$, followed by $\delta\boldsymbol{v} = [0, 1, 0]^T$ and $\delta\boldsymbol{v} = [0, 0, 1]^T$, the three components of the equation $\boldsymbol{r} = \boldsymbol{0}$ are retrieved. We can now use Equation (5.16) for the residual vector and integrate over the volume of the body to give a weak statement of the static equilibrium of the body as

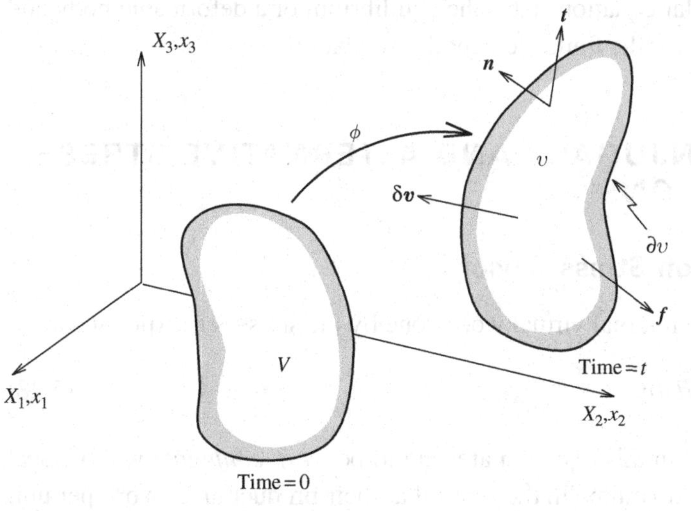


**FIGURE 5.6** Principle of Virtual Work.

$$\delta W = \int_v (\operatorname{div} \boldsymbol{\sigma} + \boldsymbol{f}) \cdot \delta \boldsymbol{v} \, dv = 0. \tag{5.23}$$

The virtual work term $\delta W$ is a function of the mapping $\boldsymbol{x} = \boldsymbol{\phi}(\boldsymbol{X}, t)$ between the initial and current particle positions given by Equation (4.1) and should strictly be written as $\delta W(\boldsymbol{\phi}, \delta \boldsymbol{v})$. While essential for the development of variational methods discussed in Chapter 8, $\delta W$ suffices for this chapter.

A more common and useful expression can be derived by recalling property (2.135e) to give the divergence of the vector $\boldsymbol{\sigma}\delta\boldsymbol{v}$ as

$$\operatorname{div}(\boldsymbol{\sigma}\delta\boldsymbol{v}) = (\operatorname{div}\boldsymbol{\sigma}) \cdot \delta\boldsymbol{v} + \boldsymbol{\sigma} : \boldsymbol{\nabla}\delta\boldsymbol{v}. \tag{5.24}$$

Using this equation together with the Gauss theorem enables Equation (5.23) to be rewritten as

$$\int_{\partial v} \boldsymbol{n} \cdot \boldsymbol{\sigma}\delta\boldsymbol{v} \, da - \int_v \boldsymbol{\sigma} : \boldsymbol{\nabla}\delta\boldsymbol{v} \, dv + \int_v \boldsymbol{f} \cdot \delta\boldsymbol{v} \, dv = 0. \tag{5.25}$$

The gradient of $\delta\boldsymbol{v}$ is, by definition, the virtual velocity gradient $\delta\boldsymbol{l}$. Additionally, we can use Equation (5.7a,b)$_a$ for the traction vector and the symmetry of $\boldsymbol{\sigma}$ to rewrite $\boldsymbol{n} \cdot \boldsymbol{\sigma}\delta\boldsymbol{v}$ as $\delta\boldsymbol{v} \cdot \boldsymbol{t}$, and consequently Equation (5.24) becomes

$$\int_v \boldsymbol{\sigma} : \delta\boldsymbol{l} \, dv = \int_v \boldsymbol{f} \cdot \delta\boldsymbol{v} \, dv + \int_{\partial v} \boldsymbol{t} \cdot \delta\boldsymbol{v} \, da. \tag{5.26}$$

Finally, expressing the virtual velocity gradient in terms of the symmetric virtual rate of deformation $\delta\boldsymbol{d}$ and the antisymmetric virtual spin tensor $\delta\boldsymbol{w}$, and taking into account again the symmetry of $\boldsymbol{\sigma}$, gives the *spatial virtual work equation* as

$$\delta W = \int_v \boldsymbol{\sigma} : \delta\boldsymbol{d} \, dv - \int_v \boldsymbol{f} \cdot \delta\boldsymbol{v} \, dv - \int_{\partial v} \boldsymbol{t} \cdot \delta\boldsymbol{v} \, da = 0. \tag{5.27}$$

This fundamental scalar equation states the equilibrium of a deformable body and will become the basis for the finite element discretization.

## 5.5 WORK CONJUGACY AND ALTERNATIVE STRESS REPRESENTATIONS

### 5.5.1 The Kirchhoff Stress Tensor

In Equation (5.27) the internal virtual work done by the stresses is expressed as

$$\delta W_{\text{int}} = \int_v \boldsymbol{\sigma} : \delta\boldsymbol{d} \, dv. \tag{5.28}$$

Pairs such as $\boldsymbol{\sigma}$ and $\boldsymbol{d}$ in this equation are said to be *work conjugate* with respect to the current deformed volume in the sense that their product gives work per unit current volume. By expressing the virtual work equation in the material coordinate system, alternative work conjugate pairs of stresses and strain rates will emerge. To achieve this objective, the spatial virtual work Equation (5.27) is first expressed with respect to the initial volume and area by transforming the integrals using Equation (4.57) for $dv$ to give

$$\int_V J\boldsymbol{\sigma} : \delta\boldsymbol{d}\, dV = \int_V \boldsymbol{f}_0 \cdot \delta\boldsymbol{v}\, dV + \int_{\partial V} \boldsymbol{t}_0 \cdot \delta\boldsymbol{v}\, dA, \tag{5.29}$$

where $\boldsymbol{f}_0 = J\boldsymbol{f}$ is the body force per unit undeformed volume and $\boldsymbol{t}_0 = \boldsymbol{t}(da/dA)$ is the traction vector per unit initial area, where the area ratio can be obtained after some algebra from Equation (4.68) or Exercise 6 in Chapter 4 as

$$\frac{da}{dA} = \frac{J}{\sqrt{\boldsymbol{n} \cdot \boldsymbol{b}\boldsymbol{n}}} = J\sqrt{\boldsymbol{N} \cdot \boldsymbol{C}^{-1}\boldsymbol{N}}. \tag{5.30}$$

The internal virtual work given by the left-hand side of Equation (5.29) can be expressed in terms of the *Kirchhoff stress tensor* $\boldsymbol{\tau}$ as

$$\delta W_{\text{int}} = \int_V \boldsymbol{\tau} : \delta\boldsymbol{d}\, dV; \quad \boldsymbol{\tau} = J\boldsymbol{\sigma}. \tag{5.31a,b}$$

This equation reveals that the Kirchhoff stress tensor $\boldsymbol{\tau}$ is work conjugate to the rate of deformation tensor with respect to the initial volume. Note that the work per unit current volume is not equal to the work per unit initial volume. However, Equation $(5.31\text{a,b})_b$ and the relationship $\rho = \rho_0/J$ ensure that the work per unit mass is invariant and can be equally written in the current or initial configuration as

$$\frac{1}{\rho}\boldsymbol{\sigma} : \boldsymbol{d} = \frac{1}{\rho_0}\boldsymbol{\tau} : \boldsymbol{d}. \tag{5.32}$$

### 5.5.2 The First Piola–Kirchhoff Stress Tensor

The crude transformation that resulted in the internal virtual work given above is not entirely satisfactory because it still relies on the spatial quantities $\boldsymbol{\tau}$ and $\boldsymbol{d}$. To alleviate this lack of consistency, note that the symmetry of $\boldsymbol{\sigma}$ together with Equation (4.93) for $\boldsymbol{l}$ in terms of $\dot{\boldsymbol{F}}$ and the properties of the trace gives

$$\begin{aligned}
\delta W_{\text{int}} &= \int_V J\boldsymbol{\sigma} : \delta\boldsymbol{l}\, dV \\
&= \int_V J\boldsymbol{\sigma} : (\delta\dot{\boldsymbol{F}}\boldsymbol{F}^{-1})\, dV \\
&= \int_V \text{tr}(J\boldsymbol{F}^{-1}\boldsymbol{\sigma}\delta\dot{\boldsymbol{F}})\, dV \\
&= \int_V (J\boldsymbol{\sigma}\boldsymbol{F}^{-T}) : \delta\dot{\boldsymbol{F}}\, dV.
\end{aligned} \tag{5.33}$$

We observe from this equality that the stress tensor work conjugate to the rate of the deformation gradient $\dot{\boldsymbol{F}}$ is the so-called *first Piola–Kirchhoff stress tensor* given as

$$\boldsymbol{P} = J\boldsymbol{\sigma}\boldsymbol{F}^{-T}. \tag{5.34a}$$

Note that, like $\boldsymbol{F}$, the first Piola–Kirchhoff tensor is an unsymmetric two-point tensor with components given as

$$\boldsymbol{P} = \sum_{i,I=1}^{3} P_{iI}\, \boldsymbol{e}_i \otimes \boldsymbol{E}_I; \quad P_{iI} = \sum_{j=1}^{3} J\sigma_{ij}(\boldsymbol{F}^{-1})_{Ij}. \tag{5.34b,c}$$

We can now rewrite the equation for the Principle of Virtual Work in terms of the first Piola–Kirchhoff tensor as

$$\int_V \boldsymbol{P} : \delta\dot{\boldsymbol{F}}\, dV = \int_V \boldsymbol{f}_0 \cdot \delta\boldsymbol{v}\, dV + \int_{\partial V} \boldsymbol{t}_0 \cdot \delta\boldsymbol{v}\, dA. \tag{5.35}$$

Additionally, if the procedure employed to obtain the virtual work Equation (5.27) from the spatial differential equilibrium Equation (5.24) is reversed, an equivalent version of the differential equilibrium equation is obtained in terms of the first Piola–Kirchhoff stress tensor as

$$\boldsymbol{r}_0 = J\boldsymbol{r} = \text{DIV}\boldsymbol{P} + \boldsymbol{f}_0 = \boldsymbol{0}, \tag{5.36}$$

where DIV$\boldsymbol{P}$ is the divergence of $\boldsymbol{P}$ with respect to the initial coordinate system, given as

$$\text{DIV}\boldsymbol{P} = \nabla_0\boldsymbol{P} : \boldsymbol{I}; \quad \nabla_0\boldsymbol{P} = \frac{\partial\boldsymbol{P}}{\partial\boldsymbol{X}}. \tag{5.37}$$

**Remark 5.1**: It is instructive to re-examine the physical meaning of the Cauchy stresses and thence the first Piola–Kirchhoff stress tensor. An element of force $d\boldsymbol{p}$ acting on an element of area $d\boldsymbol{a} = \boldsymbol{n}\,da$ in the spatial configuration can be written as

$$d\boldsymbol{p} = \boldsymbol{t}da = \boldsymbol{\sigma}d\boldsymbol{a}. \tag{5.38}$$

Broadly speaking, the Cauchy stresses give the current force per unit deformed area, which is the familiar description of stress. Using Equation (4.68) for the spatial area vector, $d\boldsymbol{p}$ can be rewritten in terms of the undeformed area corresponding to $d\boldsymbol{a}$ to give an expression involving the first Piola–Kirchhoff stresses as

$$d\boldsymbol{p} = J\boldsymbol{\sigma}\boldsymbol{F}^{-T}d\boldsymbol{A} = \boldsymbol{P}d\boldsymbol{A}. \tag{5.39}$$

This equation reveals that $\boldsymbol{P}$, like $\boldsymbol{F}$, is a two-point tensor that relates an area vector in the initial configuration to the corresponding force vector in the current configuration, as shown in Figure 5.7. Consequently, the first Piola–Kirchhoff stresses can be loosely interpreted as the current force per unit of undeformed area.

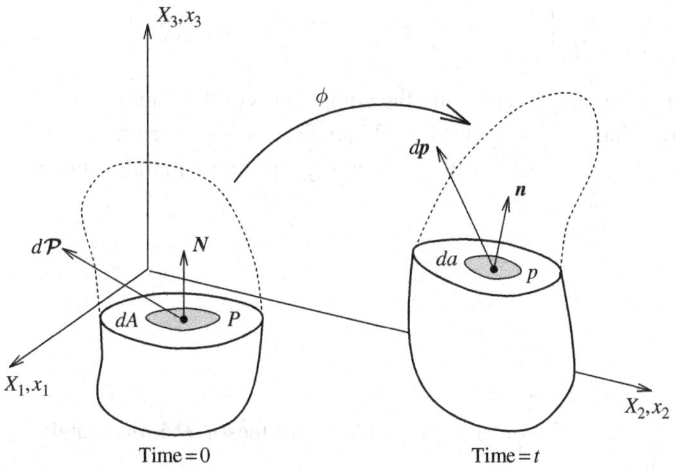


**FIGURE 5.7** Interpretation of stress tensors.

**EXAMPLE 5.3: Rectangular block under self-weight (iii)**

Using the physical interpretation for $\boldsymbol{P}$ given in Remark 5.1 we can find the first Piola–Kirchhoff tensor corresponding to the state of stresses described in Example 5.1. For this purpose note first that dividing Equation (5.39) by the current area element $da$ gives the traction vector associated with a unit normal $\boldsymbol{N}$ in the initial configuration as

$$\boldsymbol{t}(\boldsymbol{N}) = \boldsymbol{P}\boldsymbol{N}\frac{dA}{da}.$$

Using this equation with $\boldsymbol{N} = \boldsymbol{E}_2$ for the case described in Example 5.1 where the lack of lateral deformation implies $da = dA$ gives

$$\begin{aligned}\boldsymbol{t}(\boldsymbol{E}_2) &= \boldsymbol{P}\boldsymbol{E}_2\\ &= \sum_{i,I=1}^{2} P_{iI}(\boldsymbol{e}_i \otimes \boldsymbol{E}_I)\boldsymbol{E}_2\\ &= P_{12}\boldsymbol{e}_1 + P_{22}\boldsymbol{e}_2.\end{aligned}$$

*(continued)*

**EXAMPLE 5.3:** *(cont.)*

Combining the final equation with the fact that $\boldsymbol{t}(\boldsymbol{E}_2) = \boldsymbol{t}(\boldsymbol{e}_2) = -\rho_0 g(H - X_2)\boldsymbol{e}_2$ as explained in Example 5.1, we can identify $P_{12} = 0$ and $P_{22} = \rho_0 g(X_2 - H)$. Using a similar analysis for $\boldsymbol{t}(\boldsymbol{E}_1)$ eventually yields the components of $\boldsymbol{P}$ as

$$[\boldsymbol{P}] = \begin{bmatrix} 0 & 0 \\ 0 & \rho_0 g(X_2 - H) \end{bmatrix},$$

which for this particular example coincide with the components of the Cauchy stress tensor. In order to show that the above tensor $\boldsymbol{P}$ satisfies the equilibrium Equation (5.37), we first need to evaluate the force vector $\boldsymbol{f}_0$ per unit initial volume as

$$\begin{aligned} \boldsymbol{f}_0 &= \boldsymbol{f}\frac{dv}{dV} \\ &= -\rho\frac{dv}{dV}g\boldsymbol{e}_2 \\ &= -\rho_0 g\boldsymbol{e}_2. \end{aligned}$$

Combining this expression with the divergence of the above tensor $\boldsymbol{P}$ immediately leads to the desired result.

### 5.5.3 The Second Piola–Kirchhoff Stress Tensor

The first Piola-Kirchhoff tensor $\boldsymbol{P}$ is an unsymmetric two-point tensor and as such is not completely related to the material configuration. It is possible to contrive a totally material symmetric stress tensor, known as the *second Piola–Kirchhoff stress* $\boldsymbol{S}$, by pulling back the spatial element of force $d\boldsymbol{p}$ from Equation (5.39) to give a material force vector $d\boldsymbol{\mathcal{P}}$ (Figure 5.7) as

$$d\boldsymbol{\mathcal{P}} = \phi_*^{-1}[d\boldsymbol{p}] = \boldsymbol{F}^{-1}d\boldsymbol{p}. \tag{5.40}$$

Substituting from Equation (5.39) for $d\boldsymbol{p}$ gives the transformed force in terms of the *second Piola–Kirchhoff stress tensor* $\boldsymbol{S}$ and the material element of area $d\boldsymbol{A}$ as

$$d\boldsymbol{\mathcal{P}} = \boldsymbol{S}\,d\boldsymbol{A}; \quad \boldsymbol{S} = J\boldsymbol{F}^{-1}\boldsymbol{\sigma}\boldsymbol{F}^{-T}. \tag{5.41a,b}$$

It is now necessary to derive the strain rate work conjugate to the second Piola–Kirchhoff stress in the following manner. From Equation (4.100a,b)$_a$ it follows that the material and spatial virtual rates of deformation are related as

$$\delta\boldsymbol{d} = \boldsymbol{F}^{-T}\delta\dot{\boldsymbol{E}}\boldsymbol{F}^{-1}. \tag{5.42}$$

Substituting this relationship into the internal virtual work Equation (5.28) gives

$$\begin{aligned}\delta W_{\text{int}} &= \int_v \boldsymbol{\sigma} : \delta \boldsymbol{d}\, dv \\ &= \int_V J\boldsymbol{\sigma} : (\boldsymbol{F}^{-T} \delta \dot{\boldsymbol{E}} \boldsymbol{F}^{-1})\, dV \\ &= \int_V \text{tr}(\boldsymbol{F}^{-1} J \boldsymbol{\sigma} \boldsymbol{F}^{-T} \delta \dot{\boldsymbol{E}})\, dV \\ &= \int_V \boldsymbol{S} : \delta \dot{\boldsymbol{E}}\, dV, \end{aligned} \tag{5.43}$$

which shows that $\boldsymbol{S}$ is work conjugate to $\dot{\boldsymbol{E}}$ and enables the *material virtual work equation* to be alternatively written in terms of the second Piola–Kirchhoff tensor as

$$\int_V \boldsymbol{S} : \delta \dot{\boldsymbol{E}}\, dV = \int_V \boldsymbol{f}_0 \cdot \delta \boldsymbol{v}\, dV + \int_{\partial V} \boldsymbol{t}_0 \cdot \delta \boldsymbol{v}\, dA. \tag{5.44}$$

For completeness the inverse of Equations (5.34a) and (5.41b) is given as

$$\boldsymbol{\sigma} = J^{-1} \boldsymbol{P} \boldsymbol{F}^T; \quad \boldsymbol{\sigma} = J^{-1} \boldsymbol{F} \boldsymbol{S} \boldsymbol{F}^T. \tag{5.45a,b}$$

**Remark 5.2**: Applying the pull-back and push-forward concepts to the Kirchhoff and second Piola–Kirchhoff tensors yields

$$\boldsymbol{S} = \boldsymbol{F}^{-1} \boldsymbol{\tau} \boldsymbol{F}^{-T} = \phi_*^{-1}[\boldsymbol{\tau}]; \quad \boldsymbol{\tau} = \boldsymbol{F} \boldsymbol{S} \boldsymbol{F}^T = \phi_*[\boldsymbol{S}]; \tag{5.46a,b}$$

from which the second Piola–Kirchhoff and the Cauchy stresses are related as

$$\boldsymbol{S} = J \phi_*^{-1}[\boldsymbol{\sigma}]; \quad \boldsymbol{\sigma} = J^{-1} \phi_*[\boldsymbol{S}]. \tag{5.47a,b}$$

In the above equation $\boldsymbol{S}$ and $\boldsymbol{\sigma}$ are related by the so-called *Piola transformation*, which involves a push-forward or pull-back operation combined with the volume scaling $J$.

**Remark 5.3**: A useful interpretation of the second Piola–Kirchhoff stress can be obtained by observing that in the case of rigid body motion the polar decomposition given by Equation (4.27) indicates that $\boldsymbol{F} = \boldsymbol{R}$ and $J = 1$. Consequently, the second Piola–Kirchhoff stress tensor becomes

$$\boldsymbol{S} = \boldsymbol{R}^T \boldsymbol{\sigma} \boldsymbol{R}. \tag{5.48}$$

Comparing this equation with the transformation Equations (2.42) given in Section 2.2.2, it transpires that the second Piola–Kirchhoff stress components coincide with the components of the Cauchy stress tensor expressed in the local set of orthogonal axes that results from rotating the global Cartesian directions according to $\boldsymbol{R}$.

**EXAMPLE 5.4: Independence of $S$ from $Q$**

A useful property of the second Piola–Kirchhoff tensor $\boldsymbol{S}$ is its independence from possible superimposed rotations $\boldsymbol{Q}$ on the current body configuration. To prove this, note first that because $\tilde{\phi} = \boldsymbol{Q}\phi$, then $\tilde{\boldsymbol{F}} = \boldsymbol{QF}$ and $\tilde{J} = J$. Using these equations in conjunction with the objectivity of $\boldsymbol{\sigma}$ as given by Equation (5.11) gives

$$\begin{aligned}\tilde{\boldsymbol{S}} &= \tilde{J}\tilde{\boldsymbol{F}}^{-1}\tilde{\boldsymbol{\sigma}}\tilde{\boldsymbol{F}}^{-T}\\ &= J\boldsymbol{F}^{-1}\boldsymbol{Q}^T\boldsymbol{Q}\boldsymbol{\sigma}\boldsymbol{Q}^T\boldsymbol{Q}\boldsymbol{F}^{-T}\\ &= \boldsymbol{S}.\end{aligned}$$

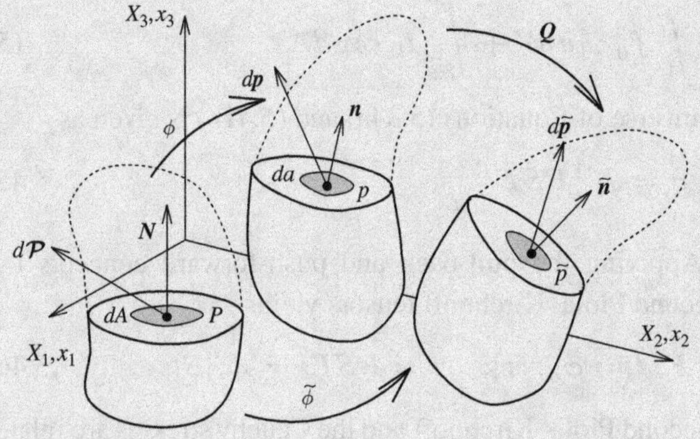


**EXAMPLE 5.5: Biot stress tensor**

Alternative stress tensors work conjugate to other strain measures can be contrived. For instance, the material stress tensor $\boldsymbol{T}$ work conjugate to the rate of the stretch tensor $\dot{\boldsymbol{U}}$ is associated with the name of Biot. In order to derive a relationship between $\boldsymbol{T}$ and $\boldsymbol{S}$, note first that differentiating the equations $\boldsymbol{UU} = \boldsymbol{C}$ and $2\boldsymbol{E} = \boldsymbol{C} - \boldsymbol{I}$ with respect to time gives

$$\dot{\boldsymbol{E}} = \tfrac{1}{2}(\boldsymbol{U}\dot{\boldsymbol{U}} + \dot{\boldsymbol{U}}\boldsymbol{U}).$$

With the help of this relationship we can express the internal work per unit of initial volume as

$$\begin{aligned}\boldsymbol{S} : \dot{\boldsymbol{E}} &= \boldsymbol{S} : \tfrac{1}{2}(\boldsymbol{U}\dot{\boldsymbol{U}} + \dot{\boldsymbol{U}}\boldsymbol{U})\\ &= \tfrac{1}{2}\mathrm{tr}(\boldsymbol{SU}\dot{\boldsymbol{U}} + \boldsymbol{S}\dot{\boldsymbol{U}}\boldsymbol{U})\\ &= \tfrac{1}{2}\mathrm{tr}(\boldsymbol{SU}\dot{\boldsymbol{U}} + \boldsymbol{US}\dot{\boldsymbol{U}})\\ &= \tfrac{1}{2}(\boldsymbol{SU} + \boldsymbol{US}) : \dot{\boldsymbol{U}},\end{aligned}$$

*(continued)*

**EXAMPLE 5.5:** *(cont.)*

and therefore the Biot tensor work conjugate to the stretch tensor is

$$T = \tfrac{1}{2}(SU + US).$$

Using the polar decomposition and the relationship between $S$ and $P$, namely, $P = FS$, an alternative equation for $T$ emerges as

$$T = \tfrac{1}{2}(R^T P + P^T R).$$

### 5.5.4 Deviatoric and Pressure Components

In many practical applications such as metal plasticity, soil mechanics, and biomechanics, it is physically relevant to isolate the hydrostatic pressure component $p$ from the deviatoric component $\sigma'$ of the Cauchy stress tensor as

$$\sigma = \sigma' + pI; \quad p = \frac{1}{3}\mathrm{tr}\sigma = \frac{1}{3}\sigma : I; \tag{5.49a,b}$$

where the deviatoric Cauchy stress tensor $\sigma'$ satisfies $\mathrm{tr}\sigma' = 0$.

Similar decompositions can be established in terms of the first and second Piola–Kirchhoff stress tensors. For this purpose, we simply substitute the above decomposition into Equations (5.34a) for $P$ and $(5.41\mathrm{a,b})_b$ for $S$ to give

$$P = P' + pJF^{-T}; \quad P' = J\sigma' F^{-T}; \tag{5.50a}$$

$$S = S' + pJC^{-1}; \quad S' = JF^{-1}\sigma' F^{-T}. \tag{5.50b}$$

The tensors $S'$ and $P'$ are often referred to as the *true* deviatoric components of $S$ and $P$. Note that, although the trace of $\sigma'$ is zero, it does not follow that the traces of $S'$ and $P'$ must also vanish. In fact, the corresponding equations can be obtained from Equations (5.50a–b) and properties (2.50) and (2.51) of the trace and double contractions as

$$S' : C = 0; \tag{5.51a}$$

$$P' : F = 0. \tag{5.51b}$$

The above equations are important as they enable the hydrostatic pressure $p$ to be evaluated directly from either $S$ or $P$ as

$$p = \frac{1}{3}J^{-1} P : F; \tag{5.52a}$$

$$p = \frac{1}{3}J^{-1} S : C. \tag{5.52b}$$

Proof of the above equations follows rapidly by taking the double contractions of (5.50a) by $\boldsymbol{F}$ and (5.50b) by $\boldsymbol{C}$.

**EXAMPLE 5.6: Proof of Equation (5.51a)**

Equation (5.51a) is easily proved as follows:

$$\begin{aligned}\boldsymbol{S}' : \boldsymbol{C} &= (J\boldsymbol{F}^{-1}\boldsymbol{\sigma}'\boldsymbol{F}^{-T}) : \boldsymbol{C} \\ &= J\mathrm{tr}(\boldsymbol{F}^{-1}\boldsymbol{\sigma}'\boldsymbol{F}^{-T}\boldsymbol{C}) \\ &= J\mathrm{tr}(\boldsymbol{\sigma}'\boldsymbol{F}^{-T}\boldsymbol{F}^{T}\boldsymbol{F}\boldsymbol{F}^{-1}) \\ &= J\mathrm{tr}\boldsymbol{\sigma}' \\ &= 0.\end{aligned}$$

A similar procedure can be used for (5.51b).

## 5.6 STRESS RATES

In Section 4.15 objective tensors were defined by imposing that under rigid body motions they transform according to Equation (4.135). Unfortunately, time differentiation of Equation (5.11) shows that the material time derivative $\dot{\boldsymbol{\sigma}}$ of the stress tensor fails to satisfy this condition as

$$\dot{\tilde{\boldsymbol{\sigma}}} = \boldsymbol{Q}\dot{\boldsymbol{\sigma}}\boldsymbol{Q}^{T} + \dot{\boldsymbol{Q}}\boldsymbol{\sigma}\boldsymbol{Q}^{T} + \boldsymbol{Q}\boldsymbol{\sigma}\dot{\boldsymbol{Q}}^{T}. \tag{5.53}$$

Consequently, $\dot{\tilde{\boldsymbol{\sigma}}} \neq \boldsymbol{Q}\dot{\boldsymbol{\sigma}}\boldsymbol{Q}^{T}$ unless the rigid body rotation is not a time-dependent transformation. Many rate-dependent materials, however, must be described in terms of stress rates and the resulting constitutive models must be frame-indifferent. It is therefore essential to derive stress rate measures that are objective. This can be achieved in several ways, each leading to a different objective stress rate tensor. The simplest of these tensors is due to Truesdell and is based on the fact that the second Piola–Kirchhoff tensor is intrinsically independent of any possible rigid body motion. The *Truesdell stress rate* $\boldsymbol{\sigma}^{\circ}$ is thus defined in terms of the Piola transformation of the time derivative of the second Piola–Kirchhoff stress as

$$\boldsymbol{\sigma}^{\circ} = J^{-1}\phi_{*}[\dot{\boldsymbol{S}}] = J^{-1}\boldsymbol{F}\left[\frac{d}{dt}(J\boldsymbol{F}^{-1}\boldsymbol{\sigma}\boldsymbol{F}^{-T})\right]\boldsymbol{F}^{T}. \tag{5.54}$$

The time derivatives of $\boldsymbol{F}^{-1}$ in the above equation can be obtained by differentiating the expression $\boldsymbol{F}\boldsymbol{F}^{-1} = \boldsymbol{I}$ and using Equation (4.93) to give

$$\frac{d}{dt}\boldsymbol{F}^{-1} = -\boldsymbol{F}^{-1}\boldsymbol{l}, \tag{5.55}$$

which combined with Equation (4.127) for $\dot{J}$ gives the Truesdell rate of stress as

$$\boldsymbol{\sigma}^{\circ} = \dot{\boldsymbol{\sigma}} - \boldsymbol{l}\boldsymbol{\sigma} - \boldsymbol{\sigma}\boldsymbol{l}^T + (\mathrm{tr}\boldsymbol{l})\boldsymbol{\sigma}. \tag{5.56}$$

The Truesdell stress rate tensor can be reinterpreted in terms of the Lie derivative of the Kirchhoff stresses as

$$J\boldsymbol{\sigma}^{\circ} = \mathcal{L}_{\phi}[\boldsymbol{\tau}]. \tag{5.57}$$

In fact, this expression defines what is known as the *Truesdell rate* of the Kirchhoff tensor $\boldsymbol{\tau}^{\circ} = J\boldsymbol{\sigma}^{\circ}$, which can be shown from Equation (5.56) or Equation (5.57) to be

$$\boldsymbol{\tau}^{\circ} = \dot{\boldsymbol{\tau}} - \boldsymbol{l}\boldsymbol{\tau} - \boldsymbol{\tau}\boldsymbol{l}^T. \tag{5.58}$$

Alternative objective stress rates can be derived in terms of the Lie derivative of the Cauchy stress tensor to give the *Oldroyd stress rate* $\boldsymbol{\sigma}^{\bullet}$ as

$$\begin{aligned}\boldsymbol{\sigma}^{\bullet} &= \mathcal{L}_{\phi}[\boldsymbol{\sigma}] \\ &= \boldsymbol{F}\left[\frac{d}{dt}(\boldsymbol{F}^{-1}\boldsymbol{\sigma}\boldsymbol{F}^{-T})\right]\boldsymbol{F}^T \\ &= \dot{\boldsymbol{\sigma}} - \boldsymbol{l}\boldsymbol{\sigma} - \boldsymbol{\sigma}\boldsymbol{l}^T.\end{aligned} \tag{5.59}$$

If the pull-back–push-forward operations are performed with $\boldsymbol{F}^T$ and $\boldsymbol{F}^{-T}$ respectively, the resulting objective stress rate tensor is the *convective stress rate* $\boldsymbol{\sigma}^{\diamond}$, given as

$$\begin{aligned}\boldsymbol{\sigma}^{\diamond} &= \boldsymbol{F}^{-T}\left[\frac{d}{dt}(\boldsymbol{F}^T\boldsymbol{\sigma}\boldsymbol{F})\right]\boldsymbol{F}^{-1} \\ &= \dot{\boldsymbol{\sigma}} + \boldsymbol{l}^T\boldsymbol{\sigma} + \boldsymbol{\sigma}\boldsymbol{l}.\end{aligned} \tag{5.60}$$

A simplified objective stress rate can be obtained by ignoring the stretch component of $\boldsymbol{F}$ in Equations (5.54), (5.59), or (5.60), thus performing the pull-back and push-forward operations using only the rotation tensor $\boldsymbol{R}$. This defines the so-called *Green–Naghdi stress rate* $\boldsymbol{\sigma}^{\triangle}$, which with the help of Equation (4.112) is given as

$$\begin{aligned}\boldsymbol{\sigma}^{\triangle} &= \boldsymbol{R}\left[\frac{d}{dt}(\boldsymbol{R}^T\boldsymbol{\sigma}\boldsymbol{R})\right]\boldsymbol{R}^T \\ &= \dot{\boldsymbol{\sigma}} + \boldsymbol{\sigma}\dot{\boldsymbol{R}}\boldsymbol{R}^T - \dot{\boldsymbol{R}}\boldsymbol{R}^T\boldsymbol{\sigma}.\end{aligned} \tag{5.61}$$

Finally, if the antisymmetric tensor $\dot{\boldsymbol{R}}\boldsymbol{R}^T$ is approximated by the spin tensor $\boldsymbol{w}$ (see Equation (4.116)), the resulting objective stress rate is known as the *Jaumann stress rate*:

$$\boldsymbol{\sigma}^{\triangledown} = \dot{\boldsymbol{\sigma}} + \boldsymbol{\sigma}\boldsymbol{w} - \boldsymbol{w}\boldsymbol{\sigma}. \tag{5.62}$$

Irrespective of the approximations made to arrive at the above definitions of $\boldsymbol{\sigma}^{\triangle}$ and $\boldsymbol{\sigma}^{\triangledown}$, they both remain objective even when these approximations do not apply.

**EXAMPLE 5.7: Objectivity of $\sigma^{\circ}$**

The objectivity of the Truesdell stress rate given by Equation (5.56) can be proved directly without referring back to the initial configuration. For this purpose recall first Equations (5.11), (5.53), and (4.137) as

$$\begin{aligned}
\tilde{\sigma} &= Q\sigma Q^T;\\
\dot{\tilde{\sigma}} &= Q\dot{\sigma}Q^T + \dot{Q}\sigma Q^T + Q\sigma\dot{Q}^T;\\
\tilde{l} &= \dot{Q}Q^T + QlQ^T;
\end{aligned}$$

and note that because $\tilde{J} = J$ then $\mathrm{tr}\boldsymbol{l} = \mathrm{tr}\tilde{\boldsymbol{l}}$. With the help of the above equations, the Truesdell stress rate on a rotated configuration $\tilde{\boldsymbol{\sigma}}^{\circ}$ emerges as

$$\begin{aligned}
\tilde{\sigma}^{\circ} &= \dot{\tilde{\sigma}} - \tilde{l}\tilde{\sigma} - \tilde{\sigma}\tilde{l}^T + (\mathrm{tr}\tilde{l})\tilde{\sigma}\\
&= Q\dot{\sigma}Q^T + \dot{Q}\sigma Q^T + Q\sigma\dot{Q}^T - (\dot{Q}Q^T + QlQ^T)Q\sigma Q^T\\
&\quad - Q\sigma Q^T(\dot{Q}Q^T + QlQ^T) + (\mathrm{tr}l)Q\sigma Q^T\\
&= Q\dot{\sigma}Q^T - Ql\sigma Q^T - Q\sigma lQ^T + (\mathrm{tr}l)Q\sigma Q^T\\
&= Q\sigma^{\circ}Q^T,
\end{aligned}$$

and is therefore objective.

## Exercises

1. A three-dimensional finite deformation of a continuum from an initial configuration $\boldsymbol{X} = (X_1, X_2, X_3)^T$ to a final configuration $\boldsymbol{x} = (x_1, x_2, x_3)^T$ is given as

$$x_1(\boldsymbol{X}) = 5 - 3X_1 - X_2 \tag{5.63a}$$

$$x_2(\boldsymbol{X}) = 2 + \frac{5}{4}X_1 - 2X_2 x_3(\boldsymbol{X}) = X_3. \tag{5.63b}$$

In addition, the first Piola–Kirchhoff stress tensor is given by

$$\boldsymbol{P} = \begin{bmatrix} X_1 & X_1 & 0\\ \alpha & X_2 & 0\\ 0 & 0 & X_3 \end{bmatrix}. \tag{5.64}$$

(a) Determine whether the deformation is isochoric.
(b) Determine the value of $\alpha$ so that the stress tensor $\boldsymbol{P}$ satisfies rotational equilibrium.

(c) Determine the body force field $\boldsymbol{f}_0$ per unit undeformed volume such that the material differential equilibrium Equation (5.36) is satisfied.

2. The deformation of a body is described by

$$x_1 = -3X_2 \; ; \; x_2 = \frac{3}{2}x_1 \; ; \; x_3 = X_3; \tag{5.65}$$

and the Cauchy stress tensor at a ceratin point in the spatial configuration is

$$\sigma = \begin{bmatrix} 10 & 2 & 0 \\ 2 & 30 & 0 \\ 0 & 0 & 10 \end{bmatrix}. \tag{5.66}$$

Determine the Cauchy traction vector $\boldsymbol{t} = \sigma\boldsymbol{n}$ and the first Piola–Kirchhoff traction vector $\boldsymbol{t}^{PK} = \boldsymbol{PN}$ acting on a plane characterized by the spatial outward normal $\boldsymbol{n} = (0,1,0)^T$.

3. A two-dimensional Cauchy stress tensor is given as

$$\boldsymbol{\sigma} = \boldsymbol{t} \otimes \boldsymbol{n}_1 + \alpha\, \boldsymbol{n}_1 \otimes \boldsymbol{n}_2,$$

where $\boldsymbol{t}$ is an arbitrary vector and $\boldsymbol{n}_1$ and $\boldsymbol{n}_2$ are orthogonal unit vectors.
(a) Describe graphically the state of stress.
(b) Determine the value of $\alpha$ (hint: $\boldsymbol{\sigma}$ must be symmetric).

4. Using Equation (5.55) and a process similar to that employed in Example 5.5, show that, with respect to the initial volume, the stress tensor $\boldsymbol{\Pi}$ is work conjugate to the tensor $\dot{\boldsymbol{H}}$, where $\boldsymbol{H} = -\boldsymbol{F}^{-T}$ and $\boldsymbol{\Pi} = \boldsymbol{PC} = J\boldsymbol{\sigma F}$.

5. Using the time derivative of the equality $\boldsymbol{CC}^{-1} = \boldsymbol{I}$, show that the tensor $\boldsymbol{\Sigma} = \boldsymbol{CSC} = J\boldsymbol{F}^T\boldsymbol{\sigma F}$ is work conjugate to $\frac{1}{2}\dot{\boldsymbol{B}}$, where $\boldsymbol{B} = -\boldsymbol{C}^{-1}$. Find relationships between $\boldsymbol{T}$, $\boldsymbol{\Sigma}$, and $\boldsymbol{\Pi}$.

6. Prove Equation (5.51b), $\boldsymbol{P}' : \boldsymbol{F} = 0$, using a procedure similar to Example 5.6.

7. Prove directly that the Jaumann stress tensor $\boldsymbol{\sigma}^{\nabla}$ is an objective tensor, using a procedure similar to Example 5.7.

8. Prove that if $d\boldsymbol{x}_1$ and $d\boldsymbol{x}_2$ are two arbitrary elemental vectors moving with the body (see Figure 4.2) then

$$\frac{d}{dt}(d\boldsymbol{x}_1 \cdot \boldsymbol{\sigma} d\boldsymbol{x}_2) = d\boldsymbol{x}_1 \cdot \boldsymbol{\sigma}^{\diamond} d\boldsymbol{x}_2.$$

# CHAPTER SIX

# HYPERELASTICITY

## 6.1 INTRODUCTION

The equilibrium equations derived in Chapter 5 are written in terms of the stresses inside the body. These stresses result from the deformation of the material, and it is now necessary to express them in terms of some measure of this deformation such as, for instance, the strain. These relationships, known as *constitutive equations*, obviously depend on the type of material under consideration and may be dependent on or independent of time. For example, the classical small strain linear elasticity equations involving Young's modulus and Poisson's ratio are time-independent, whereas viscous fluids are clearly entirely dependent on strain rate.

Generally, constitutive equations must satisfy certain physical principles. For example, the equations must obviously be objective, that is, frame-invariant. In this chapter the constitutive equations will be established in the context of a hyperelastic material, whereby stresses are derived from a stored elastic energy function. Although there are a number of alternative material descriptions that could be introduced, hyperelasticity is a particularly convenient constitutive equation, given its simplicity, and it constitutes the basis for more complex material models such as elastoplasticity, viscoplasticity, and viscoelasticity.

## 6.2 HYPERELASTICITY

Materials for which the constitutive behavior is only a function of the current state of deformation are generally known as *elastic*. Under such conditions, any stress measure at a particle $\boldsymbol{X}$ is a function of the current deformation gradient $\boldsymbol{F}$ associated with that particle. Instead of using any of the alternative strain measures given in Chapter 4, the deformation gradient $\boldsymbol{F}$, together with its conjugate first Piola–Kirchhoff stress measure $\boldsymbol{P}$, will be retained in order to define the basic material relationships. Consequently, elasticity can be generally expressed as

$$\boldsymbol{P} = \boldsymbol{P}(\boldsymbol{F}(\boldsymbol{X}), \boldsymbol{X}), \tag{6.1}$$

where the direct dependency upon $\boldsymbol{X}$ allows for the possible inhomogeneity of the material.

In the special case when the work done by the stresses during a deformation process is dependent only on the initial state at time $t_0$ and the final configuration at time $t$, the behavior of the material is said to be path-independent and the material is termed *hyperelastic*. As a consequence of the path-independent behavior and recalling from Equation (5.33) that $\boldsymbol{P}$ is work conjugate with the rate of deformation gradient $\dot{\boldsymbol{F}}$, a *stored strain energy function* or *elastic potential* $\Psi$ per unit undeformed volume can be established as the work done by the stresses from the initial to the current position as

$$\Psi(\boldsymbol{F}(\boldsymbol{X}), \boldsymbol{X}) = \int_{t_0}^{t} \boldsymbol{P}(\boldsymbol{F}(\boldsymbol{X}), \boldsymbol{X}) : \dot{\boldsymbol{F}}\, dt; \qquad \dot{\Psi} = \boldsymbol{P} : \dot{\boldsymbol{F}}. \tag{6.2a,b}$$

Presuming that from physical experiments it is possible to construct the function $\Psi(\boldsymbol{F}, \boldsymbol{X})$ which defines a given material, then the rate of change of the potential can be alternatively expressed as

$$\dot{\Psi} = \sum_{i,J=1}^{3} \frac{\partial \Psi}{\partial F_{iJ}} \dot{F}_{iJ}. \tag{6.3}$$

Comparing this with Equation (6.2a,b)$_b$ reveals that the components of the two-point tensor $\boldsymbol{P}$ are

$$P_{iJ} = \frac{\partial \Psi}{\partial F_{iJ}}. \tag{6.4}$$

For notational convenience this expression is rewritten in a more compact form as

$$\boldsymbol{P}(\boldsymbol{F}(\boldsymbol{X}), \boldsymbol{X}) = \frac{\partial \Psi(\boldsymbol{F}(\boldsymbol{X}), \boldsymbol{X})}{\partial \boldsymbol{F}}. \tag{6.5}$$

Equation (6.5) followed by Equation (6.2a,b) is often used as a definition of a hyperelastic material.

The general constitutive Equation (6.5) can be further developed by recalling the restrictions imposed by objectivity as discussed in Section 4.15. To this end, $\Psi$ must remain invariant when the current configuration undergoes a rigid body rotation. This implies that $\Psi$ depends on $\boldsymbol{F}$ only via the stretch component $\boldsymbol{U}$ and is independent of the rotation component $\boldsymbol{R}$. For convenience, however, $\Psi$ is often expressed as a function of $\boldsymbol{C} = \boldsymbol{U}^2 = \boldsymbol{F}^T\boldsymbol{F}$ as

$$\Psi(\boldsymbol{F}(\boldsymbol{X}),\boldsymbol{X}) = \Psi(\boldsymbol{C}(\boldsymbol{X}),\boldsymbol{X}). \tag{6.6}$$

Observing that $\frac{1}{2}\dot{\boldsymbol{C}} = \dot{\boldsymbol{E}}$ is work conjugate to the second Piola–Kirchhoff stress $\boldsymbol{S}$ enables a totally Lagrangian constitutive equation to be constructed in the same manner as Equation (6.5) to give

$$\dot{\Psi} = \frac{\partial \Psi}{\partial \boldsymbol{C}} : \dot{\boldsymbol{C}} = \frac{1}{2}\boldsymbol{S} : \dot{\boldsymbol{C}}; \quad \boldsymbol{S}(\boldsymbol{C}(\boldsymbol{X}),\boldsymbol{X}) = 2\frac{\partial \Psi}{\partial \boldsymbol{C}} = \frac{\partial \Psi}{\partial \boldsymbol{E}}. \tag{6.7a,b}$$

## 6.3 ELASTICITY TENSOR

### 6.3.1 The Material or Lagrangian Elasticity Tensor

The relationship between $\boldsymbol{S}$ and $\boldsymbol{C}$ or $\boldsymbol{E} = \frac{1}{2}(\boldsymbol{C} - \boldsymbol{I})$, given by Equation (6.7a,b)$_b$, will invariably be nonlinear. Within the framework of a potential Newton–Raphson solution process, this relationship will need to be linearized with respect to an increment $\boldsymbol{u}$ in the current configuration. Using the chain rule, a linear relationship between the directional derivative of $\boldsymbol{S}$ and the linearized strain $D\boldsymbol{E}[\boldsymbol{u}]$ can be obtained, initially in a component form, as

$$\begin{aligned}
DS_{IJ}[\boldsymbol{u}] &= \frac{d}{d\epsilon}\bigg|_{\epsilon=0} S_{IJ}(E_{KL}[\boldsymbol{\phi} + \epsilon \boldsymbol{u}]) \\
&= \sum_{K,L=1}^{3} \frac{\partial S_{IJ}}{\partial E_{KL}} \frac{d}{d\epsilon}\bigg|_{\epsilon=0} E_{KL}[\boldsymbol{\phi} + \epsilon \boldsymbol{u}] \\
&= \sum_{K,L=1}^{3} \frac{\partial S_{IJ}}{\partial E_{KL}} DE_{KL}[\boldsymbol{u}].
\end{aligned} \tag{6.8}$$

This relationship between the directional derivatives of $\boldsymbol{S}$ and $\boldsymbol{E}$ is more concisely expressed as

$$D\boldsymbol{S}[\boldsymbol{u}] = \boldsymbol{\mathcal{C}} : D\boldsymbol{E}[\boldsymbol{u}], \tag{6.9}$$

where the symmetric fourth-order tensor $\boldsymbol{\mathcal{C}}$, known as the *Lagrangian* or *material elasticity tensor*, is defined by the partial derivatives as

$$\begin{aligned}
\boldsymbol{\mathcal{C}} &= \sum_{I,J,K,L=1}^{3} \mathcal{C}_{IJKL}\, \boldsymbol{E}_I \otimes \boldsymbol{E}_J \otimes \boldsymbol{E}_K \otimes \boldsymbol{E}_L; \\
\mathcal{C}_{IJKL} &= \frac{\partial S_{IJ}}{\partial E_{KL}} = \frac{4\,\partial^2 \Psi}{\partial C_{IJ} \partial C_{KL}} = \mathcal{C}_{KLIJ}.
\end{aligned} \tag{6.10}$$

For convenience these expressions are often abbreviated as

$$\mathcal{C} = \frac{\partial \boldsymbol{S}}{\partial \boldsymbol{E}} = 2\frac{\partial \boldsymbol{S}}{\partial \boldsymbol{C}} = \frac{4\,\partial^2\Psi}{\partial \boldsymbol{C}\partial \boldsymbol{C}}. \tag{6.11}$$

**EXAMPLE 6.1: St. Venant–Kirchhoff material**

The simplest example of a hyperelastic material is the St. Venant–Kirchhoff model, which is defined by a strain energy function $\Psi$ as

$$\Psi(\boldsymbol{E}) = \frac{1}{2}\lambda(\text{tr}\boldsymbol{E})^2 + \mu\boldsymbol{E} : \boldsymbol{E},$$

where $\lambda$ and $\mu$ are material coefficients. Using the second part of Equation (6.7a,b)$_b$, we can obtain the second Piola–Kirchhoff stress tensor as

$$\boldsymbol{S} = \lambda(\text{tr}\boldsymbol{E})\boldsymbol{I} + 2\mu\boldsymbol{E},$$

and using Equation (6.10), the coefficients of the Lagrangian elasticity tensor emerge as

$$\mathcal{C}_{IJKL} = \lambda\delta_{IJ}\delta_{KL} + \mu\,(\delta_{IK}\delta_{JL} + \delta_{IL}\delta_{JK}).$$

Note that these last two equations are analogous to those used in linear elasticity, where the small strain tensor has been replaced by the Green strain. Unfortunately, this St. Venant–Kirchhoff material has been found to be of little practical use beyond the small strain regime.

### 6.3.2 The Spatial or Eulerian Elasticity Tensor

It would now be pertinent to attempt to find a spatial equivalent to Equation (6.9), and it is tempting to suppose that this would involve a relationship between the linearized Cauchy stress and the linearized Almansi strain. Although, in principle, this can be achieved, the resulting expression is intractable. An easier route is to interpret Equation (6.9) in a rate form and apply the push-forward operation to the resulting equation. This is achieved by linearizing $\boldsymbol{S}$ and $\boldsymbol{E}$ in the direction of $\boldsymbol{v}$, rather than $\boldsymbol{u}$. Recalling from Section 4.11.3 that $D\boldsymbol{S}[\boldsymbol{v}] = \dot{\boldsymbol{S}}$ and $D\boldsymbol{E}[\boldsymbol{v}] = \dot{\boldsymbol{E}}$ gives

$$\dot{\boldsymbol{S}} = \mathcal{C} : \dot{\boldsymbol{E}}. \tag{6.12}$$

Because the push-forward of $\dot{\boldsymbol{S}}$ has been shown in Section 5.5 to be the Truesdell rate of the Kirchhoff stress $\boldsymbol{\tau}^\circ = J\boldsymbol{\sigma}^\circ$ and the push-forward of $\dot{\boldsymbol{E}}$ is $\boldsymbol{d}$, namely, Equation (4.100a,b)$_a$, it is now possible to obtain the spatial equivalent of the material linearized constitutive Equation (6.12) as

$$\boldsymbol{\sigma}^\circ = \mathcal{c} : \boldsymbol{d}, \tag{6.13}$$

where $\mathcal{c}$, the *Eulerian or spatial elasticity tensor*, is defined as the Piola push-forward of $\mathcal{C}$, and after some careful indicial manipulations can be obtained as*

$$\mathcal{c} = J^{-1}\phi_*[\mathcal{C}]; \quad \mathcal{c} = \sum_{\substack{i,j,k,l=1\\ I,J,K,L=1}}^{3} J^{-1}F_{iI}F_{jJ}F_{kK}F_{lL}\,\mathcal{C}_{IJKL}\,\boldsymbol{e}_i \otimes \boldsymbol{e}_j \otimes \boldsymbol{e}_k \otimes \boldsymbol{e}_l. \tag{6.14a,b}$$

Often, Equation (6.13) is used, together with convenient coefficients in $\mathcal{c}$, as the fundamental constitutive equation that defines the material behavior. Use of such an approach will in general not guarantee hyperelastic behavior, and therefore the stresses cannot be obtained directly from an elastic potential. In such cases the rate equation has to be integrated in time, and this can cause substantial difficulties in a finite element analysis because of problems associated with objectivity over a finite time increment.

**Remark 6.1**: Using Equations (4.105) and (5.57), it can be observed that Equation (6.13) can be reinterpreted in terms of Lie derivatives as

$$\mathcal{L}_{\phi}[\boldsymbol{\tau}] = J\mathcal{c} : \mathcal{L}_{\phi}[\boldsymbol{e}]. \tag{6.15}$$

## 6.4 ISOTROPIC HYPERELASTICITY

### 6.4.1 Material Description

The hyperelastic constitutive equations discussed so far are unrestricted in their application. We are now going to restrict these equations to the common and important isotropic case. Isotropy is defined by requiring the constitutive behavior to be identical in any material direction.† This implies that the relationship between $\Psi$ and $\boldsymbol{C}$ must be independent of the material axes chosen and, consequently, $\Psi$ must only be a function of the invariants of $\boldsymbol{C}$ as

$$\Psi(\boldsymbol{C}(\boldsymbol{X}), \boldsymbol{X}) = \Psi(I_C, II_C, III_C, \boldsymbol{X}), \tag{6.16}$$

where the invariants of $\boldsymbol{C}$ are defined here as

$$I_C = \text{tr}\boldsymbol{C} = \boldsymbol{C} : \boldsymbol{I}; \tag{6.17a}$$

$$II_C = \text{tr}\boldsymbol{C}\boldsymbol{C} = \boldsymbol{C} : \boldsymbol{C}; \tag{6.17b}$$

$$III_C = \det \boldsymbol{C} = J^2. \tag{6.17c}$$

* Using the standard summation convention and noting from Equation (5.54) that $\sigma^{\circ}_{ij} = J^{-1}F_{iI}F_{jJ}\dot{S}_{IJ}$ and from Equation (4.100a,b)$_b$ that $\dot{E}_{KL} = F_{kK}F_{lL}d_{kl}$ gives

$$\sigma^{\circ}_{ij} = J^{-1}\tau^{\circ}_{ij} = J^{-1}F_{iI}F_{jJ}\mathcal{C}_{IJKL}F_{kK}F_{lL}d_{kl} = \mathcal{c}_{ijkl}d_{kl},$$

and, consequently, $\mathcal{c}_{ijkl} = J^{-1}F_{iI}F_{jJ}F_{kK}F_{lL}\mathcal{C}_{IJKL}$.

† Note that the resulting spatial behavior as given by the spatial elasticity tensor may be anisotropic.

As a result of the isotropic restriction, the second Piola–Kirchhoff stress tensor can be rewritten from Equation (6.7a,b)$_b$ as

$$\boldsymbol{S} = 2\frac{\partial \Psi}{\partial \boldsymbol{C}} = 2\frac{\partial \Psi}{\partial I_C}\frac{\partial I_C}{\partial \boldsymbol{C}} + 2\frac{\partial \Psi}{\partial II_C}\frac{\partial II_C}{\partial \boldsymbol{C}} + 2\frac{\partial \Psi}{\partial III_C}\frac{\partial III_C}{\partial \boldsymbol{C}}. \tag{6.18}$$

The second-order tensors formed by the derivatives of the first two invariants with respect to $\boldsymbol{C}$ can be evaluated in component form to give

$$\frac{\partial}{\partial C_{IJ}}\sum_{K=1}^{3} C_{KK} = \delta_{IJ}; \quad \frac{\partial I_C}{\partial \boldsymbol{C}} = \boldsymbol{I}; \tag{6.19a}$$

$$\frac{\partial}{\partial C_{IJ}}\sum_{K,L=1}^{3} C_{KL}C_{KL} = 2C_{IJ}; \quad \frac{\partial II_C}{\partial \boldsymbol{C}} = 2\boldsymbol{C}. \tag{6.19b}$$

The derivative of the third invariant is more conveniently evaluated using the expression for the linearization of the determinant of a tensor given in Equation (2.119). To this end note that the directional derivative with respect to an arbitrary increment tensor $\Delta \boldsymbol{C}$ and the partial derivatives are related via

$$DIII_C[\Delta \boldsymbol{C}] = \sum_{I,J=1}^{3} \frac{\partial III_C}{\partial C_{IJ}}\Delta C_{IJ} = \frac{\partial III_C}{\partial \boldsymbol{C}} : \Delta \boldsymbol{C}. \tag{6.20}$$

Rewriting Equation (2.119) as

$$DIII_C[\Delta \boldsymbol{C}] = \det \boldsymbol{C}\,(\boldsymbol{C}^{-1} : \Delta \boldsymbol{C}), \tag{6.21}$$

comparing this equation with expression (6.20), and noting that both equations are valid for any increment $\Delta \boldsymbol{C}$ yields

$$\frac{\partial III_C}{\partial \boldsymbol{C}} = J^2 \boldsymbol{C}^{-1}. \tag{6.22}$$

Introducing expressions (6.19a,b) and (6.22) into Equation (6.18) enables the second Piola–Kirchhoff stress to be evaluated as

$$\boldsymbol{S} = 2\Psi_I \boldsymbol{I} + 4\Psi_{II}\boldsymbol{C} + 2J^2\Psi_{III}\boldsymbol{C}^{-1}, \tag{6.23}$$

where $\Psi_I = \partial\Psi/\partial I_C$, $\Psi_{II} = \partial\Psi/\partial II_C$, and $\Psi_{III} = \partial\Psi/\partial III_C$.

### 6.4.2 Spatial Description

In design practice, it is obviously the Cauchy stresses that are of engineering significance. These can be obtained indirectly from the second Piola–Kirchhoff stresses by using Equation (5.45a,b)$_b$ as

$$\boldsymbol{\sigma} = J^{-1}\boldsymbol{F}\boldsymbol{S}\boldsymbol{F}^T. \tag{6.24}$$

Substituting $\boldsymbol{S}$ from Equation (6.23) and noting that the left Cauchy–Green tensor is $\boldsymbol{b} = \boldsymbol{F}\boldsymbol{F}^T$ gives

$$\boldsymbol{\sigma} = 2J^{-1}\Psi_I \boldsymbol{b} + 4J^{-1}\Psi_{II}\boldsymbol{b}^2 + 2J\Psi_{III}\boldsymbol{I}. \tag{6.25}$$

In this equation $\Psi_I$, $\Psi_{II}$, and $\Psi_{III}$ still involve derivatives with respect to the invariants of the material tensor $\boldsymbol{C}$. Nevertheless, it is easy to show that the invariants of $\boldsymbol{b}$ are identical to the invariants of $\boldsymbol{C}$, as the following expressions demonstrate:

$$I_b = \mathrm{tr}[\boldsymbol{b}] = \mathrm{tr}[\boldsymbol{F}\boldsymbol{F}^T] = \mathrm{tr}[\boldsymbol{F}^T\boldsymbol{F}] = \mathrm{tr}[\boldsymbol{C}] = I_C; \tag{6.26a}$$
$$II_b = \mathrm{tr}[\boldsymbol{b}\boldsymbol{b}] = \mathrm{tr}[\boldsymbol{F}\boldsymbol{F}^T\boldsymbol{F}\boldsymbol{F}^T] = \mathrm{tr}[\boldsymbol{F}^T\boldsymbol{F}\boldsymbol{F}^T\boldsymbol{F}] = \mathrm{tr}[\boldsymbol{C}\boldsymbol{C}] = II_C; \tag{6.26b}$$
$$III_b = \det[\boldsymbol{b}] = \det[\boldsymbol{F}\boldsymbol{F}^T] = \det[\boldsymbol{F}^T\boldsymbol{F}] = \det[\boldsymbol{C}] = III_C. \tag{6.26c}$$

Consequently, the terms $\Psi_I$, $\Psi_{II}$, and $\Psi_{III}$ in Equation (6.25) are also the derivatives of $\Psi$ with respect to the invariants of $\boldsymbol{b}$.

**Remark 6.2**: Note that any spatially-based expression for $\Psi$ must be a function of $\boldsymbol{b}$ only via its invariants, which implies an isotropic material. This follows from the condition that $\Psi$ must remain constant under rigid body rotations and only the invariants of $\boldsymbol{b}$, not $\boldsymbol{b}$ itself, remain unchanged under such rotations.

**EXAMPLE 6.2: Cauchy stresses**

It is possible to derive an alternative equation for the Cauchy stresses directly from the strain energy. For this purpose, note first that the time derivative of $\boldsymbol{b}$ is

$$\dot{\boldsymbol{b}} = \dot{\boldsymbol{F}}\boldsymbol{F}^T + \boldsymbol{F}\dot{\boldsymbol{F}}^T = \boldsymbol{l}\boldsymbol{b} + \boldsymbol{b}\boldsymbol{l}^T,$$

and therefore the internal energy rate per unit of undeformed volume $\dot{w}_0 = \dot{\Psi}$ is

$$\begin{aligned}\dot{\Psi} &= \frac{\partial \Psi}{\partial \boldsymbol{b}} : \dot{\boldsymbol{b}} \\ &= \frac{\partial \Psi}{\partial \boldsymbol{b}} : (\boldsymbol{l}\boldsymbol{b} + \boldsymbol{b}\boldsymbol{l}^T) \\ &= 2\frac{\partial \Psi}{\partial \boldsymbol{b}}\boldsymbol{b} : \boldsymbol{l}.\end{aligned}$$

Combining this equation with the fact that $\boldsymbol{\sigma}$ is work conjugate to $\boldsymbol{l}$ with respect to the current volume, that is, $\dot{w} = J^{-1}\dot{w}_0 = \boldsymbol{\sigma} : \boldsymbol{l}$, gives

$$J\boldsymbol{\sigma} = 2\frac{\partial \Psi}{\partial \boldsymbol{b}}\boldsymbol{b}.$$

It is simple to show that this equation gives the same result as Equation (6.25) for isotropic materials where $\Psi$ is a function of the invariants of $\boldsymbol{b}$.

### 6.4.3 Compressible Neo-Hookean Material

The equations derived in the previous sections refer to a general isotropic hyperelastic material. We can now focus on a particularly simple case known as *compressible neo-Hookean* material. This material exhibits characteristics that can be identified with the familiar material parameters found in linear elastic analysis. The stored energy function of such a material is defined as

$$\Psi = \frac{\mu}{2}(I_C - 3) - \mu \ln J + \frac{\lambda}{2}(\ln J)^2, \tag{6.27}$$

where the constants $\lambda$ and $\mu$ are material coefficients and $J^2 = III_C$. Note that in the absence of deformation, that is, when $\boldsymbol{C} = \boldsymbol{I}$, the stored energy function vanishes as expected.

The second Piola–Kirchhoff stress tensor can now be obtained from Equation (6.23) as

$$\boldsymbol{S} = \mu(\boldsymbol{I} - \boldsymbol{C}^{-1}) + \lambda(\ln J)\boldsymbol{C}^{-1}. \tag{6.28}$$

Alternatively, the Cauchy stresses can be obtained using Equation (6.25) in terms of the left Cauchy–Green tensor $\boldsymbol{b}$ as

$$\boldsymbol{\sigma} = \frac{\mu}{J}(\boldsymbol{b} - \boldsymbol{I}) + \frac{\lambda}{J}(\ln J)\boldsymbol{I}. \tag{6.29}$$

The Lagrangian elasticity tensor corresponding to this neo-Hookean material can be obtained by differentiation of Equation (6.28) with respect to the components of $\boldsymbol{C}$ to give, after some algebra using Equation (6.22), $\boldsymbol{\mathcal{C}}$ as

$$\boldsymbol{\mathcal{C}} = \lambda \boldsymbol{C}^{-1} \otimes \boldsymbol{C}^{-1} + 2(\mu - \lambda \ln J)\boldsymbol{\mathcal{I}}, \tag{6.30}$$

where $\boldsymbol{C}^{-1} \otimes \boldsymbol{C}^{-1} = \sum (\boldsymbol{C}^{-1})_{IJ}(\boldsymbol{C}^{-1})_{KL}\, \boldsymbol{E}_I \otimes \boldsymbol{E}_J \otimes \boldsymbol{E}_K \otimes \boldsymbol{E}_L$ and the fourth-order tensor $\boldsymbol{\mathcal{I}}$ is defined as

$$\boldsymbol{\mathcal{I}} = -\frac{\partial \boldsymbol{C}^{-1}}{\partial \boldsymbol{C}}; \quad \mathcal{I}_{IJKL} = -\frac{\partial (\boldsymbol{C}^{-1})_{IJ}}{\partial C_{KL}}. \tag{6.31}$$

In order to obtain the coefficients of this tensor, recall from Section 2.3.4 that the directional derivative of the inverse of a tensor in the direction of an arbitrary increment $\Delta \boldsymbol{C}$ is

$$D\boldsymbol{C}^{-1}[\Delta \boldsymbol{C}] = -\boldsymbol{C}^{-1}(\Delta \boldsymbol{C})\boldsymbol{C}^{-1}. \tag{6.32}$$

Alternatively, this directional derivative can be expressed in terms of the partial derivatives as

$$D\boldsymbol{C}^{-1}[\Delta \boldsymbol{C}] = \frac{\partial \boldsymbol{C}^{-1}}{\partial \boldsymbol{C}} : \Delta \boldsymbol{C}. \tag{6.33}$$

Equating the right-hand sides of the last two equations using indicial notation gives

$$-\frac{\partial(\boldsymbol{C}^{-1})_{IJ}}{\partial C_{KL}}\Delta C_{KL} = \left(\boldsymbol{C}^{-1}\right)_{IK}\Delta C_{KL}\left(\boldsymbol{C}^{-1}\right)_{LJ}. \tag{6.34}$$

By employing the symmetry of both $\Delta\boldsymbol{C}$ and $\boldsymbol{C}^{-1}$ this equation can be rewritten as

$$-\frac{\partial(\boldsymbol{C}^{-1})_{IJ}}{\partial C_{KL}}\Delta C_{KL} = \left(\boldsymbol{C}^{-1}\right)_{IL}\Delta C_{KL}\left(\boldsymbol{C}^{-1}\right)_{KJ}. \tag{6.35}$$

Consequently, the components of the tensor $\boldsymbol{\mathcal{I}}$ can now be identified by averaging Equations (6.34) and (6.35) to give

$$\mathcal{I}_{IJKL} = \frac{1}{2}\left[\left(\boldsymbol{C}^{-1}\right)_{IK}\left(\boldsymbol{C}^{-1}\right)_{JL} + \left(\boldsymbol{C}^{-1}\right)_{IL}\left(\boldsymbol{C}^{-1}\right)_{JK}\right]. \tag{6.36}$$

By constructing this tensor in the manner described above it is easy to show that $\boldsymbol{\mathcal{I}}$ and therefore the material elasticity tensor $\boldsymbol{\mathcal{C}}$ (in Equation (6.30)) satisfy the full set of symmetries:

$$\mathcal{I}_{IJKL} = \mathcal{I}_{KLIJ} = \mathcal{I}_{JIKL} = \mathcal{I}_{IJLK}; \tag{6.37a}$$

$$\mathcal{C}_{IJKL} = \mathcal{C}_{KLIJ} = \mathcal{C}_{JIKL} = \mathcal{C}_{IJLK}. \tag{6.37b}$$

The Eulerian or spatial elasticity tensor can now be obtained by pushing forward the Lagrangian tensor using Equation (6.14a,b) to give, after tedious algebra, $\boldsymbol{c}$ as

$$\boldsymbol{c} = \frac{\lambda}{J}\boldsymbol{I}\otimes\boldsymbol{I} + \frac{2}{J}(\mu - \lambda\ln J)\,\boldsymbol{\imath}, \tag{6.38}$$

where $\boldsymbol{\imath}$ is the fourth-order identity tensor obtained by pushing forward $\boldsymbol{\mathcal{I}}$ and in component form is given in terms of the Kronecker delta as

$$\boldsymbol{\imath} = \phi_*[\boldsymbol{\mathcal{I}}];\quad \imath_{ijkl} = \sum_{I,J,K,L} F_{iI}F_{jJ}F_{kK}F_{lL}\mathcal{I}_{IJKL} = \frac{1}{2}\left(\delta_{ik}\delta_{jl} + \delta_{il}\delta_{jk}\right). \tag{6.39}$$

Note that Equation (6.39) defines an isotropic fourth-order tensor as discussed in Section 2.2.4, similar to that used in linear elasticity, which can be expressed in terms of the effective Lamé moduli $\lambda'$ and $\mu'$ as

$$c_{ijkl} = \lambda'\delta_{ij}\delta_{kl} + \mu'\left(\delta_{ik}\delta_{jl} + \delta_{il}\delta_{jk}\right), \tag{6.40}$$

where the effective coefficients $\lambda'$ and $\mu'$ are

$$\lambda' = \frac{\lambda}{J};\quad \mu' = \frac{\mu - \lambda\ln J}{J}. \tag{6.41}$$

Note that in the case of small strains when $J \approx 1$, then $\lambda' \approx \lambda$, $\mu' \approx \mu$ and the standard fourth-order tensor used in linear elastic analysis is recovered. Finally, in a similar manner to Equation $(6.37)_b$ the spatial elasticity $\mathcal{c}$ satisfies the full set of symmetries:

$$\mathcal{c}_{ijkl} = \mathcal{c}_{klij} = \mathcal{c}_{jikl} = \mathcal{c}_{ijlk}. \tag{6.42}$$

The recognition of these symmetries will facilitate the numerical implementation of the linearization of the equilibrium equations.

**EXAMPLE 6.3: Pure dilatation (i)**

The simplest possible deformation is a pure dilatation case where the deformation gradient tensor $\boldsymbol{F}$ is

$$\boldsymbol{F} = \lambda \boldsymbol{I}; \quad J = \lambda^3;$$

and the left Cauchy–Green tensor $\boldsymbol{b}$ is therefore

$$\boldsymbol{b} = \lambda^2 \boldsymbol{I} = J^{2/3} \boldsymbol{I}.$$

Under such conditions the Cauchy stress tensor for a compressible neo-Hookean material is evaluated with the help of Equation (6.29) as

$$\boldsymbol{\sigma} = \left[\frac{\mu}{J}(J^{2/3} - 1) + \frac{\lambda}{J}\ln J\right]\boldsymbol{I},$$

which represents a state of hydrostatic stress with pressure $p$ equal to

$$p = \frac{\mu}{J}(J^{2/3} - 1) + \frac{\lambda}{J}\ln J.$$

**EXAMPLE 6.4: Simple shear (i)**

The case of simple shear described in Chapter 4 is defined by a deformation gradient and left Cauchy–Green tensors as

$$\boldsymbol{F} = \begin{bmatrix} 1 & \gamma & 0 \\ 0 & 1 & 0 \\ 0 & 0 & 1 \end{bmatrix}; \quad \boldsymbol{b} = \begin{bmatrix} 1+\gamma^2 & \gamma & 0 \\ \gamma & 1 & 0 \\ 0 & 0 & 1 \end{bmatrix};$$

which imply $J = 1$ and the Cauchy stresses for a neo-Hookean material are

$$\boldsymbol{\sigma} = \mu \begin{bmatrix} \gamma^2 & \gamma & 0 \\ \gamma & 0 & 0 \\ 0 & 0 & 0 \end{bmatrix}.$$

*(continued)*

**EXAMPLE 6.4:** *(cont.)*

Note that only when $\gamma \rightarrow 0$ is a state of pure shear obtained. Note also that, despite the fact that $J = 1$, that is, there is no change in volume, the pressure $p = \text{tr}\boldsymbol{\sigma}/3 = \gamma^2/3$ is not zero. This is known as the *Kelvin effect.*

# 6.5 INCOMPRESSIBLE AND NEARLY INCOMPRESSIBLE MATERIALS

Most practical large strain processes take place under incompressible or near incompressible conditions. Hence, it is pertinent to discuss the constitutive implications of this constraint on the deformation. The terminology "near incompressibility" is used here to denote materials that are truly incompressible, but their numerical treatment invokes a small measure of volumetric deformation. Alternatively, in a large strain elasto-plastic or inelastic context, the plastic deformation is often truly incompressible and the elastic volumetric strain is comparatively small.

## 6.5.1 Incompressible Elasticity

In order to determine the constitutive equation for an incompressible hyperelastic material, recall Equation (6.7a,b)$_a$ rearranged as

$$\left(\frac{1}{2}\boldsymbol{S} - \frac{\partial \Psi}{\partial \boldsymbol{C}}\right) : \dot{\boldsymbol{C}} = 0. \tag{6.43}$$

Previously the fact that $\dot{\boldsymbol{C}}$ in this equation was arbitrary implied that $\boldsymbol{S} = 2\partial\Psi/\partial\boldsymbol{C}$. In the incompressible case, the term in brackets is not guaranteed to vanish because $\dot{\boldsymbol{C}}$ is no longer arbitrary. In fact, given that $J = 1$ throughout the deformation and therefore $\dot{J} = 0$, Equation (4.129) gives the required constraint on $\dot{\boldsymbol{C}}$ as

$$\frac{1}{2}J\boldsymbol{C}^{-1} : \dot{\boldsymbol{C}} = 0. \tag{6.44}$$

The fact that Equation (6.43) has to be satisfied for any $\dot{\boldsymbol{C}}$ that complies with condition (6.44) implies that

$$\frac{1}{2}\boldsymbol{S} - \frac{\partial \Psi}{\partial \boldsymbol{C}} = \gamma\frac{J}{2}\boldsymbol{C}^{-1}, \tag{6.45}$$

where $\gamma$ is an unknown scalar that will, under certain circumstances that we will discuss later, coincide with the hydrostatic pressure and will be determined by using the additional equation given by the incompressibility constraint $J = 1$. Equation (6.44) is symbolically illustrated in Figure 6.1, where the double contraction " : " has been interpreted as a generalized dot product. This enables $(\boldsymbol{S}/2 - \partial\Psi/\partial\boldsymbol{C})$ and $J\boldsymbol{C}^{-1}/2$ to be seen as being orthogonal to any admissible $\dot{\boldsymbol{C}}$ and therefore proportional to each other.

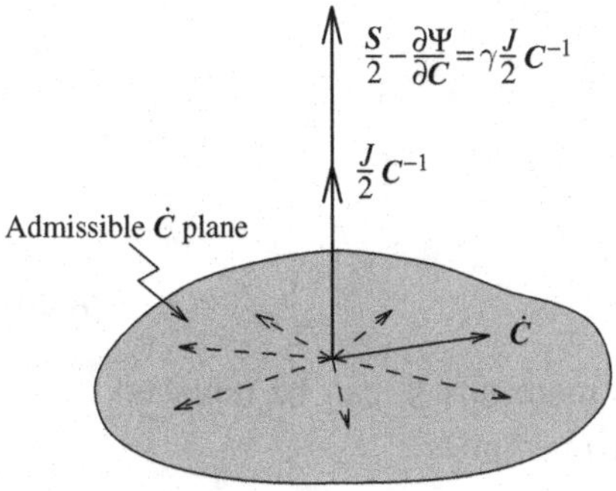


**FIGURE 6.1** Incompressibility constraint.

From Equation (6.45) the general incompressible hyperelastic constitutive equation emerges as

$$\boldsymbol{S} = 2\frac{\partial\Psi(\boldsymbol{C})}{\partial\boldsymbol{C}} + \gamma J\boldsymbol{C}^{-1}. \tag{6.46}$$

The determinant $J$ in the above equation may seem unnecessary in the case of incompressibility where $J = 1$, but retaining $J$ has the advantage that Equation (6.46) is also applicable in the nearly incompressible case. Furthermore, in practical terms, a finite element analysis rarely enforces $J = 1$ in a strict pointwise manner, and hence its retention may be important for the evaluation of stresses.

Recalling Equation (5.50b) giving the deviatoric–hydrostatic decomposition of the second Piola–Kirchhoff tensor as $\boldsymbol{S} = \boldsymbol{S}' + pJ\boldsymbol{C}^{-1}$, it would be convenient to identify the parameter $\gamma$ with the pressure $p$. With this in mind, a relationship between $p$ and $\gamma$ can be established to give

$$\begin{aligned} p &= \frac{1}{3}J^{-1}\boldsymbol{S} : \boldsymbol{C} \\ &= \frac{1}{3}J^{-1}\left[2\frac{\partial\Psi}{\partial\boldsymbol{C}} + \gamma J\boldsymbol{C}^{-1}\right] : \boldsymbol{C} \\ &= \gamma + \frac{2}{3}J^{-1}\frac{\partial\Psi}{\partial\boldsymbol{C}} : \boldsymbol{C}, \end{aligned} \tag{6.47}$$

which clearly indicates that $\gamma$ and $p$ coincide only if

$$\frac{\partial\Psi}{\partial\boldsymbol{C}} : \boldsymbol{C} = 0. \tag{6.48}$$

This implies that the function $\Psi(\boldsymbol{C})$ must be homogeneous of order 0, that is, $\Psi(\alpha\boldsymbol{C}) = \Psi(\boldsymbol{C})$ for any arbitrary constant $\alpha$.‡ This can be achieved by recognizing that for incompressible materials $III_C = \det \boldsymbol{C} = J^2 = 1$. We can therefore express the energy function $\Psi$ in terms of the distortional component of the right Cauchy–Green tensor $\hat{\boldsymbol{C}} = III_C^{-1/3}\boldsymbol{C}$ to give a formally modified (distortional) energy function $\widehat{\Psi}(\boldsymbol{C}) = \Psi(\hat{\boldsymbol{C}})$. The homogeneous properties of the resulting function $\widehat{\Psi}(\boldsymbol{C})$ are easily shown by

$$\begin{aligned}\widehat{\Psi}(\alpha\boldsymbol{C}) &= \Psi[(\det \alpha\boldsymbol{C})^{-1/3}(\alpha\boldsymbol{C})] \\ &= \Psi[(\alpha^3 \det \boldsymbol{C})^{-1/3}\alpha\boldsymbol{C}] \\ &= \Psi[(\det \boldsymbol{C})^{-1/3}\boldsymbol{C}] \\ &= \widehat{\Psi}(\boldsymbol{C}).\end{aligned} \tag{6.49}$$

Accepting that for the case of incompressible materials $\Psi$ can be replaced by $\widehat{\Psi}$, condition (6.48) is satisfied and Equation (6.46) becomes

$$\boldsymbol{S} = 2\frac{\partial\widehat{\Psi}(\boldsymbol{C})}{\partial\boldsymbol{C}} + pJ\boldsymbol{C}^{-1}. \tag{6.50}$$

It is now a trivial matter to identify the deviatoric component of the second Piola–Kirchhoff tensor by comparison of the above equation with Equation $(6.54)_b$ to give

$$\boldsymbol{S}' = 2\frac{\partial\widehat{\Psi}}{\partial\boldsymbol{C}}. \tag{6.51}$$

Note that the derivative $\partial\widehat{\Psi}(\boldsymbol{C})/\partial\boldsymbol{C}$ is not equal to the derivative $\partial\Psi(\boldsymbol{C})/\partial\boldsymbol{C}$, despite the fact that $\hat{\boldsymbol{C}} = \boldsymbol{C}$ for incompressibility. This is because $III_C$ remains a function of $\boldsymbol{C}$ while the derivative of $\hat{\boldsymbol{C}}$ is being executed. It is only after the derivative has been completed that the substitution $III_C = 1$ can be made.

### 6.5.2 Incompressible Neo-Hookean Material

In the case of incompressibility the neo-Hookean material introduced in Section 6.4.3 is defined by a hyperelastic potential $\Psi(\boldsymbol{C})$ given as

$$\Psi(\boldsymbol{C}) = \frac{1}{2}\mu(\text{tr}\boldsymbol{C} - 3). \tag{6.52}$$

‡ A scalar function $f(\boldsymbol{x})$ of a $k$-dimensional vector variable $\boldsymbol{x} = [x_1, x_2, \ldots, x_k]^T$ is said to be homogeneous of order $n$ if for any arbitrary constant $\alpha$,

$$f(\alpha\boldsymbol{x}) = \alpha^n f(\boldsymbol{x}).$$

Differentiating this expression with respect to $\alpha$ at $\alpha = 1$ gives

$$\frac{\partial f}{\partial\boldsymbol{x}} \cdot \boldsymbol{x} = nf(\boldsymbol{x}).$$

The equivalent homogeneous (distortional) potential $\widehat{\Psi}$ is established by replacing $\boldsymbol{C}$ by $\hat{\boldsymbol{C}}$ to give

$$\widehat{\Psi}(\boldsymbol{C}) = \frac{1}{2}\mu(\text{tr}\hat{\boldsymbol{C}} - 3). \tag{6.53}$$

Now using Equation (6.50) $\boldsymbol{S}$ is obtained with the help of Equations (6.19a) and (6.20) as

$$\begin{aligned}
\boldsymbol{S} &= 2\frac{\partial\widehat{\Psi}(\boldsymbol{C})}{\partial\boldsymbol{C}} + pJ\boldsymbol{C}^{-1}\\
&= \mu\frac{\partial\text{tr}\hat{\boldsymbol{C}}}{\partial\boldsymbol{C}} + pJ\boldsymbol{C}^{-1}\\
&= \mu\frac{\partial}{\partial\boldsymbol{C}}(III_{\boldsymbol{C}}^{-1/3}\boldsymbol{C} : \boldsymbol{I}) + pJ\boldsymbol{C}^{-1}\\
&= \mu[III_{\boldsymbol{C}}^{-1/3}\boldsymbol{I} - \frac{1}{3}III_{\boldsymbol{C}}^{-1/3-1}III_{\boldsymbol{C}}\boldsymbol{C}^{-1}(\boldsymbol{C} : \boldsymbol{I})] + pJ\boldsymbol{C}^{-1}\\
&= \mu III_{\boldsymbol{C}}^{-1/3}(\boldsymbol{I} - \frac{1}{3}I_{\boldsymbol{C}}\boldsymbol{C}^{-1}) + pJ\boldsymbol{C}^{-1}.
\end{aligned} \tag{6.54}$$

The corresponding Cauchy stress tensor can now be obtained by using Equation $(5.45\text{a,b})_b$ to give $\boldsymbol{\sigma}$ as

$$\begin{aligned}
\boldsymbol{\sigma} &= J^{-1}\boldsymbol{F}\boldsymbol{S}\boldsymbol{F}^T\\
&= \mu J^{-5/3}\boldsymbol{F}(\boldsymbol{I} - \frac{1}{3}I_{\boldsymbol{C}}\boldsymbol{C}^{-1})\boldsymbol{F}^T + p\boldsymbol{F}\boldsymbol{C}^{-1}\boldsymbol{F}^T\\
&= \boldsymbol{\sigma}' + p\boldsymbol{I}; \quad \boldsymbol{\sigma}' = \mu J^{-5/3}(\boldsymbol{b} - \frac{1}{3}I_{\boldsymbol{b}}\boldsymbol{I}),
\end{aligned} \tag{6.55a,b}$$

where the fact that $I_{\boldsymbol{b}} = I_{\boldsymbol{C}}$ has been used again.

We can now evaluate the Lagrangian elasticity tensor with the help of Equation (6.10) or (6.11). The result can be split into deviatoric and pressure components, $\widehat{\mathcal{C}}$ and $\mathcal{C}_p$ respectively, as

$$\mathcal{C} = 2\frac{\partial\boldsymbol{S}}{\partial\boldsymbol{C}} = \widehat{\mathcal{C}} + \mathcal{C}_p; \quad \widehat{\mathcal{C}} = 2\frac{\partial\boldsymbol{S}'}{\partial\boldsymbol{C}} = 4\frac{\partial^2\widehat{\Psi}}{\partial\boldsymbol{C}\partial\boldsymbol{C}}; \quad \mathcal{C}_p = 2p\frac{\partial(J\boldsymbol{C}^{-1})}{\partial\boldsymbol{C}}. \tag{6.56a,b,c}$$

With the help of Equations (6.22) and (6.31) these two components can be evaluated for the neo-Hookean case defined by Equation (6.53) after lengthy but simple algebra as

$$\widehat{\mathcal{C}} = 2\mu III_{\boldsymbol{C}}^{-1/3}\left[\frac{1}{3}I_{\boldsymbol{C}}\mathcal{I} - \frac{1}{3}\boldsymbol{I}\otimes\boldsymbol{C}^{-1} - \frac{1}{3}\boldsymbol{C}^{-1}\otimes\boldsymbol{I} + \frac{1}{9}I_{\boldsymbol{C}}\boldsymbol{C}^{-1}\otimes\boldsymbol{C}^{-1}\right]; \tag{6.57a}$$

$$\mathcal{C}_p = pJ[\boldsymbol{C}^{-1}\otimes\boldsymbol{C}^{-1} - 2\mathcal{I}]. \tag{6.57b}$$

Note that the pressure component $\mathcal{C}_p$ does not depend on the particular material definition being used.

The spatial elasticity tensor is obtained by the push-forward type of operation shown in Equation (6.14a,b) as

$$\mathcal{c} = \hat{\mathcal{c}} + \mathcal{c}_p; \quad \hat{\mathcal{c}} = J^{-1}\phi_*[\widehat{\mathcal{C}}]; \quad \mathcal{c}_p = J^{-1}\phi_*[\mathcal{C}_p]. \tag{6.58}$$

Performing this push-forward operation in Equations (6.57a,b) gives

$$\widehat{\mathcal{c}} = 2\mu J^{-5/3}\left[\frac{1}{3}I_b\boldsymbol{\imath} - \frac{1}{3}\boldsymbol{b}\otimes\boldsymbol{I} - \frac{1}{3}\boldsymbol{I}\otimes\boldsymbol{b} + \frac{1}{9}I_b\boldsymbol{I}\otimes\boldsymbol{I}\right]; \tag{6.59a}$$

$$\mathcal{c}_p = p[\boldsymbol{I}\otimes\boldsymbol{I} - 2\boldsymbol{\imath}]. \tag{6.59b}$$

**EXAMPLE 6.5: Mooney–Rivlin materials**

A general form for the strain energy function of incompressible rubbers, attributable to Mooney and Rivlin, is expressed as

$$\Psi(\boldsymbol{C}) = \sum_{r,s\geq 0} \mu_{rs}(I_C - 3)^r(II_C^* - 3)^s,$$

where $II_C^*$ is the second invariant of $\boldsymbol{C}$, defined as

$$II_C^* = \frac{1}{2}(I_C^2 - II_C); \qquad II_C = \boldsymbol{C} : \boldsymbol{C}.$$

The most frequently used of this family of equations is obtained when only $\mu_{01}$ and $\mu_{10}$ are different from zero. In this particular case we have

$$\Psi(\boldsymbol{C}) = \mu_{10}(I_C - 3) + \frac{1}{2}\mu_{01}(I_C^2 - II_C - 6).$$

The equivalent homogeneous potential is obtained by replacing $\boldsymbol{C}$ by $\hat{\boldsymbol{C}}$ in this equation to give

$$\widehat{\Psi}(\boldsymbol{C}) = \mu_{10}(\mathrm{tr}\hat{\boldsymbol{C}} - 3) + \frac{1}{2}\mu_{01}[(\mathrm{tr}\hat{\boldsymbol{C}})^2 - \hat{\boldsymbol{C}} : \hat{\boldsymbol{C}} - 6].$$

### 6.5.3 Nearly Incompressible Hyperelastic Materials

As explained at the beginning of Section 6.5, near incompressibility is often a device by which incompressibility can more readily be enforced within the context of the finite element formulation. This is facilitated by adding a volumetric energy component $U(J)$ to the distortional component $\widehat{\Psi}$ already defined to give the total strain energy function $\Psi(\boldsymbol{C})$ as

$$\Psi(\boldsymbol{C}) = \widehat{\Psi}(\boldsymbol{C}) + U(J), \tag{6.60}$$

where the simplest example of a volumetric function $U(J)$ is

$$U(J) = \frac{1}{2}\kappa(J-1)^2. \tag{6.61}$$

It will be seen in Chapter 8 that when equilibrium is expressed in a variational framework, the use of Equation (6.61) with a large so-called *penalty number* $\kappa$ will approximately enforce incompressibility. Typically, values of $\kappa$ in the region of $10^3$–$10^4$ are used for this purpose. Nevertheless, we must emphasize that $\kappa$ can represent a true material property, namely the bulk modulus, for a compressible material that happens to have a hyperelastic strain energy function in the form given by Equations (6.60) and (6.61).

The second Piola–Kirchhoff tensor for a material defined by Equation (6.60) is obtained in the standard manner with the help of Equation (6.22) and noting that $III_C = J^2$ to give

$$\begin{aligned} \boldsymbol{S} &= 2\frac{\partial \Psi}{\partial \boldsymbol{C}} \\ &= 2\frac{\partial \widehat{\Psi}}{\partial \boldsymbol{C}} + 2\frac{dU}{dJ}\frac{\partial J}{\partial \boldsymbol{C}} \\ &= 2\frac{\partial \widehat{\Psi}}{\partial \boldsymbol{C}} + pJ\boldsymbol{C}^{-1}, \end{aligned} \tag{6.62}$$

where, by comparison with Equation (6.50), we have identified the pressure as

$$p = \frac{dU}{dJ}, \tag{6.63}$$

which for the case where $U(J)$ is given by Equation (6.61) gives

$$p = k(J-1). \tag{6.64}$$

This value of the pressure can be substituted into the general Equation (6.62) or into the particular Equation (6.54) for the neo-Hookean case to yield the complete second Piola–Kirchhoff tensor. Alternatively, in the neo-Hookean case, $p$ can be substituted into Equation (6.55a,b) to give the Cauchy stress tensor.

**EXAMPLE 6.6: Simple shear (ii)**

Again we can study the case of simple shear for a nearly incompressible neo-Hookean material. Using Equation (6.55a,b) and the $\boldsymbol{b}$ tensor given in Example 6.4 we obtain

$$\boldsymbol{\sigma} = \mu \begin{bmatrix} \frac{2}{3}\gamma^2 & \gamma & 0 \\ \gamma & -\frac{1}{3}\gamma^2 & 0 \\ 0 & 0 & -\frac{1}{3}\gamma^2 \end{bmatrix},$$

*(continued)*

**EXAMPLE 6.6:** *(cont.)*

where now the pressure is zero as $J = 1$ for this type of deformation. Note that for this type of material there is no Kelvin effect in the sense that a volume-preserving motion leads to a purely deviatoric stress tensor.

**EXAMPLE 6.7: Pure dilatation (ii)**

It is also useful to examine the consequences of a pure dilatation on a nearly incompressible material. Recalling that this type of deformation has an associated left Cauchy–Green tensor $\boldsymbol{b} = J^{2/3}\boldsymbol{I}$ whose trace is $I_b = 3J^{2/3}$, Equations (6.55a,b) and (6.64) give

$$\boldsymbol{\sigma} = \kappa(J-1)\boldsymbol{I}.$$

As expected, a purely dilatational deformation leads to a hydrostatic state of stresses. Note also that the isochoric potential $\widehat{\Psi}$ plays no role in the value of the pressure $p$.

Again, to complete the description of this type of material it is necessary to derive the Lagrangian and spatial elasticity tensors. The Lagrangian tensor can be split into three components given as

$$\boldsymbol{\mathcal{C}} = 4\frac{\partial \widehat{\Psi}}{\partial \boldsymbol{C}\partial \boldsymbol{C}} + 2p\frac{\partial (J\boldsymbol{C}^{-1})}{\partial \boldsymbol{C}} + 2J\boldsymbol{C}^{-1} \otimes \frac{\partial p}{\partial \boldsymbol{C}}. \tag{6.65}$$

The first two components in this expression are $\widehat{\boldsymbol{\mathcal{C}}}$ and $\boldsymbol{\mathcal{C}}_p$ as evaluated in the previous section in Equations (6.57a,b). The final term, namely $\boldsymbol{\mathcal{C}}_\kappa$, represents a volumetric tangent component and follows from $U(J)$ and Equation (6.22) as

$$\begin{aligned}
\boldsymbol{\mathcal{C}}_\kappa &= 2J\boldsymbol{C}^{-1} \otimes \frac{\partial p}{\partial \boldsymbol{C}} \\
&= 2J\boldsymbol{C}^{-1} \otimes \frac{d^2U}{dJ^2}\frac{\partial J}{\partial \boldsymbol{C}} \\
&= J^2\frac{d^2U}{dJ^2}\left(\boldsymbol{C}^{-1} \otimes \boldsymbol{C}^{-1}\right),
\end{aligned} \tag{6.66}$$

which in the case $U(J) = \kappa(J-1)^2/2$ becomes

$$\boldsymbol{\mathcal{C}}_\kappa = \kappa J^2\,\boldsymbol{C}^{-1} \otimes \boldsymbol{C}^{-1}. \tag{6.67}$$

Finally, the spatial elasticity tensor is obtained by standard push-forward operation to yield

$$\mathbf{c} = J^{-1}\phi_*[\boldsymbol{\mathcal{C}}] = \hat{\mathbf{c}} + \mathbf{c}_p + \mathbf{c}_\kappa, \tag{6.68}$$

where the deviatoric and pressure components, $\hat{\mathbf{c}}$ and $\mathbf{c}_p$ respectively, are identical to those derived in the previous section and the volumetric component $\mathbf{c}_\kappa$ is

$$\mathcal{C}_\kappa = J^{-1}\phi_*[\boldsymbol{\mathcal{C}}_\kappa] = J\frac{d^2U}{dJ^2}\boldsymbol{I}\otimes\boldsymbol{I}, \tag{6.69}$$

which for the particular function $U(J)$ defined in Equation (6.61) gives

$$\mathcal{C}_\kappa = \kappa J\,\boldsymbol{I}\otimes\boldsymbol{I}. \tag{6.70}$$

**Remark 6.3**: At the initial configuration, $\boldsymbol{F} = \boldsymbol{C} = \boldsymbol{b} = \boldsymbol{I}$, $J = 1, p = 0$, and the above elasticity tensor becomes

$$\begin{aligned}\mathcal{C} &= \hat{\mathcal{C}} + \mathcal{C}_\kappa \\ &= 2\mu\big[\mathcal{I} - \tfrac{1}{3}\boldsymbol{I}\otimes\boldsymbol{I}\big] + \kappa\boldsymbol{I}\otimes\boldsymbol{I} \\ &= \big(\kappa - \tfrac{2}{3}\mu\big)\boldsymbol{I}\otimes\boldsymbol{I} + 2\mu\mathcal{I},\end{aligned} \tag{6.71}$$

which coincides with the standard spatially isotropic elasticity tensor (6.40) with the relationship between $\lambda$ and $\kappa$ given as

$$\lambda = \kappa - \frac{2}{3}\mu. \tag{6.72}$$

In fact, all isotropic hyperelastic materials have initial elasticity tensors as defined by Equation (6.40).

## 6.6 ISOTROPIC ELASTICITY IN PRINCIPAL DIRECTIONS

### 6.6.1 Material Description

It is often the case that the constitutive equations of a material are presented in terms of the stretches $\lambda_1$, $\lambda_2$, $\lambda_3$ in the principal directions $\boldsymbol{N}_1$, $\boldsymbol{N}_2$, and $\boldsymbol{N}_3$ as defined in Section 4.6. In the case of hyperelasticity, this assumes that the stored elastic energy function is obtainable in terms of $\lambda_\alpha$ rather than the invariants of $\boldsymbol{C}$. This is most likely to be the case in the experimental determination of the constitutive parameters.

In order to obtain the second Piola–Kirchhoff stress in terms of the principal directions and stretches, recall Equation (6.23) and note that the identity, the right Cauchy–Green tensor, and its inverse can be expressed as (see Equations $(2.30\text{a,b})_b$ and (4.30))

$$\boldsymbol{I} = \sum_{\alpha=1}^{3}\boldsymbol{N}_\alpha\otimes\boldsymbol{N}_\alpha; \tag{6.73a}$$

$$\boldsymbol{C} = \sum_{\alpha=1}^{3}\lambda_\alpha^2\,\boldsymbol{N}_\alpha\otimes\boldsymbol{N}_\alpha; \tag{6.73b}$$

$$\boldsymbol{C}^{-1} = \sum_{\alpha=1}^{3} \lambda_\alpha^{-2} \boldsymbol{N}_\alpha \otimes \boldsymbol{N}_\alpha. \tag{6.73c}$$

Substituting these equations into Equation (6.23) gives the second Piola–Kirchhoff stress $\boldsymbol{S}$ as

$$\boldsymbol{S} = \sum_{\alpha=1}^{3} (2\Psi_I + 4\Psi_{II}\lambda_\alpha^2 + 2III_C\Psi_{III}\lambda_\alpha^{-2})\boldsymbol{N}_\alpha \otimes \boldsymbol{N}_\alpha. \tag{6.74}$$

Given that the term in brackets is a scalar, it is immediately apparent that for an isotropic material the principal axes of stress coincide with the principal axes of strain. The terms $\Psi_I$, $\Psi_{II}$, and $\Psi_{III}$ in Equation (6.74) refer to the derivatives with respect to the invariants of $\boldsymbol{C}$. Hence it is necessary to transform these into derivatives with respect to the stretches. For this purpose note that the squared stretches $\lambda_\alpha^2$ are the eigenvalues of $\boldsymbol{C}$, which according to the general relationships (2.60a–c) are related to the invariants of $\boldsymbol{C}$ as

$$I_C = \lambda_1^2 + \lambda_2^2 + \lambda_3^2; \tag{6.75a}$$

$$II_C = \lambda_1^4 + \lambda_2^4 + \lambda_3^4; \tag{6.75b}$$

$$III_C = \lambda_1^2\,\lambda_2^2\,\lambda_3^2. \tag{6.75c}$$

Differentiating these equations gives

$$1 = \frac{\partial I_C}{\partial \lambda_\alpha^2}; \tag{6.76a}$$

$$2\lambda_\alpha^2 = \frac{\partial II_C}{\partial \lambda_\alpha^2}; \tag{6.76b}$$

$$\frac{III_C}{\lambda_\alpha^2} = \frac{\partial III_C}{\partial \lambda_\alpha^2}; \tag{6.76c}$$

which upon substitution into Equation (6.74) and using the chain rule gives the principal components of the second Piola–Kirchhoff tensor as derivatives of $\Psi$ with respect to the principal stretches as

$$\boldsymbol{S} = \sum_{\alpha=1}^{3} S_{\alpha\alpha} \boldsymbol{N}_\alpha \otimes \boldsymbol{N}_\alpha; \quad S_{\alpha\alpha} = 2\frac{\partial \Psi}{\partial \lambda_\alpha^2}. \tag{6.77}$$

### 6.6.2 Spatial Description

In order to obtain an equation analogous to (6.77) for the Cauchy stress, substitute this equation into Equation (5.45a,b)$_b$ to give

$$\boldsymbol{\sigma} = J^{-1}\boldsymbol{F}\boldsymbol{S}\boldsymbol{F}^T = \sum_{\alpha=1}^{3} \frac{2}{J}\frac{\partial \Psi}{\partial \lambda_\alpha^2}(\boldsymbol{F}\boldsymbol{N}_\alpha) \otimes (\boldsymbol{F}\boldsymbol{N}_\alpha). \tag{6.78}$$

Observing from Equation (4.44a) that $\boldsymbol{F}\boldsymbol{N}_\alpha = \lambda_\alpha \boldsymbol{n}_\alpha$ yields the principal components of Cauchy stress tensor, after simple algebra, as

$$\boldsymbol{\sigma} = \sum_{i=1}^{3} \sigma_{\alpha\alpha}\, \boldsymbol{n}_\alpha \otimes \boldsymbol{n}_\alpha; \qquad \sigma_{\alpha\alpha} = \frac{\lambda_\alpha}{J}\frac{\partial \Psi}{\partial \lambda_\alpha} = \frac{1}{J}\frac{\partial \Psi}{\partial \ln \lambda_\alpha}. \tag{6.79}$$

The evaluation of the Cartesian components of the Cauchy stress can be easily achieved by interpreting Equation (6.79) in a matrix form using Equation $(2.40)_d$ for the components of the tensor product to give

$$[\boldsymbol{\sigma}] = \sum_{i=1}^{3} \sigma_{\alpha\alpha} [\boldsymbol{n}_\alpha][\boldsymbol{n}_\alpha]^T, \tag{6.80}$$

where $[\boldsymbol{\sigma}]$ denotes the matrix formed by the Cartesian components of $\boldsymbol{\sigma}$, and $[\boldsymbol{n}_\alpha]$ are the column vectors containing the Cartesian components of $\boldsymbol{n}_\alpha$. Alternatively, a similar evaluation can be performed in an indicial manner by introducing $T_{\alpha j}$ as the Cartesian components of $\boldsymbol{n}_\alpha$, that is, $\boldsymbol{n}_\alpha = \sum_{j=1}^{3} T_{\alpha j} \boldsymbol{e}_j$, and substituting into Equation (6.79) to give

$$\boldsymbol{\sigma} = \sum_{j,k=1}^{3} \left( \sum_{i=1}^{3} \sigma_{\alpha\alpha} T_{\alpha j} T_{\alpha k} \right) \boldsymbol{e}_j \otimes \boldsymbol{e}_k. \tag{6.81}$$

The expression in brackets in the above equation again gives the Cartesian components of the Cauchy stress tensor.

### 6.6.3 Material Elasticity Tensor

To construct the material elasticity tensor for a material given in terms of the principal stretches it is again temporarily convenient to consider the time derivative Equation (6.12), that is, $\dot{\boldsymbol{S}} = \boldsymbol{\mathcal{C}} : \dot{\boldsymbol{E}}$. From Equation (4.123), it transpires that $\dot{\boldsymbol{E}}$ can be written in principal directions as

$$\dot{\boldsymbol{E}} = \sum_{\alpha=1}^{3} \frac{1}{2}\frac{d\lambda_\alpha^2}{dt} \boldsymbol{N}_\alpha \otimes \boldsymbol{N}_\alpha + \sum_{\substack{\alpha,\beta=1 \\ \alpha\neq\beta}}^{3} \frac{1}{2} W_{\alpha\beta}(\lambda_\alpha^2 - \lambda_\beta^2)\, \boldsymbol{N}_\alpha \otimes \boldsymbol{N}_\beta, \tag{6.82}$$

where $W_{\alpha\beta}$ are the components of the spin tensor of the Lagrangian triad, that is, $\dot{\boldsymbol{N}}_\alpha = \sum_{\beta=1}^{3} W_{\alpha\beta} \boldsymbol{N}_\beta$. A similar expression for the time derivative of $\boldsymbol{S}$ can be obtained by differentiating Equation (6.77) to give

$$\begin{aligned}\dot{\boldsymbol{S}} &= \sum_{\alpha,\beta=1}^{3} 2\frac{\partial^2\Psi}{\partial\lambda_\alpha^2\partial\lambda_\beta^2}\frac{d\lambda_\beta^2}{dt}\boldsymbol{N}_\alpha\otimes\boldsymbol{N}_\alpha \\ &\quad+\sum_{\alpha=1}^{3} 2\frac{\partial\Psi}{\partial\lambda_\alpha^2}(\dot{\boldsymbol{N}}_\alpha\otimes\boldsymbol{N}_\alpha+\boldsymbol{N}_\alpha\otimes\dot{\boldsymbol{N}}_\alpha) \\ &= \sum_{\alpha,\beta=1}^{3} 2\frac{\partial^2\Psi}{\partial\lambda_\alpha^2\partial\lambda_\beta^2}\frac{d\lambda_\beta^2}{dt}\boldsymbol{N}_\alpha\otimes\boldsymbol{N}_\alpha \\ &\quad+\sum_{\substack{\alpha,\beta=1\\ \alpha\neq\beta}}^{3}(S_{\alpha\alpha}-S_{\beta\beta})W_{\alpha\beta}\boldsymbol{N}_\alpha\otimes\boldsymbol{N}_\beta. \end{aligned} \tag{6.83}$$

Now observe from Equation (6.82) that the on-diagonal and off-diagonal terms of $\dot{\boldsymbol{E}}$ are

$$\frac{d\lambda_\alpha^2}{dt}=2\dot{E}_{\alpha\alpha}; \tag{6.84a}$$

$$W_{\alpha\beta}=\frac{2\dot{E}_{\alpha\beta}}{\lambda_\alpha^2-\lambda_\beta^2}=\frac{\dot{E}_{\alpha\beta}+\dot{E}_{\beta\alpha}}{\lambda_\alpha^2-\lambda_\beta^2};\quad(\alpha\neq\beta). \tag{6.84b}$$

Substituting Equations (6.84a,b) into (6.83) and expressing the components of $\dot{\boldsymbol{E}}$ as $\dot{E}_{\alpha\beta}=(\boldsymbol{N}_\alpha\otimes\boldsymbol{N}_\beta):\dot{\boldsymbol{E}}$ yields

$$\begin{aligned}\dot{\boldsymbol{S}} &= \sum_{\alpha=1}^{3} 4\frac{\partial^2\Psi}{\partial\lambda_\alpha^2\partial\lambda_\beta^2}\dot{E}_{\beta\beta}(\boldsymbol{N}_\alpha\otimes\boldsymbol{N}_\alpha) \\ &\quad+\sum_{\substack{\alpha,\beta=1\\ \alpha\neq\beta}}^{3}\frac{S_{\alpha\alpha}-S_{\beta\beta}}{\lambda_\alpha^2-\lambda_\beta^2}\left(\dot{E}_{\alpha\beta}+\dot{E}_{\beta\alpha}\right)\boldsymbol{N}_\alpha\otimes\boldsymbol{N}_\beta \\ &= \left[\sum_{\alpha,\beta=1}^{3} 4\frac{\partial^2\Psi}{\partial\lambda_\alpha^2\partial\lambda_\beta^2}\boldsymbol{\mathcal{N}}_{\alpha\alpha\beta\beta}\right. \\ &\quad\left.+\sum_{\substack{\alpha,\beta=1\\ \alpha\neq\beta}}^{3}\frac{S_{\alpha\alpha}-S_{\beta\beta}}{\lambda_\alpha^2-\lambda_\beta^2}(\boldsymbol{\mathcal{N}}_{\alpha\beta\alpha\beta}+\boldsymbol{\mathcal{N}}_{\alpha\beta\beta\alpha})\right]:\dot{\boldsymbol{E}}; \end{aligned} \tag{6.85a}$$

$$\boldsymbol{\mathcal{N}}_{\alpha\alpha\beta\beta}=\boldsymbol{N}_\alpha\otimes\boldsymbol{N}_\alpha\otimes\boldsymbol{N}_\beta\otimes\boldsymbol{N}_\beta; \tag{6.85b}$$

$$\boldsymbol{\mathcal{N}}_{\alpha\beta\alpha\beta}=\boldsymbol{N}_\alpha\otimes\boldsymbol{N}_\beta\otimes\boldsymbol{N}_\alpha\otimes\boldsymbol{N}_\beta; \tag{6.85c}$$

$$\boldsymbol{\mathcal{N}}_{\alpha\beta\beta\alpha}=\boldsymbol{N}_\alpha\otimes\boldsymbol{N}_\beta\otimes\boldsymbol{N}_\beta\otimes\boldsymbol{N}_\alpha. \tag{6.85d}$$

Comparing this expression with the rate equation $\dot{\boldsymbol{S}} = \boldsymbol{\mathcal{C}} : \dot{\boldsymbol{E}}$, the material or Lagrangian elasticity tensor emerges as

$$\boldsymbol{\mathcal{C}} = \sum_{\alpha,\beta=1}^{3} 4\frac{\partial^2 \Psi}{\partial \lambda_\alpha^2 \partial \lambda_\beta^2} \boldsymbol{\mathcal{N}}_{\alpha\alpha\beta\beta} + \sum_{\substack{\alpha,\beta=1\\ \alpha\neq\beta}}^{3} \frac{S_{\alpha\alpha} - S_{\beta\beta}}{\lambda_\alpha^2 - \lambda_\beta^2} \left(\boldsymbol{\mathcal{N}}_{\alpha\beta\alpha\beta} + \boldsymbol{\mathcal{N}}_{\alpha\beta\beta\alpha}\right). \tag{6.86}$$

**Remark 6.4**: In the particular case when $\lambda_\alpha = \lambda_\beta$, isotropy implies that $S_{\alpha\alpha} = S_{\beta\beta}$, and the quotient $(S_{\alpha\alpha} - S_{\beta\beta})(\lambda_\alpha^2 - \lambda_\beta^2)$ in Equation (6.86) must be evaluated using L'Hospital's rule to give

$$\lim_{\lambda_\beta \to \lambda_\alpha} \frac{S_{\alpha\alpha} - S_{\beta\beta}}{\lambda_\alpha^2 - \lambda_\beta^2} = 2\left(\frac{\partial^2 \Psi}{\partial \lambda_\beta^2 \partial \lambda_\beta^2} - \frac{\partial^2 \Psi}{\partial \lambda_\alpha^2 \partial \lambda_\beta^2}\right). \tag{6.87}$$

## 6.6.4 Spatial Elasticity Tensor

The spatial elasticity tensor is obtained by pushing the Lagrangian tensor forward to the current configuration using Equation (6.14a,b), which involves the product by $\boldsymbol{F}$ four times as

$$\begin{aligned} \boldsymbol{c} = & \sum_{\alpha,\beta=1}^{3} \frac{1}{J} \frac{\partial^2 \Psi}{\partial \lambda_\alpha^2 \partial \lambda_\beta^2} \phi_* \left[\boldsymbol{\mathcal{N}}_{\alpha\alpha\beta\beta}\right] \\ & + \sum_{\substack{\alpha,\beta=1\\ \alpha\neq\beta}}^{3} \frac{1}{J} \frac{S_{\alpha\alpha} - S_{\beta\beta}}{\lambda_\alpha^2 - \lambda_\beta^2} \left(\phi_* \left[\boldsymbol{\mathcal{N}}_{\alpha\beta\alpha\beta}\right] + \phi_* \left[\boldsymbol{\mathcal{N}}_{\alpha\beta\beta\alpha}\right]\right); \end{aligned} \tag{6.88a}$$

$$\phi_* \left[\boldsymbol{\mathcal{N}}_{\alpha\alpha\beta\beta}\right] = (\boldsymbol{F}\boldsymbol{N}_\alpha) \otimes (\boldsymbol{F}\boldsymbol{N}_\alpha) \otimes (\boldsymbol{F}\boldsymbol{N}_\beta) \otimes (\boldsymbol{F}\boldsymbol{N}_\beta); \tag{6.88b}$$

$$\phi_* \left[\boldsymbol{\mathcal{N}}_{\alpha\beta\alpha\beta}\right] = (\boldsymbol{F}\boldsymbol{N}_\alpha) \otimes (\boldsymbol{F}\boldsymbol{N}_\beta) \otimes (\boldsymbol{F}\boldsymbol{N}_\alpha) \otimes (\boldsymbol{F}\boldsymbol{N}_\beta); \tag{6.88c}$$

$$\phi_* \left[\boldsymbol{\mathcal{N}}_{\alpha\beta\beta\alpha}\right] = (\boldsymbol{F}\boldsymbol{N}_\alpha) \otimes (\boldsymbol{F}\boldsymbol{N}_\beta) \otimes (\boldsymbol{F}\boldsymbol{N}_\beta) \otimes (\boldsymbol{F}\boldsymbol{N}_\alpha). \tag{6.88d}$$

Noting again that $\boldsymbol{F}\boldsymbol{N}_\alpha = \lambda_\alpha \boldsymbol{n}_\alpha$ and after some algebraic manipulations using the standard chain rule we can eventually derive the Eulerian or spatial elasticity tensor as

$$\begin{aligned} \boldsymbol{c} = & \sum_{\alpha,\beta=1}^{3} \frac{1}{J} \frac{\partial^2 \Psi}{\partial \ln \lambda_\alpha \partial \ln \lambda_\beta} \boldsymbol{\eta}_{\alpha\alpha\beta\beta} - \sum_{\alpha=1}^{3} 2\sigma_{\alpha\alpha} \boldsymbol{\eta}_{\alpha\alpha\alpha\alpha} \\ & + \sum_{\substack{\alpha,\beta=1\\ \alpha\neq\beta}}^{3} \frac{\sigma_{\alpha\alpha}\lambda_\beta^2 - \sigma_{\beta\beta}\lambda_\alpha^2}{\lambda_\alpha^2 - \lambda_\beta^2} \left(\boldsymbol{\eta}_{\alpha\beta\alpha\beta} + \boldsymbol{\eta}_{\alpha\beta\beta\alpha}\right); \end{aligned} \tag{6.89a}$$

$$\boldsymbol{\eta}_{\alpha\alpha\beta\beta} = \boldsymbol{n}_\alpha \otimes \boldsymbol{n}_\alpha \otimes \boldsymbol{n}_\beta \otimes \boldsymbol{n}_\beta; \tag{6.89b}$$

$$\boldsymbol{\eta}_{\alpha\alpha\alpha\alpha} = \boldsymbol{n}_\alpha \otimes \boldsymbol{n}_\alpha \otimes \boldsymbol{n}_\alpha \otimes \boldsymbol{n}_\alpha; \tag{6.89c}$$

$$\boldsymbol{\eta}_{\alpha\beta\alpha\beta} = \boldsymbol{n}_\alpha \otimes \boldsymbol{n}_\beta \otimes \boldsymbol{n}_\alpha \otimes \boldsymbol{n}_\beta; \tag{6.89d}$$

$$\boldsymbol{\eta}_{\alpha\beta\beta\alpha} = \boldsymbol{n}_\alpha \otimes \boldsymbol{n}_\beta \otimes \boldsymbol{n}_\beta \otimes \boldsymbol{n}_\alpha. \tag{6.89e}$$

The evaluation of the Cartesian components of this tensor requires a similar transformation to that employed in Equation (6.81) for the Cauchy stresses. Using the same notation, the Cartesian components of the Eulerian triad $T_{\alpha j}$ are substituted into Equation (6.89) to give after simple algebra the Cartesian components of $\mathcal{C}$ as

$$\begin{aligned}\mathcal{C}_{ijkl} = & \sum_{\alpha,\beta=1}^{3} \frac{1}{J}\frac{\partial^2 \Psi}{\partial \ln \lambda_\alpha \partial \ln \lambda_\beta} T_{\alpha i}T_{\alpha j}T_{\beta k}T_{\beta l} - \sum_{\alpha=1}^{3} 2\sigma_{\alpha\alpha}T_{\alpha i}T_{\alpha j}T_{\alpha k}T_{\alpha l} \\ & + \sum_{\substack{\alpha,\beta=1\\ \alpha\neq\beta}}^{3} \frac{\sigma_{\alpha\alpha}\lambda_\beta^2 - \sigma_{\beta\beta}\lambda_\alpha^2}{\lambda_\alpha^2 - \lambda_\beta^2}\left(T_{\alpha i}T_{\beta j}T_{\alpha k}T_{\beta l} + T_{\alpha i}T_{\beta j}T_{\beta k}T_{\alpha l}\right).\end{aligned} \tag{6.90}$$

**Remark 6.5**: Again, recalling Remark 6.4, in the case when $\lambda_\alpha = \lambda_\beta$, L'Hospital's rule yields

$$\lim_{\lambda_\beta \to \lambda_\alpha} \frac{\sigma_{\alpha\alpha}\lambda_\beta^2 - \sigma_{\alpha\alpha}\lambda_\alpha^2}{\lambda_\alpha^2 - \lambda_\beta^2} = \frac{1}{2J}\left[\frac{\partial^2 \Psi}{\partial \ln \lambda_\beta \partial \ln \lambda_\beta} - \frac{\partial^2 \Psi}{\partial \ln \lambda_\alpha \partial \ln \lambda_\beta}\right] - \sigma_{\alpha\alpha}. \tag{6.91}$$

### 6.6.5 A Simple Stretch-Based Hyperelastic Material

A material frequently encountered in the literature is defined by a hyperelastic potential in terms of the logarithmic stretches and two material parameters $\lambda$ and $\mu$ as

$$\Psi(\lambda_1, \lambda_2, \lambda_3) = \mu[(\ln \lambda_1)^2 + (\ln \lambda_2)^2 + (\ln \lambda_3)^2] + \frac{\lambda}{2}(\ln J)^2, \tag{6.92}$$

where, because $J = \lambda_1\lambda_2\lambda_3$,

$$\ln J = \ln \lambda_1 + \ln \lambda_2 + \ln \lambda_3. \tag{6.93}$$

It will be shown that the potential $\Psi$ leads to a generalization of the stress–strain relationships employed in classical linear elasticity.

Using Equation (6.79) the principal Cauchy stress components emerge as

$$\sigma_{\alpha\alpha} = \frac{1}{J}\frac{\partial \Psi}{\partial \ln \lambda_\alpha} = \frac{2\mu}{J}\ln \lambda_\alpha + \frac{\lambda}{J}\ln J. \tag{6.94}$$

Furthermore, the coefficients of the elasticity tensor in (6.90) are

$$\frac{1}{J}\frac{\partial^2\Psi}{\partial \ln\lambda_\alpha \partial \ln\lambda_\beta} = \frac{\lambda}{J} + \frac{2\mu}{J}\delta_{\alpha\beta}. \tag{6.95}$$

The similarities between these equations and linear elasticity can be established if we first recall the standard small strain elastic equations as

$$\sigma_{\alpha\alpha} = \lambda(\varepsilon_{11} + \varepsilon_{22} + \varepsilon_{33}) + 2\mu\varepsilon_{\alpha\alpha}. \tag{6.96}$$

Recalling that $\ln J = \ln\lambda_1 + \ln\lambda_2 + \ln\lambda_3$, it transpires that Equations (6.94) and (6.96) are identical except for the small strains having been replaced by the logarithmic stretches and $\lambda$ and $\mu$ by $\lambda/J$ and $\mu/J$ respectively. The stress–strain equations can be inverted and expressed in terms of the more familiar material parameters $E$ and $\nu$, the Young's modulus and Poisson's ratio, as

$$\ln\lambda_\alpha = \frac{J}{E}[(1+\nu)\sigma_{\alpha\alpha} - \nu(\sigma_{11} + \sigma_{22} + \sigma_{33})]; \tag{6.97a}$$

$$E = \frac{\mu(2\mu + 3\lambda)}{\lambda + \mu}; \quad \nu = \frac{\lambda}{2\lambda + 2\mu}. \tag{6.97b,c}$$

**Remark 6.6**: At the initial unstressed configuration, $J = \lambda_\alpha = 1$, $\sigma_{\alpha\alpha} = 0$, and the principal directions coincide with the three spatial directions $\boldsymbol{n}_\alpha = \boldsymbol{e}_\alpha$ and therefore $T_{\alpha j} = \delta_{\alpha j}$. Substituting these values into Equations (6.95), (6.91), and (6.90) gives the initial elasticity tensor for this type of material as

$$\mathcal{c}_{ijkl} = \lambda\delta_{ij}\delta_{kl} + \mu\,(\delta_{ik}\delta_{jl} + \delta_{il}\delta_{jk}), \tag{6.98}$$

which again (see Remark 6.3) coincides with the standard spatially isotropic elasticity tensor.

### 6.6.6 Nearly Incompressible Material in Principal Directions

In view of the importance of nearly incompressible material behavior, coupled with the likelihood that such materials will be described naturally in terms of principal stretches, it is now logical to elaborate the formulation in preparation for the case when the material defined by Equation (6.92) becomes nearly incompressible. Once again, the distortional components of the kinematic variables being used, namely the stretches $\lambda_\alpha$, must be identified first. This is achieved by recalling Equations (4.43) and (4.61) for $\boldsymbol{F}$ and $\hat{\boldsymbol{F}}$ to give

$$\begin{aligned}\hat{\boldsymbol{F}} &= J^{-1/3}\boldsymbol{F}\\ &= J^{-1/3}\sum_{\alpha=1}^{3}\lambda_\alpha\,\boldsymbol{n}_\alpha \otimes \boldsymbol{N}_\alpha\\ &= \sum_{\alpha=1}^{3}(J^{-1/3}\lambda_\alpha)\,\boldsymbol{n}_\alpha \otimes \boldsymbol{N}_\alpha.\end{aligned} \tag{6.99}$$

This enables the distortional stretches $\hat{\lambda}_\alpha$ to be identified as

$$\hat{\lambda}_\alpha = J^{-1/3}\lambda_\alpha; \quad \lambda_\alpha = J^{1/3}\hat{\lambda}_\alpha. \tag{6.100a,b}$$

Substituting (6.100a,b)$_b$ into the hyperelastic potential defined in (6.92) yields after simple algebra a decoupled representation of this material as

$$\Psi(\lambda_1, \lambda_2, \lambda_3) = \widehat{\Psi}(\hat{\lambda}_1, \hat{\lambda}_2, \hat{\lambda}_3) + U(J), \tag{6.101}$$

where the distortional and volumetric components are

$$\widehat{\Psi}(\hat{\lambda}_1, \hat{\lambda}_2, \hat{\lambda}_3) = \mu[(\ln \hat{\lambda}_1)^2 + (\ln \hat{\lambda}_2)^2 + (\ln \hat{\lambda}_3)^2]; \tag{6.102a}$$

$$U(J) = \frac{1}{2}\kappa(\ln J)^2; \qquad \kappa = \lambda + \frac{2}{3}\mu. \tag{6.102b}$$

Note that this equation is a particular case of the decoupled Equation (6.60) with alternative definitions of $U(J)$ and $\widehat{\Psi}$. The function $U(J)$ will enforce incompressibility only when the ratio $\kappa$ to $\mu$ is sufficiently high, typically $10^3$–$10^4$. Under such conditions the value of $J$ is $J \approx 1$ and $\ln J \approx J - 1$, and therefore the value of $U$ will approximately coincide with the function defined in (6.61).

For the expression $U(J)$, the corresponding value of the hydrostatic pressure $p$ is re-evaluated using Equation (6.63) to give

$$p = \frac{dU}{dJ} = \frac{\kappa \ln J}{J}. \tag{6.103}$$

In order to complete the stress description, the additional deviatoric component must be evaluated by recalling Equation (6.79) as

$$\begin{aligned} \sigma_{\alpha\alpha} &= \frac{1}{J}\frac{\partial \Psi}{\partial \ln \lambda_\alpha} \\ &= \frac{1}{J}\frac{\partial \widehat{\Psi}}{\partial \ln \lambda_\alpha} + \frac{1}{J}\frac{\partial U}{\partial \ln \lambda_\alpha} \\ &= \frac{1}{J}\frac{\partial \widehat{\Psi}}{\partial \ln \lambda_\alpha} + \frac{\kappa \ln J}{J}. \end{aligned} \tag{6.104}$$

Observing that the second term in this equation is the pressure, the principal deviatoric stress components are, obviously,

$$\sigma'_{\alpha\alpha} = \frac{1}{J}\frac{\partial \widehat{\Psi}}{\partial \ln \lambda_\alpha}. \tag{6.105}$$

In order to obtain the derivatives of $\widehat{\Psi}$ it is convenient to rewrite this function with the help of Equation (6.100a,b)$_a$ as

$$\begin{aligned} \widehat{\Psi} &= \mu[(\ln \hat{\lambda}_1)^2 + (\ln \hat{\lambda}_2)^2 + (\ln \hat{\lambda}_3)^2] \\ &= \mu[(\ln \lambda_1)^2 + (\ln \lambda_2)^2 + (\ln \lambda_3)^2] + \frac{1}{3}\mu(\ln J)^2 \end{aligned}$$

$$
\begin{aligned}
&\quad - \frac{2}{3}\mu(\ln J)(\ln\lambda_1 + \ln\lambda_2 + \ln\lambda_3) \\
&= \mu[(\ln\lambda_1)^2 + (\ln\lambda_2)^2 + (\ln\lambda_3)^2] - \frac{1}{3}\mu(\ln J)^2. \qquad (6.106)
\end{aligned}
$$

This expression for $\widehat{\Psi}$ is formally identical to Equation (6.92) for the complete hyperelastic potential $\Psi$ with the Lamé coefficient $\lambda$ now replaced by $-2\mu/3$. Consequently, Equation (6.94) can now be recycled to give the deviatoric principal stress component as

$$\sigma'_{\alpha\alpha} = \frac{2\mu}{J}\ln\lambda_\alpha - \frac{2\mu}{3J}\ln J. \qquad (6.107)$$

The final stage in this development is the evaluation of the volumetric and deviatoric components of the spatial elasticity tensor $\mathcal{c}$. For a general decoupled hyperelastic potential this decomposition is embodied in Equation (6.68), where $\mathcal{c}$ is expressed as

$$\mathcal{c} = \hat{\mathcal{c}} + \mathcal{c}_p + \mathcal{c}_\kappa, \qquad (6.108)$$

where the origin of the pressure component $\mathcal{c}_p$ is the second term in the general equation for the Lagrangian elasticity tensor (6.65), which is entirely geometrical, that is, independent of the material being used, and therefore remains unchanged as given by Equation (6.59b). However, the volumetric component $\mathcal{c}_\kappa$ depends on the particular function $U(J)$ being used, and in the present case becomes

$$
\begin{aligned}
\mathcal{c}_\kappa &= J\frac{d^2U}{dJ^2}\boldsymbol{I}\otimes\boldsymbol{I} \\
&= \frac{\kappa - pJ}{J}\boldsymbol{I}\otimes\boldsymbol{I}. \qquad (6.109)
\end{aligned}
$$

The deviatoric component of the elasticity tensor $\hat{\mathcal{c}}$ emerges from the push-forward of the first term in Equation (6.65). Its evaluation is facilitated by again recalling that $\widehat{\Psi}$ coincides with $\Psi$ when the parameter $\lambda$ is replaced by $-2\mu/3$. A reformulation of the spatial elasticity tensor following the procedure previously described with this substitution and the corresponding replacement of $\sigma_{\alpha\alpha}$ by $\sigma'_{\alpha\alpha}$ inevitably leads to the Cartesian components of $\hat{\mathcal{c}}$ as

$$
\begin{aligned}
\hat{\mathcal{c}}_{ijkl} &= \sum_{\alpha,\beta=1}^{3} \frac{1}{J}\frac{\partial^2\widehat{\Psi}}{\partial\ln\lambda_\alpha\partial\ln\lambda_\beta}T_{\alpha i}T_{\alpha j}T_{\beta k}T_{\beta l} - \sum_{\alpha=1}^{3} 2\sigma'_{\alpha\alpha}T_{\alpha i}T_{\alpha j}T_{\alpha k}T_{\alpha l} \\
&\quad + \sum_{\substack{\alpha,\beta=1\\ \alpha\neq\beta}}^{3} \frac{\sigma'_{\alpha\alpha}\lambda_\beta^2 - \sigma'_{\beta\beta}\lambda_\alpha^2}{\lambda_\alpha^2 - \lambda_\beta^2}(T_{\alpha i}T_{\beta j}T_{\alpha k}T_{\beta l} + T_{\alpha i}T_{\beta j}T_{\beta k}T_{kl}), \qquad (6.110)
\end{aligned}
$$

where the derivatives of $\widehat{\Psi}$ for the material under consideration are

$$\frac{1}{J}\frac{\partial^2\widehat{\Psi}}{\partial \ln \lambda_\alpha \partial \ln \lambda_\beta} = \frac{2\mu}{J}\delta_{\alpha\beta} - \frac{2\mu}{3J}. \tag{6.111}$$

### 6.6.7 Plane Strain and Plane Stress Cases

The plane strain case is defined by the fact that the stretch in the third direction $\lambda_3 = 1$. Under such conditions, the stored elastic potential becomes

$$\Psi(\lambda_1, \lambda_2) = \mu[(\ln \lambda_1)^2 + (\ln \lambda_2)^2] + \frac{\lambda}{2}(\ln j)^2, \tag{6.112}$$

where $j = \det_{2\times 2} \boldsymbol{F}$ is the determinant of the components of $\boldsymbol{F}$ in the $\boldsymbol{n}_1$ and $\boldsymbol{n}_2$ plane. The three stresses are obtained using exactly Equation (6.94) with $\lambda_3 = 1$ and $J = j$.

The plane stress case is a little more complicated, in that it is the stress in the $\boldsymbol{n}_3$ direction rather than the stretch that is constrained, that is, $\sigma_{33} = 0$. Imposing this condition in Equation (6.94) gives

$$\sigma_{33} = 0 = \frac{\lambda}{J}\ln J + \frac{2\mu}{J}\ln \lambda_3, \tag{6.113}$$

from which the logarithmic stretch in the third direction emerges as

$$\ln \lambda_3 = -\frac{\lambda}{\lambda + 2\mu}\ln j. \tag{6.114}$$

Substituting this expression into Equation (6.92) and noting that $\ln J = \ln \lambda_3 + \ln j$ gives

$$\Psi(\lambda_1, \lambda_2) = \mu[(\ln \lambda_1)^2 + (\ln \lambda_2)^2] + \frac{\bar{\lambda}}{2}(\ln j)^2, \tag{6.115}$$

where the effective Lamé coefficient $\bar{\lambda}$ is

$$\bar{\lambda} = \gamma\lambda; \gamma = \frac{2\mu}{\lambda + 2\mu}. \tag{6.116a,b}$$

Additionally, using Equation (6.114) the three-dimensional volume ratio $J$ can be found as a function of the planar component $j$ as

$$J = j^\gamma. \tag{6.117}$$

By either substituting Equation (6.114) into Equation (6.94) or differentiating Equation (6.115) the principal Cauchy stress components are obtained as

$$\sigma_{\alpha\alpha} = \frac{\bar{\lambda}}{j^\gamma}\ln j + \frac{2\mu}{j^\gamma}\ln \lambda_\alpha, \tag{6.118}$$

and the coefficients of the Eulerian elasticity tensor become

$$\frac{1}{J}\frac{\partial^2 \Psi}{\partial \ln \lambda_\alpha \partial \ln \lambda_\beta} = \frac{\bar{\lambda}}{j^\gamma} + \frac{2\mu}{j^\gamma}\delta_{\alpha\beta}. \tag{6.119}$$

### 6.6.8 Uniaxial Rod Case

In a uniaxial rod case, the stresses in directions orthogonal to the rod, $\sigma_{22}$ and $\sigma_{33}$, vanish. Imposing this condition in Equation (6.94) gives two equations as

$$\lambda \ln J + 2\mu \ln \lambda_2 = 0; \tag{6.120a}$$

$$\lambda \ln J + 2\mu \ln \lambda_3 = 0; \tag{6.120b}$$

from which it easily follows that the stretches in the second and third directions are equal and related to the main stretch via Poisson's ratio $\nu$ as

$$\ln \lambda_2 = \ln \lambda_3 = -\nu \ln \lambda_1; \quad \nu = \frac{\lambda}{2\lambda + 2\mu}. \tag{6.121}$$

Using Equations (6.93), (6.94), and (6.121), we can find a one-dimensional constitutive equation involving the rod stress $\sigma_{11}$, the logarithmic strain $\ln \lambda_1$, and Young's modulus $E$ as

$$\sigma_{11} = \frac{E}{J} \ln \lambda_1; \quad E = \frac{\mu(2\mu + 3\lambda)}{\lambda + \mu}, \tag{6.122}$$

where $J$ can be obtained with the help of Equation (6.121)[§] in terms of $\lambda_1$ and $\nu$ as

$$J = \lambda_1^{(1-2\nu)}. \tag{6.123}$$

Observe that Equation (6.122) is precisely that used in Chapter 3, Equation (3.16), and that for the incompressible case where $J = 1$, Equation (6.122) coincides with the uniaxial constitutive equation employed in Chapter 1.

Finally, the stored elastic energy given by Equation (6.92) becomes

$$\Psi(\lambda_1) = \frac{E}{2}(\ln \lambda_1)^2, \tag{6.124}$$

and, choosing a local axis in the direction of the rod, the only effective term in the Eulerian tangent modulus $\mathcal{c}_{1111}$ is given by Equation (6.90) as

$$\mathcal{c}_{1111} = \frac{1}{J}\frac{\partial^2 \Psi}{\partial \ln \lambda_1 \partial \ln \lambda_1} - 2\sigma_{11} = \frac{E}{J} - 2\sigma_{11}. \tag{6.125}$$

[§] Alternatively, see Example 3.1.

Again, for the incompressible case $J = 1$, the term $E - 2\sigma_{11}$ was already apparent in Chapter 1, where the equilibrium equation of a rod was linearized in a direct manner.

## Exercises

1. A modified St. Venant–Kirchhoff constitutive behavior is defined by its corresponding strain energy functional $\Psi$ as

$$\Psi(J, \boldsymbol{E}) = \frac{\kappa}{2}(\ln J)^2 + \mu II_{\boldsymbol{E}},$$

where $II_{\boldsymbol{E}} = \text{tr}(\boldsymbol{E}^2)$ denotes the second invariant of the Green's strain tensor $\boldsymbol{E}$, $J$ is the Jacobian of the deformation gradient, and $\kappa$ and $\mu$ are positive material constants.

(a) Obtain an expression for the second Piola–Kirchhoff stress tensor $\boldsymbol{S}$ as a function of the right Cauchy–Green strain tensor $\boldsymbol{C}$.

(b) Obtain an expression for the Kirchhoff stress tensor $\boldsymbol{\tau}$ as a function of the left Cauchy–Green strain tensor $\boldsymbol{b}$.

(c) Calculate the material elasticity tensor.

2. In a plane stress situation, the right Cauchy–Green tensor $\boldsymbol{C}$ is

$$\boldsymbol{C} = \begin{bmatrix} C_{11} & C_{12} & 0 \\ C_{21} & C_{22} & 0 \\ 0 & 0 & C_{33} \end{bmatrix}; \quad C_{33} = \frac{h^2}{H^2};$$

where $H$ and $h$ are the initial and current thicknesses respectively. Show that incompressibility implies

$$C_{33} = III_{\overline{\boldsymbol{C}}}^{-1}; \quad (\boldsymbol{C}^{-1})_{33} = III_{\overline{\boldsymbol{C}}}; \quad \overline{\boldsymbol{C}} = \begin{bmatrix} C_{11} & C_{12} \\ C_{21} & C_{22} \end{bmatrix}.$$

Using these equations, show that for an incompressible neo-Hookean material the plane stress condition $S_{33} = 0$ enables the pressure in Equation (6.54) to be explicitly evaluated as

$$p = \frac{1}{3}\mu(I_{\overline{\boldsymbol{C}}} - 2III_{\overline{\boldsymbol{C}}}^{-1}),$$

and therefore the in-plane components of the second Piola–Kirchhoff and Cauchy tensors are

$$\overline{\boldsymbol{S}} = \mu(\bar{\boldsymbol{I}} - III_{\overline{\boldsymbol{C}}}^{-1}\,\overline{\boldsymbol{C}}^{-1});$$
$$\bar{\boldsymbol{\sigma}} = \mu(\bar{\boldsymbol{b}} - III_{\bar{\boldsymbol{b}}}^{-1}\bar{\boldsymbol{I}});$$

where the overline indicates the $2 \times 2$ components of a tensor.

3. Show that the equations in Exercise 2 can also be derived by imposing the condition $C_{33} = III_{\overline{C}}^{-1}$ in the neo-Hookean elastic function $\Psi$ to give

$$\Psi(\overline{\boldsymbol{C}}) = \frac{1}{2}\mu(I_{\overline{\boldsymbol{C}}} + III_{\overline{\boldsymbol{C}}}^{-1} - 3),$$

from which $\overline{\boldsymbol{S}}$ is obtained by differentiation with respect to the in-plane tensor $\overline{\boldsymbol{C}}$. Finally, prove that the Lagrangian and Eulerian in-plane elasticity tensors are

$$\overline{\mathcal{C}} = 2\mu III_{\overline{\boldsymbol{C}}}^{-1}(\overline{\boldsymbol{C}}^{-1} \otimes \overline{\boldsymbol{C}}^{-1} + \overline{\mathcal{I}});$$

$$\bar{\mathfrak{c}} = 2\mu III_{\bar{\boldsymbol{b}}}^{-1}(\bar{\boldsymbol{I}} \otimes \bar{\boldsymbol{I}} + \bar{\boldsymbol{\imath}}).$$

4. Using the push back–pull-forward relationships between $\dot{\boldsymbol{E}}$ and $\boldsymbol{d}$ and between $\mathcal{C}$ and $\mathfrak{c}$, show that

$$\dot{\boldsymbol{E}} : \mathcal{C} : \dot{\boldsymbol{E}} = J\boldsymbol{d} : \mathfrak{c} : \boldsymbol{d}$$

for any arbitrary motion.

5. Using the simple stretch-based hyperelastic equations discussed in Section 6.6.5, show that the principal stresses for a simple shear test are

$$\sigma_{11} = -\sigma_{22} = 2\mu \sinh^{-1}\frac{\gamma}{2}.$$

Find the Cartesian stress components.

6. A general type of incompressible hyperelastic material proposed by Ogden is defined by the following strain energy function:

$$\Psi = \sum_{p=1}^{N} \frac{\mu_p}{\alpha_p}\left(\lambda_1^{\alpha_p} + \lambda_2^{\alpha_p} + \lambda_3^{\alpha_p} - 3\right).$$

Derive the homogeneous counterpart of this functional. Obtain expressions for the principal components of the deviatoric stresses and elasticity tensor.

CHAPTER SEVEN

# LARGE ELASTO-PLASTIC DEFORMATIONS

## 7.1 INTRODUCTION

Many materials of practical importance, such as metals, do not behave in a hyperelastic manner at high levels of stress. This lack of elasticity is manifested by the fact that when the material is freed from stress it fails to return to the initial undeformed configuration, and instead permanent deformations are observed. Different constitutive theories or models such as plasticity, viscoplasticity, and others are commonly used to describe such permanent effects. Although the mathematics of these material models is well understood in the small strain case, the same is not necessarily true for finite deformation.

A complete and coherent discussion of these inelastic constitutive models is well beyond the scope of this chapter. However, because practical applications of nonlinear continuum mechanics often include some permanent inelastic deformations, it is pertinent to give a brief introduction to the basic equations used in such applications. The aim of this introduction is simply to familiarize the reader with the fundamental kinematic concepts required to deal with large strains in inelastic materials. In particular, only the simplest possible case of Von Mises plasticity with isotropic hardening will be fully considered, although the kinematic equations described and the overall procedure will be applicable to more general materials.

We will assume that the reader has some familiarity with small strain inelastic constitutive models such as plasticity, because several of the key equations to be introduced will not be fully justified but loosely based on similar expressions that are known to apply to small strain theory. More in-depth discussions can be found in the Bibliography.

## 7.2 THE MULTIPLICATIVE DECOMPOSITION

Consider the deformation of a given initial volume $V$ into the current volume $v$ as shown in Figure 7.1. An elemental vector $dX$ in the local neighborhood of a

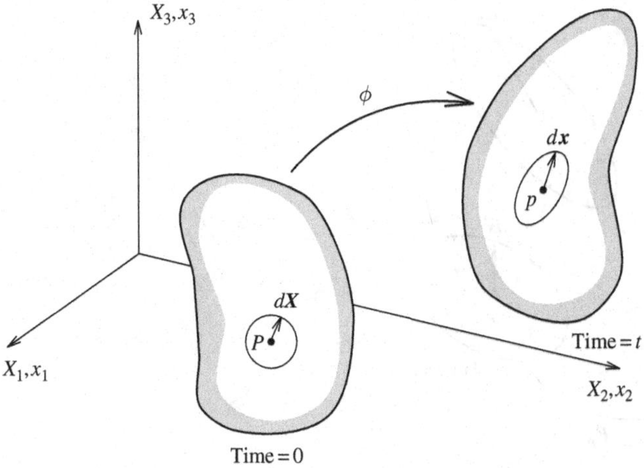


**FIGURE 7.1** Deformation in a small neighborhood.

given initial particle $P$ will deform into the spatial vector $d\boldsymbol{x}$ in the neighborhood of $p$ shown in the figure. If the neighborhood of $p$ could be isolated and freed from all forces, the material in that neighborhood would reach a new unloaded configuration characterized by the spatial vector $d\tilde{\boldsymbol{x}}$ (Figure 7.2). Observe that insofar as this neighborhood is conceptually isolated from the surrounding material it can be arbitrarily rotated without changing the intrinsic nature of the deformation of the material in the neighborhood. This potential indeterminacy will have implications on the choice of kinematic variables used in the subsequent formulation. Of course if the material is elastic this unloaded configuration will differ from the initial undeformed state only by a rigid body rotation. In the case of inelastic materials, however, this is not true and a certain amount of permanent deformation is possible. Note also that the unloaded state can only be defined locally, as the removal of all forces acting on $v$ may not lead to a global stress-free state but to a complex self-equilibrating stress distribution.

As a result of the local elastic unloading the spatial vector $d\boldsymbol{x}$ becomes $d\tilde{\boldsymbol{x}}$. The relationship between $d\boldsymbol{X}$ and $d\boldsymbol{x}$ is given by the deformation gradient $\boldsymbol{F}$ as explained in Section 4.4. Similarly, the relationship between $d\tilde{\boldsymbol{x}}$ and $d\boldsymbol{x}$ is given by the elastic component of the deformation gradient $\boldsymbol{F}_e$, and $d\boldsymbol{X}$ and $d\tilde{\boldsymbol{x}}$ are related by the permanent or inelastic component $\boldsymbol{F}_p$. These relationships are summarized as

$$d\boldsymbol{x} = \boldsymbol{F}\, d\boldsymbol{X}; \tag{7.1a}$$

$$d\boldsymbol{x} = \boldsymbol{F}_e\, d\tilde{\boldsymbol{x}}; \tag{7.1b}$$

$$d\tilde{\boldsymbol{x}} = \boldsymbol{F}_p\, d\boldsymbol{X}; \tag{7.1c}$$

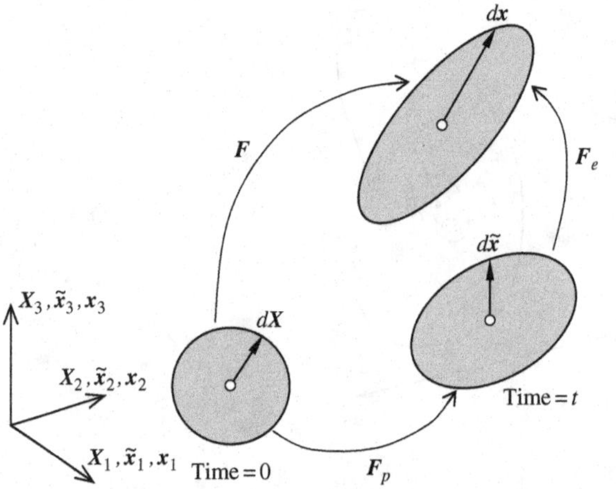


**FIGURE 7.2** Multiplicative decomposition in the small neighborhood of a particle.

for any arbitrary elemental vector $d\boldsymbol{X}$. Combining Equations (7.1b) and (7.1c) and comparing with (7.1a) gives

$$\boldsymbol{F} = \boldsymbol{F}_e \boldsymbol{F}_p. \tag{7.2}$$

This equation is known as the *multiplicative decomposition* of the deformation gradient into elastic and permanent components and constitutes the kinematic foundation for the theory that follows. Recalling that $\boldsymbol{F} = \partial \boldsymbol{x}/\partial \boldsymbol{X}$, it would be tempting to assume that it is possible to find an overall stress-free plastic state $\tilde{\boldsymbol{x}}$ such that $\boldsymbol{F}_p = \partial \tilde{\boldsymbol{x}}/\partial \boldsymbol{X}$ and $\boldsymbol{F}_e = \partial \boldsymbol{x}/\partial \tilde{\boldsymbol{x}}$. This is, unfortunately, not possible because the unloaded state can only be defined locally insofar as the locally unloaded neighborhoods cannot be re-assembled to give an overall stress-free configuration because they will not necessarily be geometrically compatible with each other.*

Strain measures that are independent of rigid body rotations can now be derived from $\boldsymbol{F}$ and its elastic and inelastic or permanent components. For instance, total right Cauchy–Green tensors given as

$$\boldsymbol{C} = \boldsymbol{F}^T \boldsymbol{F}; \quad \boldsymbol{C}_p = \boldsymbol{F}_p^T \boldsymbol{F}_p; \quad \boldsymbol{C}_e = \boldsymbol{F}_e^T \boldsymbol{F}_e; \tag{7.3a,b,c}$$

are often used for the development of inelastic constitutive equations. Observe that, like $\boldsymbol{C}$, $\boldsymbol{C}_p$ is a tensor defined in the reference configuration $\boldsymbol{X}$, whereas $\boldsymbol{C}_e$ is a tensor defined with reference to the permanent plastic state $\tilde{\boldsymbol{x}}$.

* This is analogous to a statically indeterminate truss in which the external loading is such that some members reach the yield stress but upon complete removal of the external loading the same members may not revert to their original unstressed condition. That is, if these members were completely individually unstressed then joint compatibility would not be achieved.

The stored elastic energy function $\Psi$ is now a function of the elastic right Cauchy–Green tensor $\boldsymbol{C}_e$ as

$$\Psi = \Psi(\boldsymbol{C}_e, \boldsymbol{X}), \tag{7.4}$$

and following similar arguments to those used in relation to Equations (6.7a,b), equations for the second Piola–Kirchhoff and Kirchhoff stresses are derived from $\Psi$ as

$$\widetilde{\boldsymbol{S}} = 2\frac{\partial \Psi}{\partial \boldsymbol{C}_e}; \quad \boldsymbol{\sigma} = \boldsymbol{F}_e \widetilde{\boldsymbol{S}} \boldsymbol{F}_e^T; \quad \boldsymbol{\tau} = J\boldsymbol{\sigma}. \tag{7.5a,b,c}$$

For isotropic materials it is possible and often simpler to formulate the constitutive equations in the current configuration by using the elastic left Cauchy–Green tensor $\boldsymbol{b}_e$ given as

$$\begin{aligned} \boldsymbol{b}_e &= \boldsymbol{F}_e \boldsymbol{F}_e^T \\ &= \boldsymbol{F}\boldsymbol{F}_p^{-1}\boldsymbol{F}_p^{-T}\boldsymbol{F}^T \\ &= \boldsymbol{F}\boldsymbol{C}_p^{-1}\boldsymbol{F}^T. \end{aligned} \tag{7.6}$$

Given that the invariants of $\boldsymbol{b}_e$ contain all the information needed to evaluate the stored elastic energy function and recalling Equation (6.25) for the Cauchy stress, a direct relationship between the Kirchhoff stress and strain energy is given as

$$\Psi(\boldsymbol{b}_e, \boldsymbol{X}) = \Psi\left(I_{\boldsymbol{b}_e}, II_{\boldsymbol{b}_e}, III_{\boldsymbol{b}_e}, \boldsymbol{X}\right); \tag{7.7a}$$

$$\boldsymbol{\tau} = J\boldsymbol{\sigma} = 2\Psi_I \boldsymbol{b}_e + 4\Psi_{II}\boldsymbol{b}_e^2 + 2III_{\boldsymbol{b}_e}\Psi_{III}\boldsymbol{I}; \tag{7.7b}$$

where the Kirchhoff stress tensor $\boldsymbol{\tau}$ has been introduced in this equation and will be used in the following equations in order to avoid the repeated appearance of the term $J^{-1}$. Recall that, even though the stress may be such that the material undergoes a permanent deformation, the stress is still determined from the stored *elastic* energy. For later developments it is useful to note that the tensors $\boldsymbol{\tau}$ and $\boldsymbol{b}_e$ or $\boldsymbol{b}_e^{-1}$ commute, that is, $\boldsymbol{\tau}\boldsymbol{b}_e = \boldsymbol{b}_e\boldsymbol{\tau}$ and $\boldsymbol{\tau}\boldsymbol{b}_e^{-1} = \boldsymbol{b}_e^{-1}\boldsymbol{\tau}$.

As previously mentioned, an important aspect of the locally defined unloaded state is that it can be arbitrarily rotated by a rotation matrix $\boldsymbol{Q}$ (Figure 7.3). It is therefore essential to use kinematic variables that are not only invariant with respect to overall rotation of the body but are also invariant with respect to rotations $\boldsymbol{Q}$ of the unloaded state. Fortunately, it is easy to show that the permanent strain measure $\boldsymbol{C}_p$ and the elastic left Cauchy–Green tensor $\boldsymbol{b}_e$ (but not $\boldsymbol{C}_e$, $\boldsymbol{F}_e$, and $\boldsymbol{F}_p$) satisfy these requirements.

This is explained by noting that the unit tensor $\boldsymbol{I}$ can be written as $\boldsymbol{I} = \boldsymbol{Q}^T\boldsymbol{Q} = \boldsymbol{Q}\boldsymbol{Q}^T$, which allows the multiplicative decomposition given by Equation (7.2) to be rewritten as

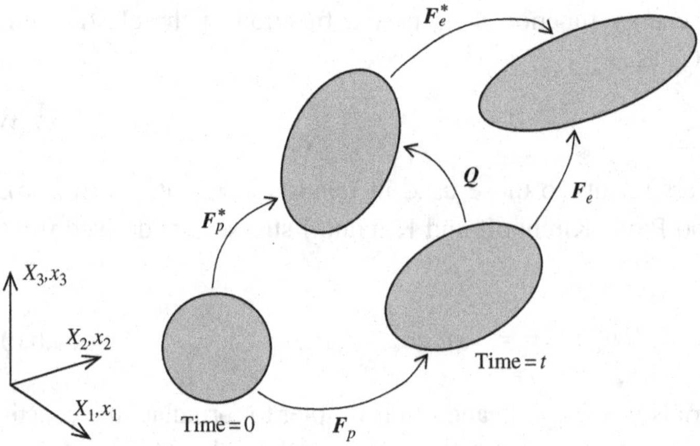


**FIGURE 7.3** Arbitrary rotation of the elastic deformation gradient.

$$F = {F^*}_e{F^*}_p; \quad {F^*}_e = F_eQ^T; \quad {F^*}_p = QF_p. \tag{7.8a,b,c}$$

Clearly, the right Cauchy–Green tensor $C$ given in Equation (7.3a,b,c)$_a$ remains unchanged. Rearranging the above equations, the right Cauchy–Green tensors $C_p$ and $C_e$ can be re-evaluated as

$$ {C^*}_p = F^{*T}_p{F^*}_p = F_p^TQ^TQF_p = F_p^TF_p = C_p; \tag{7.9a}$$

$$ {C^*}_e = F^{*T}_e{F^*}_e = QF_e^TF_eQ^T = QC_eQ^T. \tag{7.9b}$$

This reveals that the local permanent strain measure, as given by the right Cauchy–Green tensor $C_p$, remains unchanged but that the strain measure $C_e$ associated with the recovery of the local stress-free (permanent) configuration from the deformed position arbitrarily depends upon the rotation $Q$. It is also easy to show that the left Cauchy–Green tensor remains unchanged, given that

$$\begin{aligned} b_e^* &= {F^*}_eF^{*T}_e \\ &= F_eQ^TQF_e^T \\ &= b_e. \end{aligned} \tag{7.10}$$

Consequently, the stored elastic energy and the Kirchhoff stress $\tau$, given in the hyperelastic constitutive Equation (7.7), also remain invariant to the same rotation $Q$.

The fact that $C_p$ and $b_e$, unlike $F_p$ and $F_e$, are invariant with respect to an arbitrary rotation of the unloaded state suggests that $C_p$ and $b_e$ are appropriate kinematic measures to be employed in the development of the constitutive equations.

## 7.3 RATE KINEMATICS

It is clear from Section 7.2 that the state of stress in the hyperelastic–plastic material is determined, through $\boldsymbol{b}_e$, by the deformation gradient $\boldsymbol{F}$ and the plastic right Cauchy–Green tensor $\boldsymbol{C}_p$ given by Equation (7.6) as $\boldsymbol{b}_e = \boldsymbol{F}\boldsymbol{C}_p^{-1}\boldsymbol{F}^T$.

Generally, the solution process will provide the deformation gradient $\boldsymbol{F}$ and, consequently, to find the resulting $\boldsymbol{b}_e$ and thence the stresses it is necessary to derive a procedure for the evaluation of the right Cauchy–Green tensor $\boldsymbol{C}_p$. Recall that the evolution of the permanent configuration is path-dependent in the sense that $\boldsymbol{C}_p$ cannot be obtained directly from the current configuration alone. This implies that $\boldsymbol{b}_e$ is also path-dependent and must be obtained by integrating a rate form over a time or pseudo-time parameter. For elasto-plastic materials this rate expression is obtained from the *flow rule*, which enables the evaluation of $\boldsymbol{C}_p$ such that Equation (7.6) provides $\boldsymbol{b}_e$. Kirchhoff stresses can then be found that satisfy the elasto-plastic constitutive equations, that is, conform to the yield criterion. Of course such stresses may not additionally satisfy equilibrium and, computationally, the Newton–Raphson procedure will need to be employed to ensure that both constitutive equations and equilibrium are satisfied.

In order to move toward an appropriate expression of the flow rule, the rate of $\boldsymbol{b}_e$ is expressed as

$$\dot{\boldsymbol{b}}_e = \frac{d}{dt}\boldsymbol{b}_e\left(\boldsymbol{F}(t), \boldsymbol{C}_p(t)\right) = \frac{\partial \boldsymbol{b}_e}{\partial \boldsymbol{F}} : \dot{\boldsymbol{F}} + \frac{\partial \boldsymbol{b}_e}{\partial \boldsymbol{C}_p} : \dot{\boldsymbol{C}}_p. \tag{7.11}$$

The two terms on the right-hand side of the above expression can be re-expressed to provide a physical interpretation of $\dot{\boldsymbol{b}}_e$ as follows:

$$\dot{\boldsymbol{b}}_e = \frac{d}{dt}\boldsymbol{b}_e\left(\boldsymbol{F}(t), \boldsymbol{C}_p(t)\right) = \left.\frac{d\boldsymbol{b}_e}{dt}\right|_{\boldsymbol{C}_p=\text{const}} + \left.\frac{d\boldsymbol{b}_e}{dt}\right|_{\boldsymbol{F}=\text{const}}. \tag{7.12}$$

The first term measures the change in $\boldsymbol{b}_e$ that results from the overall change in deformation under the assumption that there is no further change in the permanent strain. For example, this would be correct in the case of local elastic unloading. Generally, however, further permanent deformation will occur in order to accommodate the inelastic constitutive requirements. Consequently, the second term in Equation (7.12) measures the change in $\boldsymbol{b}_e$ that results from a change in permanent deformation, with the overall deformation, as given by $\boldsymbol{F}$, remaining constant.

When recast later in incremental form suitable for computation it will be seen that the first term in Equation (7.12) provides the so-called *trial state of stress*, whereas the second term leads to the so-called *return-mapping* procedure which modifies the trial state of stress to ensure satisfaction of the inelastic constitutive requirements.

In order to develop a physically meaningful flow rule using the terms in Equation (7.12), it is necessary to employ the concept of work conjugacy. Physically, it is reasonable to assert that the internal rate of work per unit initial volume, $\dot{w}$, done by the stresses $\boldsymbol{\tau}$ can be decomposed into elastic recoverable and permanent nonrecoverable components, $\dot{w}_e$ and $\dot{w}_p$ respectively, the latter usually being called the *rate of plastic dissipation*. This can be expressed as

$$\dot{w} = \dot{w}_e + \dot{w}_p. \tag{7.13}$$

These work-rate components will now be related to the terms in Equation (7.12) and in this manner enable the development of the flow rule.

For this purpose, observe first that the total rate of work per unit undeformed volume is given in terms of the Kirchhoff stress $\boldsymbol{\tau}$ and velocity gradient $\boldsymbol{l}$ as

$$\dot{w} = \boldsymbol{\tau} : \boldsymbol{l} = \boldsymbol{\tau} : \left( \frac{1}{2} \frac{d\boldsymbol{b}_e}{dt}\bigg|_{\boldsymbol{C}_p = \text{const}} \boldsymbol{b}_e^{-1} \right). \tag{7.14}$$

The proof of this expression is given in Example 7.1. Similarly, it is shown in Example 7.2 that the elastic component of the total work rate is

$$\dot{w}_e = \boldsymbol{\tau} : \boldsymbol{l}_e = \boldsymbol{\tau} : \left( \frac{1}{2} \frac{d\boldsymbol{b}_e}{dt} \boldsymbol{b}_e^{-1} \right), \tag{7.15}$$

where the elastic velocity gradient $\boldsymbol{l}_e$ is defined by analogy with the velocity gradient, $\boldsymbol{l} = \dot{\boldsymbol{F}}\boldsymbol{F}^{-1}$, as

$$\boldsymbol{l}_e = \dot{\boldsymbol{F}}_e \boldsymbol{F}_e^{-1}. \tag{7.16}$$

**EXAMPLE 7.1: Total work rate**

Proof of Equation (7.14) relating the total rate of work, $\dot{w}$, to the time derivative of the left Cauchy–Green tensor $\boldsymbol{b}_e$. Recall the commutative property $\boldsymbol{b}_e^{-1}\boldsymbol{\tau} = \boldsymbol{\tau}\boldsymbol{b}_e^{-1}$, that is, $\boldsymbol{F}_e^{-T}\boldsymbol{F}_e^{-1}\boldsymbol{\tau} = \boldsymbol{\tau}\boldsymbol{F}_e^{-T}\boldsymbol{F}_e^{-1}$.

$$\begin{aligned}
\dot{w} &= \frac{1}{2}\boldsymbol{\tau} : \left( \frac{d\boldsymbol{b}_e}{dt}\bigg|_{\boldsymbol{C}_p = \text{const}} \boldsymbol{b}_e^{-1} \right) \\
&= \frac{1}{2}\text{tr}\left(\boldsymbol{\tau}\left(\dot{\boldsymbol{F}}\boldsymbol{C}_p^{-1}\boldsymbol{F}^T + \boldsymbol{F}\boldsymbol{C}_p^{-1}\dot{\boldsymbol{F}}^T\right)\boldsymbol{F}_e^{-T}\boldsymbol{F}_e^{-1}\right) \\
&= \frac{1}{2}\text{tr}\left(\boldsymbol{\tau}\dot{\boldsymbol{F}}\boldsymbol{F}_p^{-1}\boldsymbol{F}_p^{-T}\boldsymbol{F}_p^{T}\boldsymbol{F}_e^{T}\boldsymbol{F}_e^{-T}\boldsymbol{F}_e^{-1} + \boldsymbol{\tau}\boldsymbol{F}\boldsymbol{F}_p^{-1}\boldsymbol{F}_p^{-T}\dot{\boldsymbol{F}}^T\boldsymbol{F}_e^{-T}\boldsymbol{F}_e^{-1}\right) \\
&= \frac{1}{2}\text{tr}\left(\boldsymbol{\tau}\dot{\boldsymbol{F}}\boldsymbol{F}^{-1}\right) + \frac{1}{2}\text{tr}\left(\boldsymbol{\tau}\boldsymbol{F}_e\boldsymbol{F}_p\boldsymbol{F}_p^{-1}\boldsymbol{F}_p^{-T}\dot{\boldsymbol{F}}^T\boldsymbol{F}_e^{-T}\boldsymbol{F}_e^{-1}\right)
\end{aligned}$$

*(continued)*

**EXAMPLE 7.1:** *(cont.)*

$$
\begin{aligned}
&= \frac{1}{2}\boldsymbol{\tau} : \boldsymbol{l} + \frac{1}{2}\mathrm{tr}\left(\boldsymbol{F}_e^{-T}\boldsymbol{F}_e^{-1}\boldsymbol{\tau}\boldsymbol{F}_e\boldsymbol{F}_p^{-T}\dot{\boldsymbol{F}}^T\right)\\
&= \frac{1}{2}\boldsymbol{\tau} : \boldsymbol{l} + \frac{1}{2}\mathrm{tr}\left(\boldsymbol{\tau}\boldsymbol{F}_e^{-T}\boldsymbol{F}_e^{-1}\boldsymbol{F}_e\boldsymbol{F}_p^{-T}\dot{\boldsymbol{F}}^T\right)\\
&= \frac{1}{2}\boldsymbol{\tau} : \boldsymbol{l} + \frac{1}{2}\mathrm{tr}\left(\boldsymbol{\tau}\boldsymbol{F}^{-T}\dot{\boldsymbol{F}}^T\right)\\
&= \frac{1}{2}\boldsymbol{\tau} : \boldsymbol{l} + \frac{1}{2}\boldsymbol{\tau} : \boldsymbol{l} = \boldsymbol{\tau} : \boldsymbol{l}.
\end{aligned}
$$

The rate of plastic dissipation can now be calculated from Equations (7.12)–(7.15) as

$$
\begin{aligned}
\dot{w}_p &= \dot{w} - \dot{w}_e\\
&= \boldsymbol{\tau} : \left(\frac{1}{2}\left.\frac{d\boldsymbol{b}_e}{dt}\right|_{\boldsymbol{C}_p=\mathrm{const}}\boldsymbol{b}_e^{-1}\right) - \boldsymbol{\tau} : \left(\frac{1}{2}\frac{d\boldsymbol{b}_e}{dt}\boldsymbol{b}_e^{-1}\right)\\
&= \boldsymbol{\tau} : \left(\frac{1}{2}\left(\left.\frac{d\boldsymbol{b}_e}{dt}\right|_{\boldsymbol{C}_p=\mathrm{const}} - \frac{d\boldsymbol{b}_e}{dt}\right)\boldsymbol{b}_e^{-1}\right)\\
&= \boldsymbol{\tau} : \left(-\frac{1}{2}\left.\frac{d\boldsymbol{b}_e}{dt}\right|_{\boldsymbol{F}=\mathrm{const}}\boldsymbol{b}_e^{-1}\right)\\
&= \boldsymbol{\tau} : \boldsymbol{l}_p, \qquad (7.17)
\end{aligned}
$$

from which the definition of the so-called *plastic rate of deformation* emerges as

$$
\boldsymbol{l}_p = -\frac{1}{2}\left.\frac{d\boldsymbol{b}_e}{dt}\right|_{\boldsymbol{F}=\mathrm{const}}\boldsymbol{b}_e^{-1}. \qquad (7.18)
$$

**EXAMPLE 7.2: Elastic work rate**

Proof of Equation (7.15) relating the total rate of elastic work, $\dot{w}_e$, to the time derivative of the left Cauchy–Green tensor $\boldsymbol{b}_e$. Recall again the commutative property $\boldsymbol{b}_e^{-1}\boldsymbol{\tau} = \boldsymbol{\tau}\boldsymbol{b}_e^{-1}$, that is, $\boldsymbol{F}_e^{-T}\boldsymbol{F}_e^{-1}\boldsymbol{\tau} = \boldsymbol{\tau}\boldsymbol{F}_e^{-T}\boldsymbol{F}_e^{-1}$.

$$
\begin{aligned}
\dot{w}_e &= \frac{1}{2}\boldsymbol{\tau} : \left(\dot{\boldsymbol{b}}_e\boldsymbol{b}_e^{-1}\right)\\
&= \frac{1}{2}\mathrm{tr}\left(\boldsymbol{\tau}\left(\dot{\boldsymbol{F}}_e\boldsymbol{F}_e^T + \boldsymbol{F}_e\dot{\boldsymbol{F}}_e^T\right)\boldsymbol{F}_e^{-T}\boldsymbol{F}_e^{-1}\right)\\
&= \frac{1}{2}\mathrm{tr}\left(\boldsymbol{\tau}\dot{\boldsymbol{F}}_e\boldsymbol{F}_e^{-1}\right) + \frac{1}{2}\mathrm{tr}\left(\boldsymbol{\tau}\boldsymbol{F}_e\dot{\boldsymbol{F}}_e^T\boldsymbol{F}_e^{-T}\boldsymbol{F}_e^{-1}\right)
\end{aligned}
$$

*(continued)*

**EXAMPLE 7.2:** *(cont.)*

$$
\begin{aligned}
&= \frac{1}{2}\boldsymbol{\tau} : \boldsymbol{l}_e + \frac{1}{2}\mathrm{tr}\left(\boldsymbol{F}_e^{-T}\boldsymbol{F}_e^{-1}\boldsymbol{\tau}\boldsymbol{F}_e\dot{\boldsymbol{F}}_e^T\right) \\
&= \frac{1}{2}\boldsymbol{\tau} : \boldsymbol{l}_e + \frac{1}{2}\mathrm{tr}\left(\boldsymbol{\tau}\boldsymbol{F}_e^{-T}\boldsymbol{F}_e^{-1}\boldsymbol{F}_e\dot{\boldsymbol{F}}_e^T\right) \\
&= \frac{1}{2}\boldsymbol{\tau} : \boldsymbol{l}_e + \frac{1}{2}\mathrm{tr}\left(\boldsymbol{\tau}\boldsymbol{F}_e^{-T}\dot{\boldsymbol{F}}_e^T\right) \\
&= \frac{1}{2}\boldsymbol{\tau} : \boldsymbol{l}_e + \frac{1}{2}\boldsymbol{\tau} : \boldsymbol{l}_e = \boldsymbol{\tau} : \boldsymbol{l}_e.
\end{aligned}
$$

It will be explained in the next section that the use of this rate tensor to define the flow rule ensures that the material dissipates energy due to plastic deformation as efficiently as possible. This phenomenon is formally stated as the *principle of maximum plastic dissipation*, which is a fundamental principle underpinning plasticity theory.

It is important to observe that Equation (7.18) implicitly contains the rate of change of the inelastic right Cauchy–Green tensor. This is shown in Example 7.3 as

$$
\boldsymbol{l}_p = -\frac{1}{2}\left.\frac{d\boldsymbol{b}_e}{dt}\right|_{\boldsymbol{F}=\mathrm{const}} \boldsymbol{b}_e^{-1} = \frac{1}{2}\boldsymbol{F}\boldsymbol{C}_p^{-1}\dot{\boldsymbol{C}}_p\boldsymbol{F}^{-1}. \tag{7.19}
$$

Algorithmically, however, it will prove more convenient to use expression (7.18) for $\boldsymbol{l}_p$ rather than (7.19) and thereby calculate $\boldsymbol{C}_p$ indirectly from a time-integrated $\boldsymbol{b}_e$ as $\boldsymbol{C}_p^{-1} = \boldsymbol{F}^{-1}\boldsymbol{b}_e\boldsymbol{F}^{-T}$.

Before moving on to consider rate-independent plasticity it is worthwhile recalling the main thread of the developments achieved so far. Since large deformations and finite strains are involved, the basic kinematic description requires the overall deformation gradient $\boldsymbol{F}$, a permanent right Cauchy–Green tensor $\boldsymbol{C}_p$, and the elastic left Cauchy–Green tensor $\boldsymbol{b}_e$, all functions of a time-like parameter and related by $\boldsymbol{b}_e = \boldsymbol{F}\boldsymbol{C}_p^{-1}\boldsymbol{F}^T$. Insofar as $\boldsymbol{C}_p$ is path-dependent, $\boldsymbol{b}_e$ is also implicity path-dependent, and to determine $\boldsymbol{b}_e$ requires the integration over time of the rate of $\boldsymbol{b}_e$. Once $\boldsymbol{b}_e$ is determined, stresses can be evaluated from the hyperelastic energy function given in Equation (7.7). In addition to satisfying the equilibrium conditions the stresses must conform to the constraints of the inelastic constitutive equations (considered later in Section 7.4).

The following sections will introduce the means by which $\boldsymbol{l}_p$ can be determined and by implication how $\boldsymbol{b}_e$ can be evaluated.

**EXAMPLE 7.3: Plastic rate of deformation**

Proof of Equation (7.19) relating the plastic rate of deformation $\boldsymbol{l}_p$ to the time derivative of the right Cauchy–Green tensor, $\dot{\boldsymbol{C}}_p^{-1}$. Observe that the time derivative of $\boldsymbol{C}_p$ can be found from the time derivative of $\boldsymbol{C}_p^{-1}\boldsymbol{C}_p = \boldsymbol{I}$ as

$$\dot{\boldsymbol{C}}_p^{-1}\boldsymbol{C}_p + \boldsymbol{C}_p^{-1}\dot{\boldsymbol{C}}_p = \boldsymbol{0};$$

$$\begin{aligned}\boldsymbol{l}_p &= -\frac{1}{2}\left.\frac{d\boldsymbol{b}_e}{dt}\right|_{\boldsymbol{F}=\text{const}}\boldsymbol{b}_e^{-1} = -\frac{1}{2}\boldsymbol{F}\frac{d\boldsymbol{C}_p^{-1}}{dt}\boldsymbol{F}^T\boldsymbol{b}_e^{-1}\\ &= -\frac{1}{2}\boldsymbol{F}\frac{d\boldsymbol{C}_p^{-1}}{dt}\boldsymbol{F}^T\boldsymbol{F}^{-T}\boldsymbol{C}_p\boldsymbol{F}^{-1}\\ &= -\frac{1}{2}\boldsymbol{F}\frac{d\boldsymbol{C}_p^{-1}}{dt}\boldsymbol{C}_p\boldsymbol{F}^{-1}\\ &= \frac{1}{2}\boldsymbol{F}\boldsymbol{C}_p^{-1}\dot{\boldsymbol{C}}_p\boldsymbol{F}^{-1}.\end{aligned}$$

## 7.4 RATE-INDEPENDENT PLASTICITY

Attention is now focused on the evaluation of the plastic rate of deformation, $\boldsymbol{l}_p$. For elasto-plastic materials this is given by an equation known as the *flow rule* which relates $\boldsymbol{l}_p$ to the current state of stress in the material. The simplest case, and the only one considered in this text, is Von Mises plasticity with linear isotropic hardening, which is defined by a *yield surface* function of the Kirchhoff stress $\boldsymbol{\tau}$, a yield stress $\bar{\tau}_y$, and a hardening variable $\bar{\varepsilon}_p$ as

$$f(\boldsymbol{\tau}, \bar{\varepsilon}_p) = \sqrt{\frac{3}{2}(\boldsymbol{\tau}' : \boldsymbol{\tau}')} - \bar{\tau}_y \leq 0; \quad \bar{\tau}_y = \bar{\tau}_y^0 + H\bar{\varepsilon}_p, \tag{7.20a,b}$$

where, in terms of the mean stress, $p$, the deviatoric component of the Kirchhoff stress is

$$\boldsymbol{\tau}' = \boldsymbol{\tau} - pJ\boldsymbol{I}. \tag{7.21}$$

The constant $H$ is a material-hardening parameter and $\bar{\tau}_y^0$ is the initial yield stress. The yield surface, often called the *yield function*, defines an elastic limit as determined by the generalized scalar Von Mises equivalent stress $\sqrt{\frac{3}{2}(\boldsymbol{\tau}' : \boldsymbol{\tau}')}$. If $f(\boldsymbol{\tau}, \bar{\varepsilon}_p) < 0$ then the material behaves elastically; if $f(\boldsymbol{\tau}, \bar{\varepsilon}_p) = 0$ then the Kirchhoff stress tensor $\boldsymbol{\tau}$ must be such that the Von Mises equivalent stress equals the current yield stress $\bar{\tau}_y^0 + H\bar{\varepsilon}_p$ and elasto-plastic behavior ensues.

The flow rule can now be defined by

$$l_p = -\frac{1}{2}\left.\frac{d\boldsymbol{b}_e}{dt}\right|_{\boldsymbol{F}=\text{const}} \boldsymbol{b}_e^{-1} = \dot{\gamma}\frac{\partial f(\boldsymbol{\tau}, \bar{\varepsilon}_p)}{\partial \boldsymbol{\tau}}, \tag{7.22}$$

where $\dot{\gamma}$, a proportionality factor, is called the *consistency parameter* or *plastic multiplier*. It is couched as a rate for dimensional consistency with the plastic rate of deformation $\boldsymbol{l}_p$.

The flow rule given in Equation (7.22), in which the direction of the plastic strain rate coincides with the gradient of the yield surface, is known as *associative*.

**Remark 7.1**: The associative type of flow rule given in Equation (7.22) is a consequence of the postulate of maximum dissipation of plastic work. The total rate of work per unit initial volume, $\boldsymbol{\tau} : \boldsymbol{l}$, can be expressed using Equations (7.13)–(7.15) and (7.17) as

$$\boldsymbol{\tau} : \boldsymbol{l} = \boldsymbol{\tau} : \boldsymbol{l}_e + \boldsymbol{\tau} : \boldsymbol{l}_p, \tag{7.23}$$

where $\boldsymbol{\tau} : \boldsymbol{l}_e$ is the rate of change of elastic energy and $\boldsymbol{\tau} : \boldsymbol{l}_p$ is the plastic dissipation rate. The postulate of maximum plastic dissipation implies that the rate of the work put into the deformation that cannot be accommodated elastically is being dissipated plastically by $\boldsymbol{\tau} : \boldsymbol{l}_p$ as efficiently as possible. It is possible to show that maximizing the term $\boldsymbol{\tau} : \boldsymbol{l}_p$ subject to the constraint $f(\boldsymbol{\tau}, \bar{\varepsilon}_p) \leq 0$ leads to Equation (7.22), where $\dot{\gamma}$ can then be interpreted as a Lagrange multiplier.

Substituting the yield function given by Equation (7.20a,b) into the flow rule of Equation (7.22) gives

$$\boldsymbol{l}_p = \dot{\gamma}\frac{\boldsymbol{\tau}'}{\sqrt{\frac{2}{3}(\boldsymbol{\tau}' : \boldsymbol{\tau}')}} = \dot{\gamma}\boldsymbol{\nu}\,; \quad \boldsymbol{\nu} = \frac{\boldsymbol{\tau}'}{\sqrt{\frac{2}{3}(\boldsymbol{\tau}' : \boldsymbol{\tau}')}}. \tag{7.24a,b}$$

In order to define the evolution of the hardening variable $\bar{\varepsilon}_p$, a traditional work-hardening approach can be adopted. Given the Von Mises equivalent stress $\bar{\tau}$, defined as

$$\bar{\tau} = \sqrt{\frac{3}{2}(\boldsymbol{\tau}' : \boldsymbol{\tau}')}, \tag{7.25}$$

Where the rate $\dot{\bar{\varepsilon}}_p$ is defined to be the work conjugate to $\bar{\tau}$ as

$$\dot{w}_p = \bar{\tau}\dot{\bar{\varepsilon}}_p \tag{7.26a}$$

$$= \boldsymbol{\tau} : \boldsymbol{l}_p \tag{7.26b}$$

$$= \boldsymbol{\tau} : \dot{\gamma}\frac{\boldsymbol{\tau}'}{\sqrt{\frac{2}{3}(\boldsymbol{\tau}' : \boldsymbol{\tau}')}} \tag{7.26c}$$

$$= \dot{\gamma}\bar{\tau}, \tag{7.26d}$$

from which

$$\dot{\bar{\varepsilon}}_p = \dot{\gamma}. \tag{7.27}$$

Insofar as $\dot{\bar{\varepsilon}}_p$ has been defined as work conjugate to the Von Mises equivalent stress $\bar{\tau}$, the work-hardening variable $\bar{\varepsilon}_p$ can be designated as the Von Mises equivalent plastic strain.

**Remark 7.2**: Note that the volumetric component of $\boldsymbol{l}_p$, that is, $\text{tr}(\boldsymbol{l}_p)$, is zero. This can be shown by noting that by construction $\text{tr}(\boldsymbol{\tau}') = 0$, to give

$$\text{tr}(\boldsymbol{l}_p) = \dot{\gamma}\frac{\text{tr}(\boldsymbol{\tau}')}{\sqrt{\frac{2}{3}(\boldsymbol{\tau}' : \boldsymbol{\tau}')}} = 0. \tag{7.28}$$

As a consequence the plastic deformation is incompressible and the determinant of the inelastic (or plastic) deformation gradient is unity, that is,

$$\det \boldsymbol{F}_p = 1. \tag{7.29}$$

## 7.5 PRINCIPAL DIRECTIONS

The formulations that follow are greatly simplified by operating in principal directions. Obviously, this is essential if the hyperelastic energy is given in terms of elastic principal stretches $\lambda_{e,\alpha}$, such as for Ogden materials. But perhaps more importantly in the present context, employing principal directions will greatly facilitate the integration in time of the flow rule.†

Insofar as evaluation of the flow rule will eventually yield the left Cauchy–Green tensor $\boldsymbol{b}_e$, the elastic stretches can be obtained in the usual manner by evaluating the principal directions of $\boldsymbol{b}_e$ to give

$$\boldsymbol{b}_e = \sum_{\alpha=1}^{3} \lambda_{e,\alpha}^2 \, \boldsymbol{n}_\alpha \otimes \boldsymbol{n}_\alpha. \tag{7.30}$$

Expressing the hyperelastic energy function in terms of the elastic stretches and using algebra similar to that employed in Section 6.6 enables Equation (7.7b) to be rewritten as

$$\boldsymbol{\tau} = \sum_{\alpha=1}^{3} \tau_{\alpha\alpha} \boldsymbol{n}_\alpha \otimes \boldsymbol{n}_\alpha; \quad \tau_{\alpha\alpha} = \frac{\partial \Psi}{\partial \ln \lambda_{e,\alpha}}; \quad \alpha = 1, 2, 3. \tag{7.31a,b}$$

† In fact, working in principal directions will disguise the need to employ the so-called *exponential map* process favored by many authors.

It is crucial to observe at this point that the principal directions of $\boldsymbol{b}_e$ and $\boldsymbol{\tau}$ are in no way related to the principal directions that would be obtained from the total deformation gradient $\boldsymbol{F}$ nor to the inelastic component $\boldsymbol{F}_p$.

To proceed, it is now necessary to focus on a particular elastic material description. The fact that the formulation is expressed in principal directions implies that the strain energy function should be a function of the principal stretches $\lambda_{e,\alpha}$, such as those given in Section 6.6. Furthermore, for elasto-plastic behavior the nearly incompressible nature of the overall deformation needs to be taken into account. Even when elastic deformation is compressible the plastic strain is incompressible and much larger than the elastic component, leading to an overall behavior which is nearly incompressible. These two requirements are met by the simple stretch-based nearly incompressible material described in Section 6.6.6.

The principal components of the deviatoric Kirchhoff stress tensor are given for this material as

$$\tau'_{\alpha\alpha} = 2\mu \ln \lambda_{e,\alpha} - \frac{2}{3}\mu \ln J\,; \quad \tau'_{\alpha\alpha} = \tau_{\alpha\alpha} - p\,; \quad \alpha = 1,2,3; \qquad (7.32\text{a,b})$$

where the hydrostatic pressure $p$ is defined in Equation (6.103).

The plastic rate of deformation given in Equation (7.22) can be re-expressed in principal directions by introducing Equation (7.30) as

$$\begin{aligned} \boldsymbol{l}_p &= -\frac{1}{2}\left.\frac{d\boldsymbol{b}_e}{dt}\right|_{\boldsymbol{F}} \boldsymbol{b}_e^{-1} \\ &= -\frac{1}{2}\left[\left.\frac{d}{dt}\right|_{\boldsymbol{F}} \left(\sum_{\alpha=1}^{3} \lambda_{e,\alpha}^2\, \boldsymbol{n}_\alpha \otimes \boldsymbol{n}_\alpha\right)\right]\left(\sum_{\beta=1}^{3} \lambda_{e,\beta}^{-2}\, \boldsymbol{n}_\beta \otimes \boldsymbol{n}_\beta\right). \end{aligned} \qquad (7.33)$$

Due to the possibility of plastic deformation, both the stretches $\lambda_{e,\alpha}$ and the direction vectors $\boldsymbol{n}_\alpha$ in the above equation are potentially subject to change with time at constant $\boldsymbol{F}$. As shown in Example 7.4, the time derivatives of the principal directions $\boldsymbol{n}_\alpha$ at constant $\boldsymbol{F}$ can be expressed in terms of the components of a skew symmetric tensor $\boldsymbol{W}$ as

$$\left.\frac{d\boldsymbol{n}_\alpha}{dt}\right|_{\boldsymbol{F}} = \sum_{\beta=1}^{3} W_{\alpha\beta}\, \boldsymbol{n}_\beta\,; \quad W_{\alpha\beta} = -W_{\beta\alpha}\,; \quad W_{\alpha\alpha} = 0. \qquad (7.34\text{a,b,c})$$

This enables the plastic rate of deformation tensor to be expressed in principal directions (see Example 7.4 for proof) as

$$\boldsymbol{l}_p = \sum_{\alpha=1}^{3} -\frac{1}{2}\frac{d\lambda_{e,\alpha}^2}{dt}\lambda_{e,\alpha}^{-2}\, \boldsymbol{n}_\alpha \otimes \boldsymbol{n}_\alpha - \sum_{\substack{\alpha,\beta=1 \\ \alpha\neq\beta}}^{3} \frac{1}{2} W_{\alpha\beta}\left(\frac{\lambda_{e,\alpha}^2 - \lambda_{e,\beta}^2}{\lambda_{e,\beta}^2}\right) \boldsymbol{n}_\alpha \otimes \boldsymbol{n}_\beta. \qquad (7.35)$$

In addition, for isotropic materials the yield function can always be expressed in principal directions of $\boldsymbol{\tau}$, that is, $f(\boldsymbol{\tau}, \bar{\varepsilon}_p) = f(\tau_{\alpha\alpha}, \bar{\varepsilon}_p)$, enabling the flow rule Equation (7.22) to be expressed in principal directions as

$$\boldsymbol{l}_p = \dot{\gamma} \sum_{\alpha=1}^{3} \frac{\partial f(\tau_{\alpha\alpha}, \bar{\varepsilon}_p)}{\partial \tau_{\alpha\alpha}} \boldsymbol{n}_\alpha \otimes \boldsymbol{n}_\alpha. \tag{7.36}$$

The above isotropic flow rule shows that, when expressed in principal directions, $\boldsymbol{l}_p$ can only contain diagonal entries. That is, there are no terms in the above equation with a tensorial basis $\boldsymbol{n}_\alpha \otimes \boldsymbol{n}_\beta$ where $\alpha \neq \beta$. Consequently, the off-diagonal components in Equation (7.35) must be zero, which reveals that $W_{\alpha\beta} = 0$ and hence, from Equation (7.34a,b,c)$_a$,

$$\left.\frac{d\boldsymbol{n}_\alpha}{dt}\right|_{\boldsymbol{F}=\text{const}} = 0. \tag{7.37}$$

This statement enables the flow rule to be written succinctly in principal directions as

$$l_{p,\alpha\alpha} = -\frac{1}{2}\left.\frac{d\lambda_{e,\alpha}^2}{dt}\right|_{\boldsymbol{F}=\text{const}} \lambda_{e,\alpha}^{-2} = \dot{\gamma}\frac{\partial f(\tau_{\alpha\alpha}, \bar{\varepsilon}_p)}{\partial \tau_{\alpha\alpha}}; \quad \alpha = 1, 2, 3; \tag{7.38}$$

which can be conveniently rearranged to give

$$l_{p,\alpha\alpha} = \left.\frac{d\varepsilon_{e,\alpha}}{dt}\right|_{\boldsymbol{F}=\text{const}} = -\dot{\gamma}\frac{\partial f(\tau_{\alpha\alpha}, \bar{\varepsilon}_p)}{\partial \tau_{\alpha\alpha}}; \quad \varepsilon_{e,\alpha} = \ln \lambda_{e,\alpha}. \tag{7.39a,b}$$

The fact that the rotation term $W_{\alpha\beta} = 0$, in Equation (7.35), implies that $\boldsymbol{l}_p$ is symmetric, which justifies the terminology "plastic rate of deformation," $\boldsymbol{d}_p$, since $\boldsymbol{d}_p = \frac{1}{2}\left(\boldsymbol{l}_p + \boldsymbol{l}_p^T\right)$.

**Remark 7.3**: Note that, for the simple one-dimensional truss case described in Chapter 3, the term $-\left.\frac{d\varepsilon_{e,\alpha}}{dt}\right|_{\boldsymbol{F}=\text{const}}$ coincides with $\dot{\varepsilon}_p$ used in Equation (3.53). This is easily shown by rearranging Equation (3.49a,b,c,d)$_d$ as $\varepsilon_e = \varepsilon - \varepsilon_p$ and taking the time derivative, at constant $\boldsymbol{F}$ equivalent to constant $\varepsilon$.

The partial derivative of the yield function with respect to the principal directions of $\boldsymbol{\tau}$ emerges as

$$\frac{\partial f(\tau_{\alpha\alpha}, \bar{\varepsilon}_p)}{\partial \tau_{\alpha\alpha}} = \frac{\tau'_{\alpha\alpha}}{\sqrt{\frac{2}{3}\boldsymbol{\tau}' : \boldsymbol{\tau}'}} = \nu_\alpha\,; \quad \sqrt{\frac{2}{3}\boldsymbol{\tau}' : \boldsymbol{\tau}'} = \left(\frac{2}{3}\sum_{\alpha=1}^{3}(\tau'_{\alpha\alpha})^2\right)^{\frac{1}{2}}$$

(7.40a,b)

where $\boldsymbol{\nu}$ is a dimensionless direction vector normal to the yield surface described with respect to the principal directions of $\boldsymbol{\tau}$ (Figure 7.4).

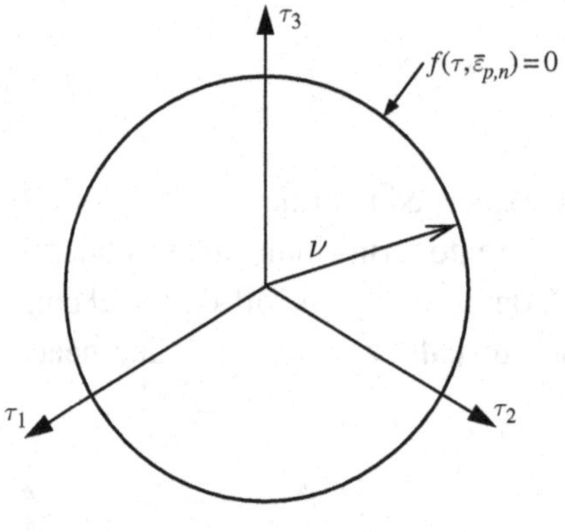


**FIGURE 7.4** Dimensionless direction vector $\boldsymbol{\nu}$ normal to yield surface $f(\boldsymbol{\tau},\bar{\varepsilon}_p)$.

**EXAMPLE 7.4: Plastic rate of deformation in principal directions**

Proof of Equation (7.35) for the plastic rate of deformation $\boldsymbol{l}_p$. Recall Equation (7.34a,b,c)$_a$ as

$$\left.\frac{d\boldsymbol{n}_\alpha}{dt}\right|_{\boldsymbol{F}} = \sum_{\beta=1}^{3} W_{\alpha\beta}\,\boldsymbol{n}_\beta,$$

where $W_{\alpha\beta}$ describes the rotation of the principal directions $\boldsymbol{n}_\alpha$ at constant $\boldsymbol{F}$. Now, observe that using the above equation the time derivative of $\boldsymbol{n}_\alpha \cdot \boldsymbol{n}_\beta = \delta_{\alpha\beta}$ at $\boldsymbol{F}$ constant yields Equations (7.34a,b,c)$_{b,c}$, that is,

$$W_{\alpha\beta} = -W_{\beta\alpha}\,; \quad W_{\alpha\alpha} = 0.$$

The plastic rate of deformation $\boldsymbol{l}_p$ can now be developed as follows:

$$\begin{aligned}
\boldsymbol{l}_p &= -\frac{1}{2}\left.\frac{d\boldsymbol{b}_e}{dt}\right|_{\boldsymbol{F}}\boldsymbol{b}_e^{-1} = -\frac{1}{2}\left.\frac{d}{dt}\right|_{\boldsymbol{F}}\left(\sum_{\alpha=1}^{3}\lambda_{e,\alpha}^2\,\boldsymbol{n}_\alpha\otimes\boldsymbol{n}_\alpha\right)\left(\sum_{\gamma=1}^{3}\lambda_{e,\gamma}^{-2}\,\boldsymbol{n}_\gamma\otimes\boldsymbol{n}_\gamma\right)\\
&= -\frac{1}{2}\left(\sum_{\alpha=1}^{3}\left.\frac{d\lambda_{e,\alpha}^2}{dt}\right|_{\boldsymbol{F}}\boldsymbol{n}_\alpha\otimes\boldsymbol{n}_\alpha + \sum_{\alpha=1}^{3}\lambda_{e,\alpha}^2\left.\frac{d\boldsymbol{n}_\alpha}{dt}\right|_{\boldsymbol{F}}\otimes\boldsymbol{n}_\alpha\right.\\
&\qquad\left.+\sum_{\alpha=1}^{3}\lambda_{e,\alpha}^2\boldsymbol{n}_\alpha\otimes\left.\frac{d\boldsymbol{n}_\alpha}{dt}\right|_{\boldsymbol{F}}\right)\left(\sum_{\gamma=1}^{3}\lambda_{e,\gamma}^{-2}\,\boldsymbol{n}_\gamma\otimes\boldsymbol{n}_\gamma\right)
\end{aligned}$$

*(continued)*

**EXAMPLE 7.4:** *(cont.)*

$$= -\frac{1}{2}\left(\sum_{\alpha=1}^{3} \left.\frac{d\lambda_{e,\alpha}^2}{dt}\right|_{\boldsymbol{F}} \boldsymbol{n}_\alpha \otimes \boldsymbol{n}_\alpha + \sum_{\alpha,\beta=1}^{3} \lambda_{e,\alpha}^2 W_{\alpha\beta}\, \boldsymbol{n}_\beta \otimes \boldsymbol{n}_\alpha \right.$$

$$\left. + \sum_{\alpha,\beta=1}^{3} \lambda_{e,\alpha}^2 W_{\alpha\beta}\, \boldsymbol{n}_\alpha \otimes \boldsymbol{n}_\beta \right) \left(\sum_{\gamma=1}^{3} \lambda_{e,\gamma}^{-2} \boldsymbol{n}_\gamma \otimes \boldsymbol{n}_\gamma\right)$$

$$= -\frac{1}{2}\left(\sum_{\alpha=1}^{3} \left.\frac{d\lambda_{e,\alpha}^2}{dt}\right|_{\boldsymbol{F}} \boldsymbol{n}_\alpha \otimes \boldsymbol{n}_\alpha\right)$$

$$-\frac{1}{2}\left(\sum_{\substack{\alpha,\beta=1\\ \alpha\neq\beta}}^{3} W_{\alpha\beta}\left(\lambda_{e,\alpha}^2 - \lambda_{e,\beta}^2\right) \boldsymbol{n}_\alpha \otimes \boldsymbol{n}_\beta\right)\left(\sum_{\gamma=1}^{3} \lambda_{e,\gamma}^{-2} \boldsymbol{n}_\gamma \otimes \boldsymbol{n}_\gamma\right);$$

hence

$$\boldsymbol{l}_p = \sum_{\alpha=1}^{3} -\frac{1}{2}\frac{d\lambda_{e,\alpha}^2}{dt} \lambda_{e,\alpha}^{-2} \boldsymbol{n}_\alpha \otimes \boldsymbol{n}_\alpha - \sum_{\substack{\alpha,\beta=1\\ \alpha\neq\beta}}^{3} \frac{1}{2} W_{\alpha\beta} \left(\frac{\lambda_{e,\alpha}^2 - \lambda_{e,\beta}^2}{\lambda_{e,\beta}^2}\right) \boldsymbol{n}_\alpha \otimes \boldsymbol{n}_\beta.$$

## 7.6 INCREMENTAL KINEMATICS

Computationally, as explained in Chapters 1 and 3, a solution to the nonlinear equilibrium equations that are obtained from consideration of geometric or combined geometric and material nonlinearity is found by taking a sufficient number of load increments, which may be related to an artificial or real-time parameter. Consequently, it is now necessary to revisit many of the above equations in which differentiability in "time" was implied and consider equivalent expressions in an incremental setting.

Consider the motion between two arbitrary consecutive increments as shown in Figure 7.5. At increment $n$ the deformation gradient $\boldsymbol{F}_n$ has known elastic and permanent components, $\boldsymbol{F}_{e,n}$ and $\boldsymbol{F}_{p,n}$ respectively, that determine the state of stress at this configuration. In order to proceed to the next configuration at $n+1$, a standard Newton–Raphson process is employed. At each iteration the deformation gradient $\boldsymbol{F}_{n+1}$ can be obtained from the current geometry. However, in order to obtain the corresponding stresses at this increment $n+1$ and thus check for equilibrium, it is first necessary to determine the elastic and permanent components

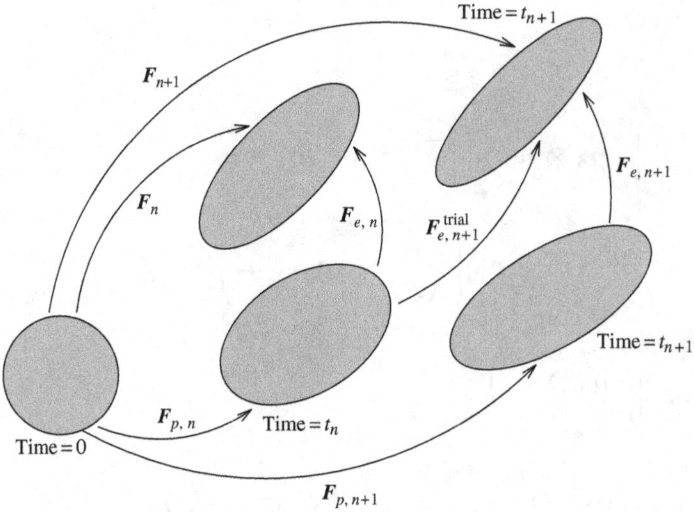


**FIGURE 7.5** Multiplicative decomposition at times $t$ and $t + \Delta t$.

of the current deformation gradient $\boldsymbol{F}_{n+1} = \boldsymbol{F}_{e,n+1}\boldsymbol{F}_{p,n+1}$. Clearly, this is not an obvious process because during the increment an as yet unknown amount of additional inelastic deformation may take place.

It is, however, possible that during the motion from $n$ to $n+1$ no further permanent deformation takes place. Making this preliminary assumption, $\boldsymbol{F}_{p,n+1} = \boldsymbol{F}_{p,n}$ and therefore $\boldsymbol{C}_{p,n+1} = \boldsymbol{C}_{p,n}$, and a trial left Cauchy–Green tensor can be found as

$$\boldsymbol{b}^{\text{trial}}_{\text{e},n+1} = \boldsymbol{F}_{n+1}\boldsymbol{C}^{-1}_{p,n}\boldsymbol{F}^{T}_{n+1}. \tag{7.41}$$

Since $\boldsymbol{C}_p$ is now assumed temporarily constant, the above expression conveniently represents the exact time integration of the first term in Equation (7.12).

Using this trial strain tensor, principal directions and a preliminary state of stress can now be evaluated as

$$\boldsymbol{b}^{\text{trial}}_{\text{e},n+1} = \sum_{\alpha=1}^{3} \left(\lambda^{\text{trial}}_{\text{e},\alpha}\right)^2 \boldsymbol{n}^{\text{trial}}_{\alpha} \otimes \boldsymbol{n}^{\text{trial}}_{\alpha}; \tag{7.42a}$$

$$\boldsymbol{\tau}^{\text{trial}} = \sum_{\alpha=1}^{3} \frac{\partial \Psi}{\partial \ln \lambda^{\text{trial}}_{\text{e},\alpha}} \boldsymbol{n}^{\text{trial}}_{\alpha} \otimes \boldsymbol{n}^{\text{trial}}_{\alpha}; \tag{7.42b}$$

from which the deviatoric Kirchhoff stress tensor is expressed in component form, using Equation (7.32a,b), as

$$\boldsymbol{\tau}'^{\,\text{trial}} = \sum_{\alpha=1}^{3} \tau'^{\,\text{trial}}_{\alpha\alpha} \boldsymbol{n}^{\text{trial}}_{\alpha} \otimes \boldsymbol{n}^{\text{trial}}_{\alpha}; \quad \tau'^{\,\text{trial}}_{\alpha\alpha} = 2\mu \ln \lambda^{\text{trial}}_{\text{e},\alpha} - \frac{2}{3}\mu \ln J^{n+1}. \tag{7.43a,b}$$

Note that $\boldsymbol{\tau}'^{\text{trial}}$ is the state of stress that is obtained directly from $\boldsymbol{b}_{e,n+1}^{\text{trial}}$ under the assumption that no further inelastic strain takes place during the increment. Invariably, this trial state of stress will not be compatible with the assumption that no further permanent deformation takes place during the increment and consequently the state of stress requires modification occasioned by further permanent deformation.

To effect this, stress modification requires a further update to $\boldsymbol{b}_e$ as indicated by the second term in Equation (7.12). As previously explained, this is achieved by integrating the flow rule over the time increment to give this additional change in $\boldsymbol{b}_e$ due to plastic deformation.

In Von Mises plasticity theory, as in many other metal plasticity models, the plastic deformation is isochoric, that is, $\det \boldsymbol{F}_p = 1$; see Equation (7.29). Under such conditions, $J^{n+1} = J_e^{n+1}$ in Equation (7.43a,b), where $J_e^{n+1}$ is the hyperelastic elemental volume ratio, and the hydrostatic pressure $p$ can be evaluated directly from $J^{n+1}$ as in standard hyperelasticity; see, for instance, Equation (6.103).

The time derivative at constant $\boldsymbol{F}$ in the flow rule Equation (7.39a,b) can now be approximated incrementally as

$$\left.\frac{d\ln\lambda_{e,\alpha}}{dt}\right|_{\boldsymbol{F}=\text{const}} \simeq \frac{\ln\lambda_{e,\alpha}^{n+1} - \ln\lambda_{e,\alpha}^{\text{trial}}}{\Delta t}, \tag{7.44}$$

and therefore the flow rule in principal directions can now be expressed in incremental terms as

$$\ln\lambda_{e,\alpha}^{n+1} - \ln\lambda_{e,\alpha}^{\text{trial}} = -\Delta\gamma\nu_{\alpha}^{n+1}; \quad \Delta\gamma = \dot{\gamma}^{n+1}\Delta t. \tag{7.45a,b}$$

The above equation represents a backward Euler time integration of the second term in Equation (7.12).

**Remark 7.4**: Equation (7.45a,b)$_a$ provides a generalization to large strains of the small strain incremental plastic update rule, typically given as

$$\Delta\varepsilon_{p,\alpha} = \Delta\gamma\frac{\partial f}{\partial\sigma_\alpha}. \tag{7.46}$$

To show this, simply note that the small strain additive decomposition $\varepsilon = \varepsilon_e + \varepsilon_p$ taken at increments $n$ and $n+1$ implies

$$\begin{aligned}\Delta\varepsilon_{p,\alpha} &= \left(\varepsilon_{\alpha}^{n+1} - \varepsilon_{e,\alpha}^{n+1}\right) - \varepsilon_{p,\alpha}^{n}\\ &= \left(\varepsilon_{\alpha}^{n+1} - \varepsilon_{p,\alpha}^{n}\right) - \varepsilon_{e,\alpha}^{n+1}\\ &= \varepsilon_{e,\alpha}^{\text{trial}} - \varepsilon_{e,\alpha}^{n+1}.\end{aligned} \tag{7.47}$$

This enables the flow rule to be written as

$$\varepsilon_{e,\alpha}^{n+1} - \varepsilon_{e,\alpha}^{\text{trial}} = -\Delta\gamma\frac{\partial f}{\partial\sigma_\alpha}, \tag{7.48}$$

which, in the case of small strains, coincides with Equation (7.45a,b)$_a$.

This observation can be employed to extend the theory developed in this chapter to more general plasticity models using widely formulated return-mapping algorithms derived in the small strain context.

The evaluation of $\Delta\gamma$, $\nu_\alpha^{n+1}$, and the elastic stretches $\ln \lambda_{e,\alpha}^{n+1}$ that ensure that $\boldsymbol{\tau}$ *radial* lies on the yield surface is known as the, *return-mapping* algorithm and is considered in the following section. Assuming this has been accomplished and the elastic stretches have been obtained, the updated left Cauchy–Green tensor $\boldsymbol{b}_e^{n+1}$ can be determined as

$$\boldsymbol{b}_{e,n+1} = \sum_{\alpha=1}^{3} \left(\lambda_{e,\alpha}^{n+1}\right)^2 \boldsymbol{n}_\alpha^{n+1} \otimes \boldsymbol{n}_\alpha^{n+1}, \tag{7.49}$$

where $\boldsymbol{n}_\alpha^{n+1}$ can be obtained by approximating the time integration of Equation (7.37) as

$$\left.\frac{d\boldsymbol{n}_\alpha}{dt}\right|_{\boldsymbol{F}=\text{const}} = \frac{\boldsymbol{n}_\alpha^{n+1} - \boldsymbol{n}_\alpha^{\text{trial}}}{\Delta t} = 0 \tag{7.50}$$

to give $\boldsymbol{n}_\alpha^{n+1} = \boldsymbol{n}_\alpha^{\text{trial}}$.

The updated plastic right Cauchy–Green tensor and its inverse are finally found from Equation (7.6) as

$$\boldsymbol{C}_{p,n+1} = \boldsymbol{F}_{n+1}^T \boldsymbol{b}_{e,n+1}^{-1} \boldsymbol{F}_{n+1}; \quad \boldsymbol{C}_{p,n+1}^{-1} = \boldsymbol{F}_{n+1}^{-1} \boldsymbol{b}_{e,n+1} \boldsymbol{F}_{n+1}^{-T}. \tag{7.51a,b}$$

### 7.6.1 The Radial Return Mapping

From Equations (7.32a,b), (7.43a,b), and (7.45a,b) the deviatoric Kirchhoff stress in principal directions can be evaluated as follows:‡

$$\begin{aligned} \tau'_{\alpha\alpha} - \tau'^{\text{trial}}_{\alpha\alpha} &= 2\mu(\ln \lambda_{e,\alpha}^{n+1} - \ln \lambda_{e,\alpha}^{\text{trial}}) \\ &= -2\mu\Delta\gamma\nu_\alpha^{n+1} \end{aligned} \tag{7.52}$$

to give

$$\tau'_{\alpha\alpha} = \tau'^{\text{trial}}_{\alpha\alpha} - 2\mu\Delta\gamma\nu_\alpha^{n+1}. \tag{7.53}$$

By virtue of Equation (7.40a,b) the above equation indicates that $\boldsymbol{\tau}'$ is proportional to $\boldsymbol{\tau}'^{\text{trial}}$ and is therefore known as the *radial return mapping* (Figure 7.6). As a

‡ Strictly in (7.52) $\tau'_{\alpha\alpha}$ should be denoted as $(\tau'_{\alpha\alpha})_{n+1}^k$, where $n+1$ refers to the load increment (or pseudo time step) and $k$ refers to the Newton–Raphson iteration that is driving the solution toward equilibrium. However, to simplify, the notation $\tau'_{\alpha\alpha}$ is retained.

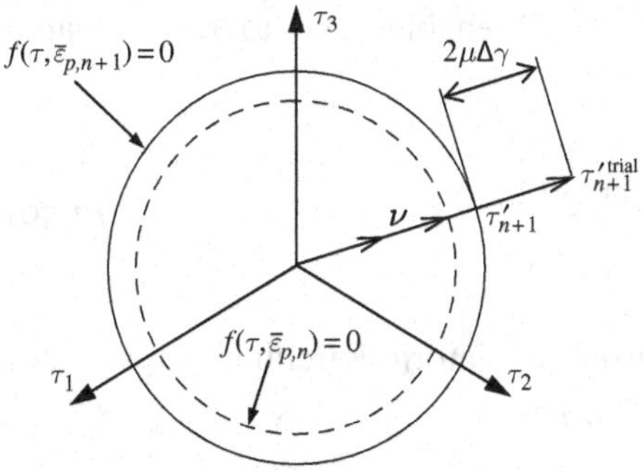


**FIGURE 7.6** Radial return.

consequence of this proportionality, the nondimensional direction vector $\boldsymbol{\nu}$ can be equally obtained from $\boldsymbol{\tau}'^{\,\text{trial}}$, that is,

$$\nu_\alpha^{n+1} = \frac{\tau'^{\,\text{trial}}_{\alpha\alpha}}{\sqrt{\frac{2}{3}}\|\boldsymbol{\tau}'^{\,trial}\|} = \frac{\tau'_{\alpha\alpha}}{\sqrt{\frac{2}{3}}\|\boldsymbol{\tau}'\|}; \quad \text{where, generally,} \quad \|\boldsymbol{\tau}\| = \sqrt{\boldsymbol{\tau} : \boldsymbol{\tau}}, \tag{7.54a,b}$$

and therefore the only unknown in Equation (7.45a,b) is now $\Delta\gamma$.

In order to evaluate $\Delta\gamma$, multiply Equation (7.53) by $\nu_\alpha^{n+1}$ and sum over $\alpha$ to give

$$\sum_{\alpha=1}^{3} \tau'_{\alpha\alpha}\nu_\alpha^{n+1} = \sum_{\alpha=1}^{3} \tau'^{\,\text{trial}}_{\alpha\alpha}\nu_\alpha^{n+1} - 2\mu\Delta\gamma \sum_{\alpha=1}^{3} \left(\nu_\alpha^{n+1}\right)^2. \tag{7.55}$$

Substituting from Equation (7.54a,b), the appropriate expression for $\nu_\alpha^{n+1}$ gives

$$\frac{\|\boldsymbol{\tau}'\|^2}{\sqrt{\frac{2}{3}}\|\boldsymbol{\tau}'\|} = \frac{\|\boldsymbol{\tau}'^{\,\text{trial}}\|^2}{\sqrt{\frac{2}{3}}\|\boldsymbol{\tau}'^{\,\text{trial}}\|} - 2\mu\Delta\gamma\left(\frac{3}{2}\right). \tag{7.56}$$

Simplifying the first two terms of the above equation together with the enforcement of the yield condition (7.20a,b) gives

$$\sqrt{\frac{3}{2}\boldsymbol{\tau}' : \boldsymbol{\tau}'} = \sqrt{\frac{3}{2}\boldsymbol{\tau}'^{\,\text{trial}} : \boldsymbol{\tau}'^{\,\text{trial}}} - 3\mu\Delta\gamma = (\bar{\tau}_y^0 + H\bar{\varepsilon}_{p,n} + H\Delta\bar{\varepsilon}_p). \tag{7.57}$$

In addition, integrating Equation (7.27) in time gives

$$\Delta\bar{\varepsilon}_p = \Delta\gamma. \tag{7.58}$$

Substituting this expression into Equation (7.57) enables $\Delta\gamma$ to be evaluated explicitly as

$$\Delta\gamma = \begin{cases} \dfrac{f(\boldsymbol{\tau}^{\text{trial}}, \bar{\varepsilon}_{p,n})}{3\mu + H} & \text{if } f(\boldsymbol{\tau}^{\text{trial}}, \bar{\varepsilon}_{p,n}) > 0; \\ 0 & \text{if } f(\boldsymbol{\tau}^{\text{trial}}, \bar{\varepsilon}_{p,n}) \leq 0. \end{cases} \tag{7.59}$$

Once the value of $\Delta\gamma$ is known, the current deviatoric Kirchhoff stresses are easily obtained by re-expressing Equation (7.53) as

$$\tau'_{\alpha\alpha} = \left(1 - \frac{2\mu\Delta\gamma}{\sqrt{2/3}\|\boldsymbol{\tau}'^{\,\text{trial}}\|}\right)\tau'^{\,\text{trial}}_{\alpha\alpha}. \tag{7.60}$$

In order to be able to move on to the next increment, it is necessary to record the current state of permanent or plastic deformation. In particular, the new value of the Von Mises equivalent plastic strain emerges from (7.58) as

$$\bar{\varepsilon}_{p,n+1} = \bar{\varepsilon}_{p,n} + \Delta\gamma. \tag{7.61}$$

### 7.6.2 Algorithmic Tangent Modulus

The derivation of the deviatoric component of the tangent modulus for this constitutive model follows the same process employed in Section 6.6, with the only difference being that the fixed reference configuration is now the unloaded state at increment $n$ rather than the initial configuration. Similar algebra thus leads to

$$\begin{aligned} \hat{\boldsymbol{c}} = &\sum_{\alpha,\beta=1}^{3} \frac{1}{J}\frac{\partial \tau'_{\alpha\alpha}}{\partial \ln \lambda^{\text{trial}}_{e,\beta}}\boldsymbol{\eta}_{\alpha\alpha\beta\beta} - \sum_{\alpha=1}^{3} 2\sigma'_{\alpha\alpha}\boldsymbol{\eta}_{\alpha\alpha\alpha\alpha} \\ &+ \sum_{\substack{\alpha,\beta=1\\ \alpha\neq\beta}}^{3} \frac{\sigma'_{\alpha\alpha}\left(\lambda^{\text{trial}}_{e,\beta}\right)^2 - \sigma'_{\beta\beta}\left(\lambda^{\text{trial}}_{e,\alpha}\right)^2}{\left(\lambda^{\text{trial}}_{e,\alpha}\right)^2 - \left(\lambda^{\text{trial}}_{e,\beta}\right)^2}\left(\boldsymbol{\eta}_{\alpha\beta\alpha\beta} + \boldsymbol{\eta}_{\alpha\beta\beta\alpha}\right), \end{aligned} \tag{7.62a}$$

where for any $\alpha, \beta, \gamma, \delta$ the fourth-order tensor $\boldsymbol{\eta}_{\alpha\beta\gamma\delta}$ is defined as

$$\boldsymbol{\eta}_{\alpha\beta\gamma\delta} = \boldsymbol{n}_\alpha \otimes \boldsymbol{n}_\beta \otimes \boldsymbol{n}_\gamma \otimes \boldsymbol{n}_\delta. \tag{7.62b}$$

In order to implement this equation the derivatives in the first term must first be evaluated from Equation (7.60) as

$$\begin{aligned} \frac{\partial \tau'_{\alpha\alpha}}{\partial \ln \lambda^{\text{trial}}_{e,\beta}} = &\left(1 - \frac{2\mu\,\Delta\gamma}{\sqrt{2/3}\,\|\boldsymbol{\tau}'^{\,\text{trial}}\|}\right)\frac{\partial \tau'^{\,\text{trial}}_{\alpha\alpha}}{\partial \ln \lambda^{\text{trial}}_{e,\beta}} \\ &- \frac{2\mu\tau'^{\,\text{trial}}_{\alpha\alpha}}{\sqrt{2/3}}\frac{\partial}{\partial \ln \lambda^{\text{trial}}_{e,\beta}}\left(\frac{\Delta\gamma}{\|\boldsymbol{\tau}'^{\,\text{trial}}\|}\right). \end{aligned} \tag{7.63}$$

If the material is in the elastic regime, $\Delta\gamma = 0$; hence the second term in Equation (7.63) vanishes and Equation (7.43a,b)$_b$ shows that the derivatives of the trial stresses are given by the elastic modulus as

$$\frac{\partial \tau'^{\text{trial}}_{\alpha\alpha}}{\partial \ln \lambda^{\text{trial}}_{e,\beta}} = 2\mu\delta_{\alpha\beta} - \frac{2}{3}\mu. \tag{7.64}$$

In the elasto-plastic regime, $\Delta\gamma \neq 0$, and the second derivative of Equation (7.63) must be evaluated with the help of Equation (7.59) and the chain rule as

$$\frac{\partial}{\partial \ln \lambda^{\text{trial}}_{e,\beta}}\left(\frac{\Delta\gamma}{\|\boldsymbol{\tau}'^{\text{trial}}\|}\right) = \frac{1}{\|\boldsymbol{\tau}'^{\text{trial}}\|}\left(\frac{\sqrt{3/2}}{3\mu + H} - \frac{\Delta\gamma}{\|\boldsymbol{\tau}'^{\text{trial}}\|}\right)\frac{\partial \|\boldsymbol{\tau}'^{\text{trial}}\|}{\partial \ln \lambda^{\text{trial}}_{e,\beta}}, \tag{7.65}$$

where simple algebra shows that

$$\begin{aligned}\frac{\partial \|\boldsymbol{\tau}'^{\text{trial}}\|}{\partial \ln \lambda^{\text{trial}}_{e,\beta}} &= \frac{1}{\|\boldsymbol{\tau}'^{\text{trial}}\|}\sum_{\alpha=1}^{3} \tau'^{\text{trial}}_{\alpha\alpha}\frac{\partial \tau'^{\text{trial}}_{\alpha\alpha}}{\partial \ln \lambda^{\text{trial}}_{e,\beta}} \\ &= \frac{1}{\|\boldsymbol{\tau}'^{\text{trial}}\|}\sum_{\alpha=1}^{3} \tau'^{\text{trial}}_{\alpha\alpha}\left(2\mu\delta_{\alpha\beta} - \frac{2}{3}\mu\right) \\ &= 2\mu\sqrt{2/3}\,\nu_\beta. \end{aligned} \tag{7.66}$$

Finally, combining Equation (7.63) and Equation (7.66) gives the derivatives needed to evaluate the consistent algorithmic[§] tangent modulus as

$$\begin{aligned}\frac{\partial \tau'_{\alpha\alpha}}{\partial \ln \lambda^{\text{trial}}_{e,\beta}} &= \left(1 - \frac{2\mu\,\Delta\gamma}{\sqrt{2/3}\,\|\boldsymbol{\tau}'^{\text{trial}}\|}\right)\left(2\mu\delta_{\alpha\beta} - \frac{2}{3}\mu\right) \\ &\quad - 2\mu\,\nu_\alpha\nu_\beta\left(\frac{2\mu}{3\mu + H} - \frac{2\mu\sqrt{2/3}\,\Delta\gamma}{\|\boldsymbol{\tau}'^{\text{trial}}\|}\right).\end{aligned} \tag{7.67}$$

The algorithmic procedure for implementing rate-independent Von Mises plasticity with isotropic hardening is given in Boxes 7.1 and 7.2.

## 7.7 TWO-DIMENSIONAL CASES

As explained in Section 6.6.7, plane strain is defined by the fact that the stretch in the third direction $\lambda_3 = 1$. Insofar as $\lambda_3 = \lambda_{e,3}\lambda_{p,3}$ it is clear that neither $\lambda_{e,3}$ nor $\lambda_{p,3}$ need be equal to unity. This implies that the trial Kirchhoff stress given by

§ This is alternatively called the *consistent tangent modulus* insofar as its derivation is consistent with the chosen method of returning the trial stress to the yield surface; see also Section 3.5.6.

Equation (7.43a,b) must be evaluated in three dimensions. However, the trial left Cauchy–Green tensor given by Equation (7.41) assumes a simpler form as

$$\boldsymbol{b}^{\text{trial}}_{e,n+1} = \begin{bmatrix} \boldsymbol{F}_{n+1(2\times 2)} & \begin{matrix} 0 \\ 0 \end{matrix} \\ \begin{matrix} 0 & 0 \end{matrix} & 1 \end{bmatrix} \begin{bmatrix} \boldsymbol{C}^{-1}_{p,n(2\times 2)} & \begin{matrix} 0 \\ 0 \end{matrix} \\ \begin{matrix} 0 & 0 \end{matrix} & C^{-1}_{p,n(3,3)} \end{bmatrix} \begin{bmatrix} \boldsymbol{F}^{T}_{n+1(2\times 2)} & \begin{matrix} 0 \\ 0 \end{matrix} \\ \begin{matrix} 0 & 0 \end{matrix} & 1 \end{bmatrix}. \tag{7.68}$$

From Equations (7.3a,b,c) and (7.29), $\det \boldsymbol{C}^{-1}_{p,n} = 1$; consequently, $C^{-1}_{p,n(3,3)}$ can easily be found as

$$C^{-1}_{p,n(3,3)} = \frac{1}{\det \boldsymbol{C}^{-1}_{p,n(2\times 2)}}. \tag{7.69}$$

Observe that since $\lambda_3 \neq 0$ the calculation of the principal directions of $\boldsymbol{b}^{\text{trial}}_{e,n+1}$, the principal Kirchhoff stresses, and the evaluation of the yield function proceed as for the fully three-dimensional case with other calculations remaining two-dimensional.

For the case of plane stress the situation is not so simple, as the return algorithm is no longer radial. This is because the yield surface becomes an ellipse given by the intersection of a cylinder in principal directions and the horizontal plane $\tau_3 = 0$. Plane stress behavior is not pursued further; however, details can be found in Simo and Hughes (1997).

**BOX 7.1: Algorithm for Rate-Independent Von Mises Plasticity with Isotropic Hardening**

1. For load increment $n+1$
2. For Newton–Raphson iteration $k$
3. Find $\boldsymbol{F}^{k}_{n+1}$ and given $\boldsymbol{C}^{-1}_{p,n}$ and $\bar{\varepsilon}_{p,n}$, ($k$ now implied)

| | |
|---|---|
| $J_{n+1} = \det \boldsymbol{F}_{n+1}$ | Jacobean |
| $p = \left(\dfrac{\kappa \ln J_{n+1}}{J_{n+1}}\right) \quad \left[\approx \kappa \dfrac{\ln (v^e/V^e)}{(v^e/V^e)}\right]$ | Pressure (mean stress) [mean dilatation, Section 8.6.5] |
| $\boldsymbol{b}^{\text{trial}}_{e,n+1} = \boldsymbol{F}_{n+1}\boldsymbol{C}^{-1}_{p,n}\boldsymbol{F}^{T}_{n+1}$ | Trial left Cauchy–Green tensor |
| $\boldsymbol{b}^{\text{trial}}_{e,n+1} = \sum_{\alpha=1}^{3} \left(\lambda^{\text{trial}}_{e,\alpha}\right)^2 \boldsymbol{n}^{\text{trial}}_{\alpha} \otimes \boldsymbol{n}^{\text{trial}}_{\alpha}$ | Trial stretches and principal directions |
| $\tau'^{\text{trial}}_{\alpha\alpha} = 2\mu \ln \lambda^{\text{trial}}_{e,\alpha} - \dfrac{2}{3}\mu \ln J^{n+1}$ | Trial Kirchhoff stress |

*(continued)*

**BOX 7.1:** *(cont.)*

**4.** Check for yielding

$$f(\boldsymbol{\tau}^{\text{trial}}, \bar{\varepsilon}_{p,n}) \leq 0$$ Yield function

**5.** Radial return algorithm

**If** $f > 0$

**Then**

$$\nu_\alpha^{n+1} = \frac{\tau'^{\,\text{trial}}_{\alpha\alpha}}{\sqrt{\frac{2}{3}}\|\boldsymbol{\tau}'^{\,\text{trial}}\|}$$ Direction vector

$$\Delta\gamma = \frac{f(\boldsymbol{\tau}^{\text{trial}}, \bar{\varepsilon}_{p,n})}{3\mu + H}$$ Incremental plastic multiplier

**Else**

$$\Delta\gamma = 0$$ Elastic response

$$\ln \lambda_{e,\alpha}^{n+1} = \ln \lambda_{e,\alpha}^{\text{trial}} - \Delta\gamma \nu_\alpha^{n+1}$$ Logarithmic elastic stretch

$$\tau'_{\alpha\alpha} = \left(1 - \frac{2\mu\Delta\gamma}{\sqrt{2/3}\|\boldsymbol{\tau}'^{\,\text{trial}}\|}\right)\tau'^{\,\text{trial}}_{\alpha\alpha}$$ Return map

**6.** Update inverse of elastic left Cauchy–Green tensor

$$\boldsymbol{b}_{e,n+1} = \sum_{\alpha=1}^{3} \left(\lambda_{e,\alpha}^{n+1}\right)^2 \boldsymbol{n}_\alpha^{n+1} \otimes \boldsymbol{n}_\alpha^{n+1}$$

$$\boldsymbol{n}_\alpha = \boldsymbol{n}_\alpha^{\text{trial}}$$

**7.** Update stress

$$\sigma'_{\alpha\alpha} = \frac{1}{J_{n+1}} \tau'_{\alpha\alpha}$$ Principal deviatoric Cauchy stress

$$\sigma_{\alpha\alpha} = \sigma'_{\alpha\alpha} + p$$ Principal Cauchy stress

$$\boldsymbol{\sigma} = \sum_{\alpha=1}^{3} \sigma_{\alpha\alpha} \boldsymbol{n}_\alpha^{n+1} \otimes \boldsymbol{n}_\alpha^{n+1}$$ Cauchy stress

**8.** Update tangent modulus

$$\hat{\mathcal{c}} = \sum_{\alpha,\beta=1}^{3} \frac{1}{J} \frac{\partial \tau'_{\alpha\alpha}}{\partial \ln \lambda_{e,\beta}^{\text{trial}}} \boldsymbol{\eta}_{\alpha\alpha\beta\beta} - \sum_{\alpha=1}^{3} 2\sigma'_{\alpha\alpha} \boldsymbol{\eta}_{\alpha\alpha\alpha\alpha}$$

$$+ \sum_{\substack{\alpha,\beta=1 \\ \alpha\neq\beta}}^{3} \frac{\sigma'_{\alpha\alpha}\left(\lambda_{e,\beta}^{\text{trial}}\right)^2 - \sigma'_{\beta\beta}\left(\lambda_{e,\alpha}^{\text{trial}}\right)^2}{\left(\lambda_{e,\alpha}^{\text{trial}}\right)^2 - \left(\lambda_{e,\beta}^{\text{trial}}\right)^2} \left(\boldsymbol{\eta}_{\alpha\beta\alpha\beta} + \boldsymbol{\eta}_{\alpha\beta\beta\alpha}\right)$$

*(continued)*

**BOX 7.1:** *(cont.)*

**9.** Update state variables

$$C^{-1}_{p,n+1} = F^{-1}_{n+1} b_{e,n+1} F^{-T}_{n+1}$$ Update plastic right Cauchy–Green tensor

$$\bar{\varepsilon}_{p,n+1} = \bar{\varepsilon}_{p,n} + \Delta\gamma$$ Update equivalent plastic strain

**10.** Check equilibrium

**If R** $>$ tolerance — See **Box 1.1**

**Then**

$k \Leftarrow k+1$ GO TO 2

**Else**

$n \Leftarrow n+1$ GO TO 1

**BOX 7.2: Tangent Modulus**

1. Update tangent modulus

$$\hat{\mathcal{C}} = \sum_{\alpha,\beta=1}^{3} \frac{1}{J} \mathcal{C}_{\alpha\beta}\, \boldsymbol{\eta}_{\alpha\alpha\beta\beta} - \sum_{\alpha=1}^{3} 2\sigma'_{\alpha\alpha}\, \boldsymbol{\eta}_{\alpha\alpha\alpha\alpha} + \sum_{\substack{\alpha,\beta=1\\ \alpha\neq\beta}}^{3} \frac{\sigma'_{\alpha\alpha}\left(\lambda^{\text{trial}}_{e,\beta}\right)^2 - \sigma'_{\beta\beta}\left(\lambda^{\text{trial}}_{e,\alpha}\right)^2}{\left(\lambda^{\text{trial}}_{e,\alpha}\right)^2 - \left(\lambda^{\text{trial}}_{e,\beta}\right)^2}\left(\boldsymbol{\eta}_{\alpha\beta\alpha\beta} + \boldsymbol{\eta}_{\alpha\beta\beta\alpha}\right)$$

**If** $f > 0$

**Then**

$$\mathcal{C}_{\alpha\beta} = \left(1 - \frac{2\mu\,\Delta\gamma}{\sqrt{2/3}\,\|\boldsymbol{\tau}'^{\,\text{trial}}\|}\right)\left(2\mu\delta_{\alpha\beta} - \frac{2}{3}\mu\right) - 2\mu\,\nu_\alpha\nu_\beta\left(\frac{2\mu}{3\mu + H} - \frac{2\mu\sqrt{2/3}\,\Delta\gamma}{\|\boldsymbol{\tau}'^{\,\text{trial}}\|}\right)$$ Plastic response

**Else**

$$\mathcal{C}_{\alpha\beta} = 2\mu\delta_{\alpha\beta} - \frac{2}{3}\mu$$ Elastic response

## Exercises

1. Using the multiplicative decomposition $\boldsymbol{F} = \boldsymbol{F}_e\boldsymbol{F}_p$ and the expressions $\boldsymbol{l} = \dot{\boldsymbol{F}}\boldsymbol{F}^{-1}$ and $\boldsymbol{l}_e = \dot{\boldsymbol{F}}_e\boldsymbol{F}_e^{-1}$, show that the plastic rate of deformation $\boldsymbol{l}_p$ can be obtained as

$$l_p = F_e \dot{F}_p F_p^{-1} F_e^{-1}.$$

2. Starting from the expression $F_e = F_e(F, F_p)$ and using a decomposition similar to that shown in Equation (7.12), show that

$$l_p = -\left.\frac{dF_e}{dt}\right|_{F=\text{const}} F_e^{-1}.$$

3. Use the equation obtained in the previous exercise to derive the flow rule in principal directions, following a procedure similar to that described in Section 7.5.
4. Consider a material in which the internal elastic energy is expressed as $\Psi(C, C_p)$. Show that the plastic dissipation rate can be expressed as

$$\dot{w}_p = -\frac{\partial \Psi}{\partial C_P} : \dot{C}_p.$$

Starting from this expression and using the principle of maximum plastic dissipation, show that if the yield surface is defined in terms of $C$ and $C_p$ by $f(C, C_p) \leq 0$ then the flow rule becomes

$$\frac{\partial^2 \Psi}{\partial C \partial C_P} : \dot{C}_p = -\gamma \frac{\partial f}{\partial C}.$$

CHAPTER EIGHT

# LINEARIZED EQUILIBRIUM EQUATIONS

## 8.1 INTRODUCTION

The virtual work representation of the equilibrium equation presented in Section 5.4 is nonlinear with respect to both the geometry and the material. For a given material and loading conditions, its solution is given by a deformed configuration $\phi$ in a state of equilibrium. In order to obtain this equilibrium position using a Newton–Raphson iterative solution, it is necessary to linearize the equilibrium equations using the general directional derivative procedure discussed in Chapter 2. Two approaches are in common use: some authors prefer to discretize the equilibrium equations and then linearize with respect to the nodal positions, whereas others prefer to linearize the virtual work statement and then discretize. The latter approach is more suitable for solid continua and will be adopted herein, although in some cases where a nonstandard discretization is used this approach may not be possible.

## 8.2 LINEARIZATION AND THE NEWTON–RAPHSON PROCESS

The Principle of Virtual Work has been expressed in Chapter 5 in terms of the virtual velocity in the spatial configuration as

$$\delta W(\phi, \delta \boldsymbol{v}) = \int_v \boldsymbol{\sigma} : \delta \boldsymbol{d}\, dv - \int_v \boldsymbol{f} \cdot \delta \boldsymbol{v}\, dv - \int_{\partial v} \boldsymbol{t} \cdot \delta \boldsymbol{v}\, da = 0, \tag{8.1}$$

or alternatively in the material configuration as

$$\delta W(\phi, \delta \boldsymbol{v}) = \int_V \boldsymbol{S} : \delta \dot{\boldsymbol{E}}\, dV - \int_V \boldsymbol{f}_0 \cdot \delta \boldsymbol{v}\, dV - \int_{\partial V} \boldsymbol{t}_0 \cdot \delta \boldsymbol{v}\, dA. \tag{8.2}$$

Considering a trial solution $\phi_k$, either of the above two equations can be linearized in the direction of an increment $\boldsymbol{u}$ in $\phi_k$ as

$$\delta W(\phi_k, \delta\boldsymbol{v}) + D\delta W(\phi_k, \delta\boldsymbol{v})[\boldsymbol{u}] = 0. \tag{8.3}$$

Consequently, it is necessary to find the directional derivative of the virtual work equation at $\phi_k$ in the direction of $\boldsymbol{u}$. It is worth pausing first to ask what this means! To begin, a virtual velocity $\delta\boldsymbol{v}(\phi(\boldsymbol{X}))$ is associated with every particle labeled $\boldsymbol{X}$ in the body, and it is not allowed to alter during the incremental change $\boldsymbol{u}(\boldsymbol{x})$ (Figure 8.1). At a trial solution position $\phi_k$, $\delta W(\phi_k, \delta\boldsymbol{v})$ will have some value, probably not equal to zero as required for equilibrium. The directional derivative $D\delta W(\phi_k, \delta\boldsymbol{v})[\boldsymbol{u}]$ is simply the change in $\delta W$ due to $\phi_k$ changing to $\phi_k + \boldsymbol{u}$. Since $\delta\boldsymbol{v}$ remains constant during this change, the directional derivative must represent the change in the internal forces due to $\boldsymbol{u}$ (assuming that external forces are constant). This is precisely what is needed in the Newton–Raphson procedure to adjust the configuration $\phi_k$ in order to bring the internal forces into equilibrium with the external forces. Hence, the directional derivative of the virtual work equation will be the source of the tangent matrix.

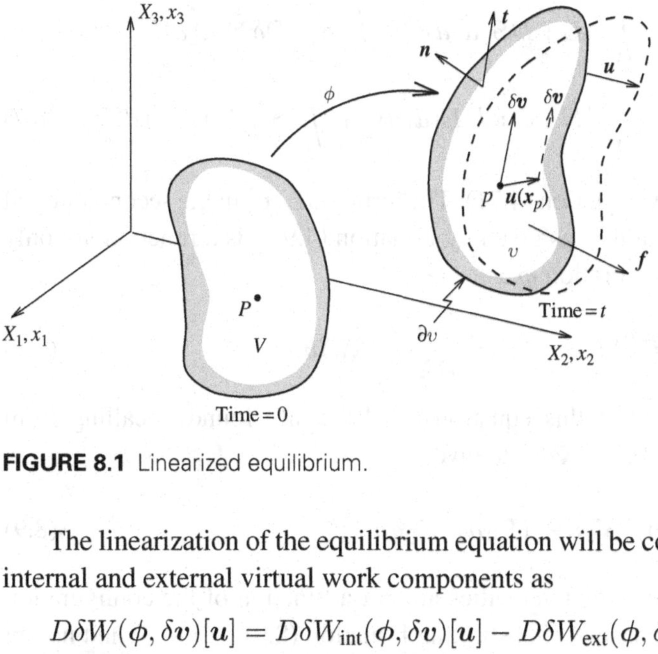


**FIGURE 8.1** Linearized equilibrium.

The linearization of the equilibrium equation will be considered in terms of the internal and external virtual work components as

$$D\delta W(\phi, \delta\boldsymbol{v})[\boldsymbol{u}] = D\delta W_{\text{int}}(\phi, \delta\boldsymbol{v})[\boldsymbol{u}] - D\delta W_{\text{ext}}(\phi, \delta\boldsymbol{v})[\boldsymbol{u}], \tag{8.4}$$

where

$$\delta W_{\text{int}}(\phi, \delta\boldsymbol{v}) = \int_v \boldsymbol{\sigma} : \delta\boldsymbol{d}\, dv; \tag{8.5a}$$

$$\delta W_{\text{ext}}(\phi, \delta\boldsymbol{v}) = \int_v \boldsymbol{f} \cdot \delta\boldsymbol{v}\, dv + \int_{\partial v} \boldsymbol{t} \cdot \delta\boldsymbol{v}\, da. \tag{8.5b}$$

## 8.3 LAGRANGIAN LINEARIZED INTERNAL VIRTUAL WORK

Although the eventual discretization of the linearized equilibrium equations will be formulated only for the Eulerian case, it is nevertheless convenient to perform the linearization with respect to the material description of the equilibrium equations, simply because the initial elemental volume $dV$ is constant during the linearization. This will then be transformed by a push-forward operation to the spatial configuration. Recall from Equation (5.43) that the internal virtual work can be expressed in a Lagrangian form as

$$\delta W_{\text{int}}(\boldsymbol{\phi}, \delta \boldsymbol{v}) = \int_V \boldsymbol{S} : \delta \dot{\boldsymbol{E}} \, dV. \tag{8.6}$$

Using the product rule for directional derivatives and the definition of the material elasticity tensor, the directional derivative is obtained as

$$\begin{aligned} D\delta W_{\text{int}}(\boldsymbol{\phi}, \delta \boldsymbol{v})[\boldsymbol{u}] &= \int_V D(\delta \dot{\boldsymbol{E}} : \boldsymbol{S})[\boldsymbol{u}] \, dV \\ &= \int_V \delta \dot{\boldsymbol{E}} : D\boldsymbol{S}[\boldsymbol{u}] \, dV + \int_V \boldsymbol{S} : D\delta \dot{\boldsymbol{E}}[\boldsymbol{u}] \, dV \\ &= \int_V \delta \dot{\boldsymbol{E}} : \boldsymbol{\mathcal{C}} : D\boldsymbol{E}[\boldsymbol{u}] \, dV + \int_V \boldsymbol{S} : D\delta \dot{\boldsymbol{E}}[\boldsymbol{u}] \, dV, \end{aligned} \tag{8.7}$$

where $D\boldsymbol{E}[\boldsymbol{u}]$ is given by Equation (4.71). The term $D\delta\dot{\boldsymbol{E}}[\boldsymbol{u}]$ in the second integral emerges from the fact that $\delta\dot{\boldsymbol{E}}$, as given by Equation (4.97), is a function not only of $\delta\boldsymbol{v}$ but also of the configuration $\boldsymbol{\phi}$ as

$$\delta \dot{\boldsymbol{E}} = \frac{1}{2}(\delta \dot{\boldsymbol{F}}^T \boldsymbol{F} + \boldsymbol{F}^T \delta \dot{\boldsymbol{F}}); \quad \delta \dot{\boldsymbol{F}} = \frac{\partial \delta \boldsymbol{v}}{\partial \boldsymbol{X}} = \boldsymbol{\nabla}_0 \delta \boldsymbol{v}. \tag{8.8}$$

The directional derivative of this equation can be easily found, recalling from Equation (4.70) that $D\boldsymbol{F}[\boldsymbol{u}] = \boldsymbol{\nabla}_0 \boldsymbol{u}$, to give

$$D\delta \dot{\boldsymbol{E}}[\boldsymbol{u}] = \frac{1}{2}[(\boldsymbol{\nabla}_0 \delta \boldsymbol{v})^T \boldsymbol{\nabla}_0 \boldsymbol{u} + (\boldsymbol{\nabla}_0 \boldsymbol{u})^T \boldsymbol{\nabla}_0 \delta \boldsymbol{v}]. \tag{8.9}$$

Observe that because the virtual velocities are not a function of the configuration the term $\boldsymbol{\nabla}_0 \delta\boldsymbol{v}$ remains constant. Substituting Equation (8.9) into (8.7) and noting the symmetry of $\boldsymbol{S}$ gives the material or Lagrangian linearized Principle of Virtual Work as

$$D\delta W_{\text{int}}(\boldsymbol{\phi}, \delta \boldsymbol{v})[\boldsymbol{u}] = \int_V \delta \dot{\boldsymbol{E}} : \boldsymbol{\mathcal{C}} : D\boldsymbol{E}[\boldsymbol{u}] \, dV + \int_V \boldsymbol{S} : [(\boldsymbol{\nabla}_0 \boldsymbol{u})^T \boldsymbol{\nabla}_0 \delta \boldsymbol{v}] \, dV. \tag{8.10}$$

Given the relationship between the directional and time derivatives as explained in Section 4.11.3, note that $\delta\dot{\boldsymbol{E}}$ can be expressed as $D\boldsymbol{E}[\delta\boldsymbol{v}]$, which enables Equation (8.10) to be written in a more obviously symmetric form as

$$D\delta W_{\text{int}}(\boldsymbol{\phi}, \delta\boldsymbol{v})[\boldsymbol{u}] = \int_V D\boldsymbol{E}[\delta\boldsymbol{v}] : \boldsymbol{\mathcal{C}} : D\boldsymbol{E}[\boldsymbol{u}]\, dV + \int_V \boldsymbol{S} : [(\boldsymbol{\nabla}_0\boldsymbol{u})^T \boldsymbol{\nabla}_0\delta\boldsymbol{v}]\, dV. \tag{8.11}$$

## 8.4 EULERIAN LINEARIZED INTERNAL VIRTUAL WORK

Equation (8.11) can perfectly well be used and may indeed be more appropriate in some cases for the development of the tangent stiffness matrix. Nevertheless, much simplification can be gained by employing the equivalent spatial alternative to give the same tangent matrix. To this end, the materially-based terms in Equation (8.11) must be expressed in terms of spatially-based quantities. These relationships are manifest in the following pull-back and push-forward operations:

$$D\boldsymbol{E}[\boldsymbol{u}] = \boldsymbol{\phi}_*^{-1}[\boldsymbol{\varepsilon}] = \boldsymbol{F}^T\boldsymbol{\varepsilon}\boldsymbol{F}; \quad 2\boldsymbol{\varepsilon} = \boldsymbol{\nabla}\boldsymbol{u} + (\boldsymbol{\nabla}\boldsymbol{u})^T; \tag{8.12a}$$

$$D\boldsymbol{E}[\delta\boldsymbol{v}] = \boldsymbol{\phi}_*^{-1}[\delta\boldsymbol{d}] = \boldsymbol{F}^T\delta\boldsymbol{d}\,\boldsymbol{F}; \quad 2\delta\boldsymbol{d} = \boldsymbol{\nabla}\delta\boldsymbol{v} + (\boldsymbol{\nabla}\delta\boldsymbol{v})^T; \tag{8.12b}$$

$$J\boldsymbol{\sigma} = \boldsymbol{\phi}_*[\boldsymbol{S}] = \boldsymbol{F}\boldsymbol{S}\boldsymbol{F}^T; \tag{8.12c}$$

$$J\boldsymbol{c} = \boldsymbol{\phi}_*[\boldsymbol{\mathcal{C}}]; \quad J c_{ijkl} = \sum_{I,J,K,L=1}^{3} F_{iI}F_{jJ}F_{kK}F_{lL}\mathcal{C}_{IJKL}; \tag{8.12d}$$

$$JdV = dv. \tag{8.12e}$$

With the help of these transformations it can be shown that the first integrand in Equation (8.11) can be re-expressed in a spatial framework as

$$D\boldsymbol{E}[\delta\boldsymbol{v}] : \boldsymbol{\mathcal{C}} : D\boldsymbol{E}[\boldsymbol{u}]dV = \delta\boldsymbol{d} : \boldsymbol{c} : \boldsymbol{\varepsilon} dv. \tag{8.13}$$

Additionally, the gradient with respect to the initial particle coordinates appearing in the second integral in Equation (8.11) can be related to the spatial gradient using the chain rule (see Equation (4.69)) to give

$$\boldsymbol{\nabla}_0\boldsymbol{u} = (\boldsymbol{\nabla}\boldsymbol{u})\boldsymbol{F}; \tag{8.14a}$$

$$\boldsymbol{\nabla}_0\delta\boldsymbol{v} = (\boldsymbol{\nabla}\delta\boldsymbol{v})\boldsymbol{F}. \tag{8.14b}$$

Substituting these expressions into the second term of Equation (8.11) and using Equation (8.12c) for the Cauchy and second Piola–Kirchhoff stresses reveals that the second integrand can be rewritten as

$$\boldsymbol{S} : [(\boldsymbol{\nabla}_0\boldsymbol{u})^T(\boldsymbol{\nabla}_0\delta\boldsymbol{v})]dV = \boldsymbol{\sigma} : [(\boldsymbol{\nabla}\boldsymbol{u})^T(\boldsymbol{\nabla}\delta\boldsymbol{v})]dv. \tag{8.15}$$

Finally, Equation (8.11) can be rewritten using Equations (8.13) and (8.15) to give the spatial or Eulerian linearized equilibrium equations as

$$D\delta W_{\text{int}}(\phi, \delta \boldsymbol{v})[\boldsymbol{u}] = \int_v \delta \boldsymbol{d} : \mathcal{c} : \boldsymbol{\varepsilon}\, dv + \int_v \boldsymbol{\sigma} : [(\boldsymbol{\nabla u})^T \boldsymbol{\nabla} \delta \boldsymbol{v}] dv. \tag{8.16}$$

This equation will be the basis for the Eulerian or spatial evaluation of the tangent matrix. Observe that the functional relationship between $\delta \boldsymbol{d}$ and $\delta \boldsymbol{v}$ is identical to that between $\boldsymbol{\varepsilon}$ and $\boldsymbol{u}$. This together with the symmetry of $\mathcal{c}$ and $\boldsymbol{\sigma}$ implies that the terms $\boldsymbol{u}$ and $\delta \boldsymbol{v}$ can be interchanged in this equation without altering the result. Consequently, the linearized virtual work equation is symmetric in $\delta \boldsymbol{v}$ and $\boldsymbol{u}$, that is,

$$D\delta W_{\text{int}}(\phi, \delta \boldsymbol{v})[\boldsymbol{u}] = D\delta W_{\text{int}}(\phi, \boldsymbol{u})[\delta \boldsymbol{v}]. \tag{8.17}$$

This symmetry will, upon discretization, yield a symmetric tangent stiffness matrix.

**EXAMPLE 8.1: Proof of Equation (8.13)**

In order to prove Equation (8.13), first rewrite (8.12a–b) in indicial notation as

$$DE_{IJ}[\delta \boldsymbol{v}] = \sum_{i,j} F_{iI} \delta d_{ij} F_{jJ}; \qquad DE_{KL}[\boldsymbol{u}] = \sum_{k,l} F_{kK} \varepsilon_{kl} F_{lL}.$$

With the help of these expressions and Equation (8.12d), the left-hand side of Equation (8.13) can be manipulated to give

$$\begin{aligned}
&D\boldsymbol{E}[\delta \boldsymbol{v}] : \boldsymbol{\mathcal{C}} : D\boldsymbol{E}[\boldsymbol{u}]\, dV \\
&\quad = \sum_{I,J,K,L} DE_{IJ}[\delta \boldsymbol{v}] \mathcal{C}_{IJKL} DE_{KL}[\boldsymbol{u}]\, dV \\
&\quad = \sum_{I,J,K,L} \Big( \sum_{i,j} F_{iI} \delta d_{ij} F_{jJ} \Big) \mathcal{C}_{IJKL} \Big( \sum_{k,l} F_{kK} \varepsilon_{kl} F_{lL} \Big) J^{-1} dv \\
&\quad = \sum_{i,j,k,l} \delta d_{ij} \Big( \sum_{I,J,K,L} F_{iI} F_{jJ} F_{kK} F_{lL} \mathcal{C}_{IJKL} J^{-1} \Big) \varepsilon_{kl}\, dv \\
&\quad = \sum_{i,j,k,l} \delta d_{ij}\; \mathcal{c}_{ijkl} \varepsilon_{kl}\, dv \\
&\quad = \delta \boldsymbol{d} : \mathcal{c} : \boldsymbol{\varepsilon}\, dv.
\end{aligned}$$

## 8.5 LINEARIZED EXTERNAL VIRTUAL WORK

The external virtual work has contributions from body forces $\boldsymbol{f}$ and surface tractions $\boldsymbol{t}$. These two cases will now be considered separately.

### 8.5.1 Body Forces

The most common example of a body force is self-weight or gravity loading, in which case $\boldsymbol{f} = \rho \boldsymbol{g}$, where $\rho$ is the current density and $\boldsymbol{g}$ is the acceleration due to gravity. By a simple pull-back of the body force component in Equation (8.5b) it is easy to show that in this simple case the loading is not deformation-dependent and therefore the corresponding directional derivative vanishes. Recall for this purpose Equation (4.59) as $\rho = \rho_0/J$, which when substituted in the first term of the external virtual work Equation (8.5b) gives

$$\delta W^{\boldsymbol{f}}_{\text{ext}}(\boldsymbol{\phi}, \delta\boldsymbol{v}) = \int_v \frac{\rho_0}{J}\boldsymbol{g} \cdot \delta\boldsymbol{v}\, dv = \int_V \rho_0 \boldsymbol{g} \cdot \delta\boldsymbol{v}\, dV. \tag{8.18}$$

It is clear that none of the terms in this expression depends on the current geometry, and hence its linearization is superfluous, that is, $D\delta W^{\boldsymbol{f}}_{\text{ext}}(\boldsymbol{\phi}, \delta\boldsymbol{v})[\boldsymbol{u}] = 0$.

### 8.5.2 Surface Forces

Although a wide variety of traction forces exist, only the important case of uniform normal pressure will be addressed. However, the techniques illustrated by this simple example are relevant to more complex situations such as frictional contact.

Figure 8.2 shows a general body with an applied uniform pressure $p$ acting on a surface $a$ having a pointwise normal $\boldsymbol{n}$. The traction force vector $\boldsymbol{t}$ is therefore $p\boldsymbol{n}$ and the corresponding virtual work component is

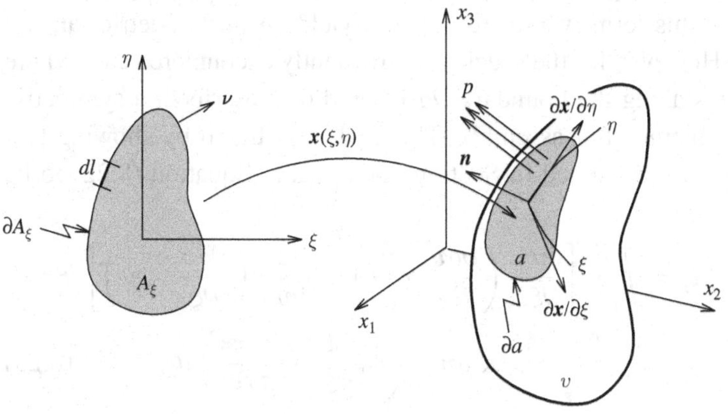


**FIGURE 8.2** Uniform surface pressure.

$$\delta W^p_{\text{ext}}(\boldsymbol{\phi}, \delta\boldsymbol{v}) = \int_a p\boldsymbol{n} \cdot \delta\boldsymbol{v}\, da. \tag{8.19}$$

In this equation the magnitude of the area element and the orientation of the normal are both displacement-dependent. Consequently, any change in geometry will result in a change in the equilibrium condition and the emergence of a stiffness term. Although it may be tempting to attempt the linearization of Equation (8.19) by a pull-back to the initial configuration in the usual manner, a more direct approach is available by using an arbitrary parameterization of the surface as shown in Figure 8.2. (An understanding of this approach is facilitated by imagining the surface area $a$ to be a single isoparametric element.) In terms of this parameterization the normal and area elements can be obtained in terms of the tangent vectors $\partial\boldsymbol{x}/\partial\xi$ and $\partial\boldsymbol{x}/\partial\eta$ as

$$\boldsymbol{n} = \frac{\frac{\partial \boldsymbol{x}}{\partial \xi} \times \frac{\partial \boldsymbol{x}}{\partial \eta}}{\left\| \frac{\partial \boldsymbol{x}}{\partial \xi} \times \frac{\partial \boldsymbol{x}}{\partial \eta} \right\|}; \quad da = \left\| \frac{\partial \boldsymbol{x}}{\partial \xi} \times \frac{\partial \boldsymbol{x}}{\partial \eta} \right\| d\xi d\eta, \tag{8.20}$$

which enables the integral in (8.19) to be expressed in the parameter plane as

$$\delta W^p_{\text{ext}}(\boldsymbol{\phi}, \delta\boldsymbol{v}) = \int_{A_\xi} p\delta\boldsymbol{v} \cdot \left( \frac{\partial \boldsymbol{x}}{\partial \xi} \times \frac{\partial \boldsymbol{x}}{\partial \eta} \right) d\xi d\eta. \tag{8.21}$$

Note that the only displacement-dependent items in this equation are the vectors $\partial\boldsymbol{x}/\partial\xi$ and $\partial\boldsymbol{x}/\partial\eta$, which linearize to $\partial\boldsymbol{u}/\partial\xi$ and $\partial\boldsymbol{u}/\partial\eta$ respectively. Hence the use of the product rule and a cyclic manipulation of the triple product gives

$$D\delta W^p_{\text{ext}}(\boldsymbol{\phi}, \delta\boldsymbol{v})[\boldsymbol{u}] = \int_{A_\xi} p \left[ \frac{\partial \boldsymbol{x}}{\partial \xi} \cdot \left( \frac{\partial \boldsymbol{u}}{\partial \eta} \times \delta\boldsymbol{v} \right) - \frac{\partial \boldsymbol{x}}{\partial \eta} \cdot \left( \frac{\partial \boldsymbol{u}}{\partial \xi} \times \delta\boldsymbol{v} \right) \right] d\xi d\eta. \tag{8.22}$$

It is clear that Equation (8.22) is unsymmetric in the sense that the terms $\boldsymbol{u}$ and $\delta\boldsymbol{v}$ cannot be interchanged without altering the result of the integral. Hence the discretization of this term would, in general, yield an unsymmetric tangent matrix component. However, for the special but frequently encountered case where the position of points along the boundary $\partial a$ is fixed or prescribed, a symmetric matrix will indeed emerge after assembly. This is demonstrated by showing that the integration theorems discussed in Section 2.4.2 enable Equation (8.22) to be rewritten as

$$\begin{aligned} D\delta W^p_{\text{ext}}(\boldsymbol{\phi}, \delta\boldsymbol{v})[\boldsymbol{u}] &= \int_{A_\xi} p \left[ \frac{\partial \boldsymbol{x}}{\partial \xi} \cdot \left( \frac{\partial \delta\boldsymbol{v}}{\partial \eta} \times \boldsymbol{u} \right) - \frac{\partial \boldsymbol{x}}{\partial \eta} \cdot \left( \frac{\partial \delta\boldsymbol{v}}{\partial \xi} \times \boldsymbol{u} \right) \right] d\xi d\eta \\ &\quad + \oint_{\partial A_\xi} p(\boldsymbol{u} \times \delta\boldsymbol{v}) \cdot \left( \nu_\eta \frac{\partial \boldsymbol{x}}{\partial \xi} - \nu_\xi \frac{\partial \boldsymbol{x}}{\partial \eta} \right) dl, \end{aligned} \tag{8.23}$$

where $\boldsymbol{\nu} = [\nu_\xi, \nu_\eta]^T$ is the vector in the parameter plane normal to $\partial A_\xi$. For the special case where the positions along $\partial a$ are fixed or prescribed, both the iterative displacement $\boldsymbol{u}$ and the virtual velocity $\delta\boldsymbol{v}$ are zero a priori along $\partial A_\xi$ and the second integral in the above expression vanishes. (Additionally, if a symmetry plane bisects the region $a$, then it is possible to show that the triple product in the second term of Equation (8.23) is also zero along this plane.) Anticipating closed boundary conditions a symmetric expression for $D\delta W^p_{\text{ext}}(\boldsymbol{\phi}, \delta\boldsymbol{v})[\boldsymbol{u}]$ can be constructed by adding half Equations (8.22) and (8.23) to give

$$\begin{aligned} D\delta W^p_{\text{ext}}(\boldsymbol{\phi}, \delta\boldsymbol{v})[\boldsymbol{u}] &= \frac{1}{2}\int_{A_\xi} p\frac{\partial \boldsymbol{x}}{\partial \xi}\cdot\left[\left(\frac{\partial \boldsymbol{u}}{\partial \eta}\times\delta\boldsymbol{v}\right)+\left(\frac{\partial \delta\boldsymbol{v}}{\partial \eta}\times\boldsymbol{u}\right)\right]d\xi d\eta \\ &-\frac{1}{2}\int_{A_\xi} p\frac{\partial \boldsymbol{x}}{\partial \eta}\cdot\left[\left(\frac{\partial \boldsymbol{u}}{\partial \xi}\times\delta\boldsymbol{v}\right)+\left(\frac{\partial \delta\boldsymbol{v}}{\partial \xi}\times\boldsymbol{u}\right)\right]d\xi d\eta. \end{aligned} \tag{8.24}$$

Discretization of this equation will obviously lead to a symmetric component of the tangent matrix.

**EXAMPLE 8.2: Proof of Equation (8.23)**

Repeated use of cyclic permutations of the triple product on Equation (8.22) and the integration theorem give

$$\begin{aligned} D\delta W^p_{\text{ext}}(\boldsymbol{\phi}, \delta\boldsymbol{v})[\boldsymbol{u}] &= \int_{A_\xi} p\left[\frac{\partial \boldsymbol{u}}{\partial \eta}\cdot\left(\delta\boldsymbol{v}\times\frac{\partial \boldsymbol{x}}{\partial \xi}\right)-\frac{\partial \boldsymbol{u}}{\partial \xi}\cdot\left(\delta\boldsymbol{v}\times\frac{\partial \boldsymbol{x}}{\partial \eta}\right)\right]d\xi d\eta \\ &= \int_{A_\xi} p\left[\frac{\partial}{\partial \eta}\left(\frac{\partial \boldsymbol{x}}{\partial \xi}\cdot(\boldsymbol{u}\times\delta\boldsymbol{v})\right)-\frac{\partial}{\partial \xi}\left(\frac{\partial \boldsymbol{x}}{\partial \eta}\cdot(\boldsymbol{u}\times\delta\boldsymbol{v})\right)\right]d\xi d\eta \\ &-\int_{A_\xi} p\left[\frac{\partial \boldsymbol{x}}{\partial \xi}\cdot\left(\boldsymbol{u}\times\frac{\partial \delta\boldsymbol{v}}{\partial \eta}\right)-\frac{\partial \boldsymbol{x}}{\partial \eta}\cdot\left(\boldsymbol{u}\times\frac{\partial \delta\boldsymbol{v}}{\partial \xi}\right)\right]d\xi d\eta \\ &= \int_{A_\xi} p\left[\frac{\partial \boldsymbol{x}}{\partial \xi}\cdot\left(\frac{\partial \delta\boldsymbol{v}}{\partial \eta}\times\boldsymbol{u}\right)-\frac{\partial \boldsymbol{x}}{\partial \eta}\cdot\left(\frac{\partial \delta\boldsymbol{v}}{\partial \xi}\times\boldsymbol{u}\right)\right]d\xi d\eta \\ &+\oint_{\partial A_\xi} p(\boldsymbol{u}\times\delta\boldsymbol{v})\cdot\left(\nu_\eta\frac{\partial \boldsymbol{x}}{\partial \xi}-\nu_\xi\frac{\partial \boldsymbol{x}}{\partial \eta}\right)dl. \end{aligned}$$

## 8.6 VARIATIONAL METHODS AND INCOMPRESSIBILITY

It is well known in small strain linear elasticity that the equilibrium equation can be derived by finding the stationary position of a total energy potential with respect to displacements. This applies equally to finite deformation situations and has the additional advantage that such a treatment provides a unified framework within which such topics as incompressibility, contact boundary conditions, and finite element technology can be formulated. In particular, in the context of incompressibility a variational approach conveniently facilitates the introduction of Lagrangian multipliers or penalty methods of constraint, where the resulting multi-field variational principles incorporate variables such as the internal pressure. The use of an independent discretization for these additional variables resolves the well-known locking problem associated with incompressible finite element formulations.

### 8.6.1 Total Potential Energy and Equilibrium

For a truss, Section 3.5 introduced the equivalence between the stationary condition of the Total Potential Energy (TPE) and the Principle of Virtual Work as expressions of equilbrium. In that section the TPE, given by Equation (3.34), was divided into internal and external components given by Equations (3.32) and (3.33) respectively. The internal energy was given by the strain energy per unit initial volume $\Psi(\lambda_e)$ and the external energy, per node, by $-\boldsymbol{F}_a \cdot \boldsymbol{x}_a$. Here $\lambda_e$ indicates that $\Psi$ is a function of strain and $\boldsymbol{x}_a$ is the position of the nodal force $\boldsymbol{F}_a$ with respect to an arbitrary datum. These concepts can be generalized to the case of a continuum by introducing a total potential energy functional whose directional derivative yields the principle of virtual work as

$$\Pi(\boldsymbol{\phi}) = \Pi_{\text{int}}(\boldsymbol{\phi}) + \Pi_{\text{ext}}(\boldsymbol{\phi}), \tag{8.25}$$

where

$$\Pi_{\text{int}}(\boldsymbol{\phi}) = \int_V \Psi(\boldsymbol{C})\, dV; \tag{8.26a}$$

$$\Pi_{\text{ext}}(\boldsymbol{\phi}) = -\int_V \boldsymbol{f}_0 \cdot \boldsymbol{x}\, dV - \int_{\partial V} \boldsymbol{t}_0 \cdot \boldsymbol{x}\, dA. \tag{8.26b}$$

Here the strain energy per unit initial volume $\Psi(\boldsymbol{C})$ is given by Equation (6.6) and the body force and surface traction external potential energy is an integrated expression of Equation (3.33).

Note that traditionally the definition of external energy given by Equation (8.26b) has an opposite sign convention to the definition of external virtual work given by Equation (8.5b). This implies that

$$D\Pi_{\text{ext}}(\boldsymbol{\phi})[\delta\boldsymbol{v}] = -\delta W_{\text{ext}}(\boldsymbol{\phi}, \delta\boldsymbol{v}). \tag{8.27}$$

To proceed, we assume that the body and traction forces are not functions of the motion. This is usually the case for body forces $\boldsymbol{f}_0$, but it is unlikely that traction forces $\boldsymbol{t}_0$ will conform to this requirement in a finite deformation context.

(Obviously both these terms are independent of deformation in the small displacement case.) Under these assumptions the stationary position of the above functional is found by equating to zero the variation of $\Pi(\phi)$ in an arbitrary direction $\delta\boldsymbol{u}$ obtained using the directional derivative as

$$D\Pi(\phi)[\delta\boldsymbol{u}] = \int_V \frac{\partial\Psi}{\partial\boldsymbol{C}} : D\boldsymbol{C}[\delta\boldsymbol{u}]\,dV - \int_V \boldsymbol{f}_0 \cdot \delta\boldsymbol{u}\,dV - \int_{\partial V} \boldsymbol{t}_0 \cdot \delta\boldsymbol{u}\,dA$$
$$= \int_V \boldsymbol{S} : D\boldsymbol{E}[\delta\boldsymbol{u}]\,dV - \int_V \boldsymbol{f}_0 \cdot \delta\boldsymbol{u}\,dV - \int_{\partial V} \boldsymbol{t}_0 \cdot \delta\boldsymbol{u}\,dA = 0, \tag{8.28}$$

where Equation (6.7a,b) for $\boldsymbol{S}$ has been used. With the help of Equation (4.90), namely $D\boldsymbol{E}[\delta\boldsymbol{v}] = \delta\dot{\boldsymbol{F}}$, it is easy to see that, similarly to the case of the truss (see Section 3.5.1), the above equation is identical to the principle of virtual work given by Equation (8.2), that is,

$$D\Pi(\phi)[\delta\boldsymbol{v}] = \delta W(\phi, \delta\boldsymbol{v}), \tag{8.29}$$

and consequently the equilibrium configuration $\phi$ renders stationary the total potential energy. The stationary condition of Equation (8.25), i.e. Equation (8.28), is also known as a *variational statement* of equilibrium.

Furthermore, the linearized equilibrium Equation (8.11) or (8.16) can be interpreted as the second derivative of the total potential energy as

$$D\delta W(\phi, \delta\boldsymbol{v})[\boldsymbol{u}] = D^2\Pi(\phi)[\delta\boldsymbol{v}, \boldsymbol{u}]. \tag{8.30}$$

### 8.6.2 Lagrange Multiplier Approach to Incompressibility

We have seen in Chapter 6 that for incompressible materials the constitutive equations are only a function of the distortional component of the deformation. In addition, the incompressibility constraint $J = 1$ has to be enforced explicitly. This is traditionally achieved by augmenting the variational functional (8.25) with a Lagrange multiplier term to give

$$\Pi_L(\phi, p) = \hat{\Pi}(\phi) + \int_V p(J-1)\,dV, \tag{8.31}$$

where $p$ has been used to denote the Lagrange multiplier in anticipation of the fact that it will be identified as the internal pressure, and the notation $\hat{\Pi}$ implies that the strain energy $\Psi$ is now a function of the distortional component $\hat{\boldsymbol{C}}$ of the right Cauchy–Green tensor (see Section 6.5.1), that is,

$$\hat{\Pi}(\phi) = \int_V \hat{\Psi}(\boldsymbol{C})\,dV - \int_V \boldsymbol{f}_0 \cdot \boldsymbol{x}\,dV - \int_{\partial V} \boldsymbol{t}_0 \cdot \boldsymbol{x}\,dA; \quad \hat{\Psi}(\boldsymbol{C}) = \Psi(\hat{\boldsymbol{C}}). \tag{8.32}$$

The stationary condition of the above functional with respect to $\phi$ and $p$ will be considered separately. First, the directional derivative of $\Pi_L$ with respect to $p$ in the arbitrary direction $\delta p$ is

$$D\Pi_L(\phi, p)[\delta p] = \int_V \delta p(J - 1)\, dV = 0. \tag{8.33}$$

Hence, if this condition is satisfied for all functions $\delta p$, the incompressibility constraint $J = 1$ is ensured.

The derivative of Equation (8.31) with respect to the motion in the direction of $\delta \boldsymbol{v}$ is given as

$$D\Pi_L(\phi, p)[\delta \boldsymbol{v}] = D\hat{\Pi}(\phi)[\delta \boldsymbol{v}] + \int_V pDJ[\delta \boldsymbol{v}]\, dV = 0. \tag{8.34}$$

Substituting Equation (8.32) and using Equation (4.76) for $DJ[\delta \boldsymbol{v}]$ and (6.51) for the derivative of the second Piola–Kirchhoff stress gives, after some algebra, the principle of virtual work as

$$D\Pi_L(\phi, p)[\delta \boldsymbol{v}] = \int_V \frac{\partial \hat{\Psi}}{\partial \boldsymbol{C}} : D\boldsymbol{C}[\delta \boldsymbol{v}]\, dV + \int_V Jp \operatorname{div} \delta \boldsymbol{v}\, dV - \delta W_{\text{ext}}(\phi, \delta \boldsymbol{v}) \tag{8.35a}$$

$$= \int_V \boldsymbol{S}' : D\boldsymbol{E}[\delta \boldsymbol{v}]\, dV + \int_v p \operatorname{div} \delta \boldsymbol{v}\, dv - \delta W_{\text{ext}}(\phi, \delta \boldsymbol{v}) \tag{8.35b}$$

$$= \int_v \boldsymbol{\sigma}' : \delta \boldsymbol{d}\, dv + \int_v p\boldsymbol{I} : \delta \boldsymbol{d}\, dv - \delta W_{\text{ext}}(\phi, \delta \boldsymbol{v}) \tag{8.35c}$$

$$= \int_v \boldsymbol{\sigma} : \delta \boldsymbol{d}\, dv - \delta W_{\text{ext}}(\phi, \delta \boldsymbol{v}) = 0, \tag{8.35d}$$

where Equation (8.12b) for the relevant push-forward operation and Equations (5.49a,b) and (5.50b) have been invoked. Observe that Equation (8.35c) clearly identifies the Lagrange multiplier $p$ as the physical internal (hydrostatic) pressure.

In the context of a Newton–Raphson solution process, the governing Equations (8.33) and (8.35d) need to be linearized with respect to both variables $p$ and $\phi$ in the direction of respective increments $\Delta p$ and $\boldsymbol{u}$. Starting with Equation (8.33) and using the linearization of $J$ as given in Equation (4.76), we have

$$D^2\Pi_L(\phi, p)[\delta p, \Delta p] = 0; \tag{8.36a}$$

$$D^2\Pi_L(\phi, p)[\delta p, \boldsymbol{u}] = \int_v \delta p \operatorname{div} \boldsymbol{u}\, dv. \tag{8.36b}$$

The linearization of Equation (8.35d) with respect to increments in $p$ is easily found using (8.35b) to give

$$D^2\Pi_L(\phi, p)[\delta\boldsymbol{v}, \Delta p] = \int_v \Delta p \,\mathrm{div}\, \delta\boldsymbol{v}\, dv. \tag{8.37}$$

The obvious symmetry between Equations (8.37) and (8.36b) leads to a symmetric tangent matrix upon discretization.

Finally, for the purpose of obtaining the derivative of (8.35d) in the direction of an increment $\boldsymbol{u}$ in the motion, it is convenient to revert to a material description and rewrite Equation (8.35d) as

$$D\Pi_L(\phi, p)[\delta\boldsymbol{v}] = \int_V \boldsymbol{S} : D\boldsymbol{E}[\delta\boldsymbol{v}]\, dV - \delta W_{\text{ext}}(\phi, \delta\boldsymbol{v}); \quad \boldsymbol{S} = \boldsymbol{S}' + pJ\boldsymbol{C}^{-1}. \tag{8.38}$$

The linearization of this expression in the direction of $\boldsymbol{u}$ is obtained in the same manner as in Section 8.3 to give

$$D^2\Pi_L(\phi, p)[\delta\boldsymbol{v}, \boldsymbol{u}] = \int_V D\boldsymbol{E}[\delta\boldsymbol{v}] : \boldsymbol{\mathcal{C}} : D\boldsymbol{E}[\boldsymbol{u}]\, dV + \int_V \boldsymbol{S} : [(\boldsymbol{\nabla}_0\boldsymbol{u})^T \boldsymbol{\nabla}_0\delta\boldsymbol{v}]\, dV, \tag{8.39}$$

where the tangent modulus is given in Section 6.5.2 as

$$\boldsymbol{\mathcal{C}} = 2\frac{\partial}{\partial \boldsymbol{C}}(\boldsymbol{S}' + pJ\boldsymbol{C}^{-1}) = \hat{\boldsymbol{\mathcal{C}}} + \boldsymbol{\mathcal{C}}_p; \tag{8.40a}$$

$$\hat{\boldsymbol{\mathcal{C}}} = 2\frac{\partial \boldsymbol{S}'}{\partial \boldsymbol{C}}; \tag{8.40b}$$

$$\boldsymbol{\mathcal{C}}_p = pJ[\boldsymbol{C}^{-1} \otimes \boldsymbol{C}^{-1} - 2\boldsymbol{\mathcal{I}}]. \tag{8.40c}$$

Similarly, in the current configuration Equation (8.39) becomes

$$D^2\Pi_L(\phi, p)[\delta\boldsymbol{v}, \boldsymbol{u}] = \int_v \delta\boldsymbol{d} : \boldsymbol{c} : \boldsymbol{\varepsilon}\, dv + \int_v \boldsymbol{\sigma} : [(\boldsymbol{\nabla}\boldsymbol{u})^T \boldsymbol{\nabla}\delta\boldsymbol{v}]\, dv, \tag{8.41}$$

where the spatial tangent modulus is

$$\boldsymbol{c} = \hat{\boldsymbol{c}} + \boldsymbol{c}_p; \qquad \hat{\boldsymbol{c}} = J^{-1}\phi_*[\hat{\boldsymbol{\mathcal{C}}}]; \qquad \boldsymbol{c}_p = p[\boldsymbol{I} \otimes \boldsymbol{I} - 2\,\boldsymbol{i}]. \tag{8.42}$$

In conclusion, the linearization of the governing Equations (8.33), (8.37), and (8.41) is the combination of Equations (8.36), (8.37), and (8.41).

The above Lagrangian multiplier approach can be the basis for a successful finite element implementation provided that the interpolations of $\phi$ and $p$ are carefully chosen so as to avoid volumetric locking. It does, however, suffer from the limitation of the presence of additional pressure variables. These two problems will be considered in the next two sections, and the Lagrange multiplier approach will not be pursued further.

### 8.6.3 Penalty Methods for Incompressibility

Penalty methods are an alternative to the Lagrangian multiplier approach to incompressibility. There are two ways in which penalty methods can be introduced in the formulation. A popular, physically-based approach, which conveniently eliminates the pressure as an independent variable, is to consider the material as being nearly incompressible, whereby a large value of the bulk modulus effectively prevents significant volumetric changes. A second, less intuitive route is to perturb the Lagrangian functional given in Equation (8.31) by the addition of a further "penalty" term that enables the pressure to eventually be artificially associated with the deformation, thereby again eliminating the pressure variable. It will transpire that these two methods lead to identical equations. The perturbed Lagrangian functional is

$$\Pi_P(\phi, p) = \Pi_L(\phi, p) - \int_V \frac{1}{2\kappa} p^2 \, dV, \tag{8.43}$$

where $\kappa$ is the penalty parameter and clearly $\Pi_P \to \Pi_L$ as $\kappa \to \infty$.

The stationary condition of the functional (8.43) with respect to $p$ now becomes

$$D\Pi_P(\phi, p)[\delta p] = \int_V \delta p \left[ (J-1) - \frac{p}{\kappa} \right] dV = 0. \tag{8.44}$$

Consequently, enforcing this equation for all $\delta p$ functions gives an equation artificially relating $p$ and $J$ as

$$p = \kappa(J-1). \tag{8.45}$$

Referring to Section 6.5.3, Equation (6.64), it is clear that this equation represents a nearly incompressible material with the penalty number $\kappa$ as the bulk modulus. The use of this equation in a finite element context, either directly as a nearly incompressible material or indirectly as a perturbed Lagrangian method, will lead to a formulation involving only kinematic unknown variables.

The stationary condition of the functional (8.43) with respect to the motion is identical to Equation (8.35d) and gives the principle of virtual work, where now the internal pressure in the Cauchy stresses can be evaluated using Equation (8.45) or, in general for a nearly incompressible material, using Equation (6.63). In conclusion, the linearized equilibrium equation is given directly by Equation (8.16), where, as shown in Section 6.5.3, the tangent modulus is now

$$\mathcal{C} = \hat{\mathcal{C}} + \mathcal{C}_p + \mathcal{C}_\kappa; \tag{8.46a}$$

$$\hat{\mathcal{C}} = J^{-1}\phi_*[\hat{\boldsymbol{\mathcal{C}}}]; \qquad \hat{\boldsymbol{\mathcal{C}}} = 4\frac{\partial^2 \hat{\Psi}}{\partial \boldsymbol{C} \partial \boldsymbol{C}} = 2\frac{\partial \boldsymbol{S}'}{\partial \boldsymbol{C}}; \tag{8.46b}$$

$$\mathcal{C}_p = p[\boldsymbol{I} \otimes \boldsymbol{I} - 2\mathcal{i}]; \tag{8.46c}$$

$$\mathcal{C}_\kappa = J\frac{dp}{dJ}\boldsymbol{I} \otimes \boldsymbol{I} = \kappa J \boldsymbol{I} \otimes \boldsymbol{I}. \tag{8.46d}$$

### 8.6.4 Hu–Washizu Variational Principle for Incompressibility

Neither the Lagrange multiplier method nor the penalty method as presented above will result in a simple and efficient finite element formulation. This is because the kinematically restricted motion implicit in the finite element discretization is, in general, unable to distort while simultaneously meeting the incompressibility requirement at each point of the body. This phenomenon manifests itself as a catastrophic artificial stiffening of the system known as *volumetric locking*. One common practical solution to this problem involves the use of different discretizations for $p$ and $\phi$ in the Lagrange multiplier approach. Another ad hoc solution, associated with the enforcement of incompressibility using nearly incompressible constitutive equations, is to first separate the volumetric and distortional strain energy as

$$\Pi(\phi) = \int_V \hat{\Psi}(\boldsymbol{C})\, dV + \int_V U(J)\, dV - \Pi_{\text{ext}}(\phi), \tag{8.47}$$

and then underintegrate the volumetric term containing $U(J)$.

A formal framework for the solution of locking problems for nearly incompressible materials that has potential for further developments is provided by a functional that permits the use of independent kinematic descriptions for the volumetric and distortional deformations. This can be achieved by introducing a three-field Hu–Washizu type of variational principle as

$$\Pi_{HW}(\phi, \bar{J}, p) = \int_V \hat{\Psi}(\boldsymbol{C})\, dV + \int_V U(\bar{J})\, dV + \int_V p(J - \bar{J})\, dV - \Pi_{\text{ext}}(\phi), \tag{8.48}$$

where $\bar{J}$ is a new kinematic variable representing the dilatation or volume change independently of the motion and $p$ is a Lagrange multiplier enforcing the condition that $\bar{J} = J$. The opportunity that this functional affords of using independent discretizations for $\phi$ and $\bar{J}$ will introduce sufficient flexibility to prevent locking.

The stationary conditions of the functional (8.48) with respect to $\phi$, $\bar{J}$, and $p$ will yield the equilibrium equation and the constitutive and kinematic relationships associated with the volumetric behavior. To this end, we find the directional derivative of $\Pi_{HW}$ in the direction $\delta\boldsymbol{v}$ as

$$\begin{aligned} D\Pi_{HW}(\phi, \bar{J}, p)[\delta\boldsymbol{v}] &= \int_V \frac{\partial \hat{\Psi}}{\partial \boldsymbol{C}} : D\boldsymbol{C}[\delta\boldsymbol{v}]dV + \int_V pDJ[\delta\boldsymbol{v}]\, dV \\ &\quad - D\Pi_{\text{ext}}(\phi)[\delta\boldsymbol{v}]. \end{aligned} \tag{8.49}$$

Repeating the algebra used in Equation (8.35), the stationary condition with respect to $\phi$ gives the principle of virtual work as

$$D\Pi_{HW}(\phi, \bar{J}, p)[\delta\boldsymbol{v}] = \int_v \boldsymbol{\sigma} : \delta\boldsymbol{d}\, dv - \delta W_{\text{ext}}(\phi, \delta\boldsymbol{v}) = 0. \tag{8.50}$$

The stationary condition of the functional (8.48) with respect to changes in $\bar{J}$ gives a constitutive equation for $p$ as

$$D\Pi_{HW}(\phi, \bar{J}, p)[\delta \bar{J}] = \int_V \left( \frac{dU}{d\bar{J}} - p \right) \delta \bar{J}\, dV = 0. \tag{8.51}$$

Similarly, the stationary condition of the functional (8.48) with respect to $p$ gives a kinematic equation for $\bar{J}$ as

$$D\Pi_{HW}(\phi, \bar{J}, p)[\delta p] = \int_V (J - \bar{J})\delta p\, dV = 0. \tag{8.52}$$

Obviously, if $\delta p$ and $\delta \bar{J}$ are arbitrary functions, Equation (8.52) gives $\bar{J} = J$ and Equation (8.51) gives $p = dU/dJ$. However, as explained, a finite element formulation based on this procedure where the discretization of $\bar{J}$ is based on the same interpolation as $\phi$ confers no advantage, and volumetric locking will emerge. This is usually prevented in a finite element context by judiciously choosing the interpolating functions used for the volumetric variables $\bar{J}$, $p$, and their variations, $\delta \bar{J}$ and $\delta p$. In particular, the simplest possible procedure involves using constant interpolations for these variables over a given volume, typically a finite element. The resulting method is known as the *mean dilatation technique*.

**EXAMPLE 8.3: Hu–Washizu variational principle**

Equation (8.48) is a Hu–Washizu type of variational equation in the sense that it incorporates three independent variables, namely, the motion, a volumetric strain field, and its corresponding volumetric stress. It is, however, a very particular case as only the volumetric stress and strain components are present in the equation. A more general Hu–Washizu variational principle involving the motion, a complete stress field such as the first Piola–Kirchhoff tensor $\boldsymbol{P}$, and its associated strain $\boldsymbol{F}$ is given as

$$\Pi_{HW}(\phi, \boldsymbol{F}, \boldsymbol{P}) = \int_V \Psi(\boldsymbol{F})\, dv + \int_V \boldsymbol{P} : (\boldsymbol{\nabla}_0 \phi - \boldsymbol{F})\, dV - \Pi_{\text{ext}}(\phi),$$

where now $\boldsymbol{F}$ is an independent variable as yet unrelated to the deformation gradient of the motion, $\boldsymbol{\nabla}_0 \phi$. The stationary condition of this functional with respect to a variation $\delta \boldsymbol{v}$ in motion $\phi$ gives the principle of virtual work as

$$D\Pi(\phi, \boldsymbol{F}, \boldsymbol{P})[\delta \boldsymbol{v}] = \int_V \boldsymbol{P} : \boldsymbol{\nabla}_0 \delta \boldsymbol{v}\, dV - \delta W_{\text{ext}}(\phi)[\delta \boldsymbol{v}] = 0,$$

whereas the stationary conditions with respect to $\boldsymbol{F}$ and $\boldsymbol{P}$ give a constitutive equation and a relationship between the strain and the motion as

*(continued)*

**EXAMPLE 8.3:** *(cont.)*

$$D\Pi(\phi, \boldsymbol{F}, \boldsymbol{P})[\delta \boldsymbol{F}] = \int_V \left( \frac{\partial \Psi}{\partial \boldsymbol{F}} - \boldsymbol{P} \right) : \delta \boldsymbol{F} \, dV = 0;$$

$$D\Pi(\phi, \boldsymbol{F}, \boldsymbol{P})[\delta \boldsymbol{P}] = \int_V (\boldsymbol{\nabla}_0 \phi - \boldsymbol{F}) : \delta \boldsymbol{P} \, dV = 0.$$

These expressions are the weak forms equivalent to the hyperelastic relationship $\boldsymbol{P} = \partial\Psi/\partial\boldsymbol{F}$ and the strain equation $\boldsymbol{F} = \boldsymbol{\nabla}_0 \phi$ respectively.

### 8.6.5 Mean Dilatation Procedure

For convenience, the following discussion is based on an arbitrary volume $V$, which after discretization will inevitably become the volume of each element.

Assuming that $p$, $\bar{J}$ and $\delta p$, $\delta \bar{J}$ are constant over the integration volume, Equations (6.48) and (6.49) yield

$$\bar{J} = \frac{1}{V} \int_V J dV = \frac{v}{V}; \tag{8.53a}$$

$$p = \left. \frac{dU}{d\bar{J}} \right|_{\bar{J}=v/V}. \tag{8.53b}$$

Observe that Equation (8.53a) shows that the surrogate Jacobian $\bar{J}$ is the integral equivalent of the pointwise Jacobian $J = dv/dV$. A typical expression for the volumetric strain energy has already been introduced in Section 6.5.3, Equation (6.61). In terms of $\bar{J}$, this equation now becomes

$$U(\bar{J}) = \frac{1}{2}\kappa(\bar{J} - 1)^2, \tag{8.54}$$

from which the mean pressure emerges as

$$p(\bar{J}) = \kappa \left( \frac{v - V}{V} \right). \tag{8.55}$$

At this juncture, it is convenient to acknowledge that the mean dilatation method will be used in the finite element formulation presented in Chapter 9. This enables us to incorporate a priori the mean pressure derived above using the Hu–Washizu functional directly into the principle of virtual work to give

$$\delta \overline{W}(\phi, \delta \boldsymbol{v}) = \int_v \boldsymbol{\sigma} : \delta \boldsymbol{d} \, dv - \delta W_{\text{ext}}(\phi, \delta \boldsymbol{v}) = 0, \qquad \boldsymbol{\sigma} = \boldsymbol{\sigma}' + p\boldsymbol{I};$$

$$p = \left. \frac{dU}{d\bar{J}} \right|_{\bar{J}=v/V}; \tag{8.56}$$

which can also be expressed in the initial configuration as

$$\delta\overline{W}(\boldsymbol{\phi},\delta\boldsymbol{v}) = \int_V \boldsymbol{S} : D\boldsymbol{E}[\delta\boldsymbol{v}]\,dV - \delta W_{\text{ext}}(\boldsymbol{\phi},\delta\boldsymbol{v}) = 0, \boldsymbol{S} = \boldsymbol{S}' + pJ\boldsymbol{C}^{-1}. \tag{8.57}$$

Given that the pressure is constant over the volume, Equation (8.56) can also be written as

$$\begin{aligned}\delta\overline{W}(\boldsymbol{\phi},\delta\boldsymbol{v}) &= \int_v \boldsymbol{\sigma}' : \delta\boldsymbol{d}\,dv + p\int_v \text{div}\,\delta\boldsymbol{v}\,dv - \delta W_{\text{ext}}(\boldsymbol{\phi},\delta\boldsymbol{v})\\ &= \int_v \boldsymbol{\sigma}' : \delta\boldsymbol{d}\,dv + pv(\overline{\text{div}}\,\delta\boldsymbol{v}) - \delta W_{\text{ext}}(\boldsymbol{\phi},\delta\boldsymbol{v}) = 0,\end{aligned} \tag{8.58}$$

where the notation $\overline{\text{div}}$ implies the average divergence over the volume $v$, for instance,

$$\overline{\text{div}}\,\delta\boldsymbol{v} = \frac{1}{v}\int_v \text{div}\,\delta\boldsymbol{v}\,dv. \tag{8.59}$$

**EXAMPLE 8.4: Mean deformation gradient method**

Using the Hu–Washizu variational principle introduced in Example 8.3, it is possible to derive a technique whereby the complete deformation gradient $\boldsymbol{F}$, rather than just its volumetric component, is taken as constant over the integration volume. Finite element discretizations based on this type of technique are sometimes used to avoid shear locking as well as volumetric locking. By assuming that $\boldsymbol{P}$, $\boldsymbol{F}$, and their variations are constant in Example 8.3, the following equations are obtained for $\boldsymbol{F}$ and $\boldsymbol{P}$:

$$\boldsymbol{F} = \frac{1}{V}\int_V \boldsymbol{\nabla}_0\boldsymbol{\phi}\,dV = \overline{\boldsymbol{\nabla}}_0\boldsymbol{\phi};$$

$$\boldsymbol{P} = \left.\frac{\partial\Psi}{\partial\boldsymbol{F}}\right|_{\boldsymbol{F}=\overline{\boldsymbol{\nabla}}_0\boldsymbol{\phi}},$$

where $\overline{\boldsymbol{\nabla}}_0$ represents the mean (or average) gradient over the integration volume $V$. With the help of these equations the principle of virtual work becomes

$$\delta\overline{W}(\boldsymbol{\phi},\delta\boldsymbol{v}) = V\boldsymbol{P} : \overline{\boldsymbol{\nabla}}_0\delta\boldsymbol{v} - \delta W_{\text{ext}}(\boldsymbol{\phi},\delta\boldsymbol{v}) = 0.$$

Note that this can be written in terms of the corresponding second Piola–Kirchhoff tensor $\boldsymbol{S} = (\overline{\boldsymbol{\nabla}}_0\boldsymbol{\phi})^{-1}\boldsymbol{P}$ as

$$\begin{aligned}\delta\overline{W}(\boldsymbol{\phi})[\delta\boldsymbol{v}] &= V\boldsymbol{S} : D\overline{\boldsymbol{E}}[\delta\boldsymbol{v}] - \delta W_{\text{ext}}(\boldsymbol{\phi},\delta\boldsymbol{v}) = 0;\\ 2D\overline{\boldsymbol{E}}[\delta\boldsymbol{v}] &= (\overline{\boldsymbol{\nabla}}_0\boldsymbol{\phi})^T\overline{\boldsymbol{\nabla}}_0\delta\boldsymbol{v} + (\overline{\boldsymbol{\nabla}}_0\delta\boldsymbol{v})^T\overline{\boldsymbol{\nabla}}_0\boldsymbol{\phi},\end{aligned}$$

or in terms of the Cauchy stress tensor $v\boldsymbol{\sigma} = V\boldsymbol{P}(\overline{\boldsymbol{\nabla}}_0\boldsymbol{\phi})^T$ as

$$\delta\overline{W}(\boldsymbol{\phi})[\delta\boldsymbol{v}] = v\boldsymbol{\sigma} : \overline{\boldsymbol{\nabla}}\delta\boldsymbol{v} - \delta W_{\text{ext}}(\boldsymbol{\phi},\delta\boldsymbol{v}) = 0;\quad \overline{\boldsymbol{\nabla}}\delta\boldsymbol{v} = (\overline{\boldsymbol{\nabla}}_0\delta\boldsymbol{v})(\overline{\boldsymbol{\nabla}}_0\boldsymbol{\phi})^{-1}.$$

As usual, it is now necessary to linearize the modified virtual work equation in preparation for a Newton–Raphson iteration and the development of a tangent matrix. Again, this linearization is first obtained using the initial configuration Equation (8.57) for which the integral limits remain constant. Disregarding the external force component, this gives

$$\begin{aligned}
D\delta\overline{W}_{\text{int}}(\boldsymbol{\phi},\delta\boldsymbol{v})[\boldsymbol{u}] &= \int_V D\boldsymbol{E}[\delta\boldsymbol{v}] : 2\frac{\partial \boldsymbol{S}}{\partial \boldsymbol{C}} : D\boldsymbol{E}[\boldsymbol{u}]\,dV + \int_V \boldsymbol{S} : D\delta\dot{\boldsymbol{E}}[\boldsymbol{u}]\,dV \\
&= \int_V D\boldsymbol{E}[\delta\boldsymbol{v}] : 2\left(\frac{\partial \boldsymbol{S}'}{\partial \boldsymbol{C}} + p\frac{\partial (J\boldsymbol{C}^{-1})}{\partial \boldsymbol{C}}\right) : D\boldsymbol{E}[\boldsymbol{u}]\,dV \\
&\quad + \int_V (D\boldsymbol{E}[\delta\boldsymbol{v}] : J\boldsymbol{C}^{-1}) Dp[\boldsymbol{u}]\,dV + \int_V \boldsymbol{S} : D\delta\dot{\boldsymbol{E}}[\boldsymbol{u}]\,dV \\
&= \int_V D\boldsymbol{E}[\delta\boldsymbol{v}] : (\hat{\boldsymbol{\mathcal{C}}} + \boldsymbol{\mathcal{C}}_p) : D\boldsymbol{E}[\boldsymbol{u}]\,dV \\
&\quad + \int_V \boldsymbol{S} : [(\boldsymbol{\nabla}_0\boldsymbol{u})^T\boldsymbol{\nabla}_0\delta\boldsymbol{v}] + Dp[\boldsymbol{u}]\int_V D\boldsymbol{E}[\delta\boldsymbol{v}] : J\boldsymbol{C}^{-1}\,dV.
\end{aligned} \tag{8.60}$$

Observe that, via Equation (8.53b), the pressure is now an explicit function of the current volume and thus of $\boldsymbol{\phi}$, and hence is subject to linearization in the direction of $\boldsymbol{u}$. With the help of Equations (4.129) and (8.56) together with Section 4.11.3 and the usual push-forward operations (8.12), this equation is rewritten in the current configuration as

$$\begin{aligned}
D\delta\overline{W}_{\text{int}}(\boldsymbol{\phi},\delta\boldsymbol{v})[\boldsymbol{u}] &= \int_v \delta\boldsymbol{d} : (\hat{\mathbb{c}} + \mathbb{c}_p) : \boldsymbol{\varepsilon}\,dv + \int_v \boldsymbol{\sigma} : [(\boldsymbol{\nabla}\boldsymbol{u})^T\boldsymbol{\nabla}\delta\boldsymbol{v}]\,dv \\
&\quad + v(\overline{\text{div}\,\delta\boldsymbol{v}})Dp[\boldsymbol{u}].
\end{aligned} \tag{8.61}$$

The linearization of the pressure term follows from Equation (8.53b) as

$$\begin{aligned}
Dp[\boldsymbol{u}] &= \left.\frac{d^2U}{d\bar{J}^2}\right|_{\bar{J}=v/V} D(v/V)[\boldsymbol{u}] \\
&= \frac{1}{V}\left.\frac{d^2U}{d\bar{J}^2}\right|_{\bar{J}=v/V}\int_V DJ[\boldsymbol{u}]\,dV \\
&= \frac{1}{V}\left.\frac{d^2U}{d\bar{J}^2}\right|_{\bar{J}=v/V}\int_V J\text{div}\,\boldsymbol{u}\,dV \\
&= \bar{\kappa}\,\overline{\text{div}\,\boldsymbol{u}}; \qquad \bar{\kappa} = \frac{v}{V}\left.\frac{d^2U}{d\bar{J}^2}\right|_{\bar{J}=v/V},
\end{aligned} \tag{8.62}$$

from which, finally,

$$D\delta\overline{W}_{\text{int}}(\phi,\delta\boldsymbol{v})[\boldsymbol{u}] = \int_v \delta\boldsymbol{d} : (\hat{\mathbb{c}} + \mathbb{c}_p) : \boldsymbol{\varepsilon}\, dv$$
$$+ \int_v \boldsymbol{\sigma} : [(\boldsymbol{\nabla}\boldsymbol{u})^T \boldsymbol{\nabla}\delta\boldsymbol{v}]\, dv + \bar{\kappa}v(\overline{\text{div}}\,\delta\boldsymbol{v})(\overline{\text{div}}\,\boldsymbol{u}), \quad (8.63)$$

where, for instance, for the volumetric potential shown in Equation (8.54),

$$\bar{\kappa} = \frac{v\kappa}{V}. \quad (8.64)$$

The discretization of the mean dilatation technique will be considered in Chapter 9.

## Exercises

1. Show that the linearized internal virtual work can also be expressed as

$$D\delta W(\phi,\delta\boldsymbol{v})[\boldsymbol{u}] = \int_V (\boldsymbol{\nabla}_0\delta\boldsymbol{v}) : \boldsymbol{\mathcal{A}} : (\boldsymbol{\nabla}_0\boldsymbol{u})\, dV; \boldsymbol{\mathcal{A}} = \frac{\partial\boldsymbol{P}}{\partial\boldsymbol{F}} = \frac{\partial^2\Psi}{\partial\boldsymbol{F}\partial\boldsymbol{F}},$$

where $\boldsymbol{P}$ is the first Piola–Kirchhoff tensor.

2. Show that, for the case of uniform pressure over an enclosed fixed boundary, the external virtual work can be derived from an associated potential as $\delta W^p_{\text{ext}}(\phi,\delta\boldsymbol{v}) = D\Pi^p_{\text{ext}}(\phi)[\delta\boldsymbol{v}]$, where

$$\Pi^p_{\text{ext}}(\phi) = \frac{1}{3}\int_a p\boldsymbol{x}\cdot\boldsymbol{n}\, da.$$

Explain the physical significance of this integral.

3. Prove that, for two-dimensional applications, Equation (8.24) becomes

$$D\delta W^p_{\text{ext}}(\phi,\delta\boldsymbol{v})[\boldsymbol{u}] = \frac{1}{2}\int_{L_\eta} p\boldsymbol{k}\cdot\left[\left(\frac{\partial\boldsymbol{u}}{\partial\eta}\times\delta\boldsymbol{v}\right) + \left(\frac{\partial\delta\boldsymbol{v}}{\partial\eta}\times\boldsymbol{u}\right)\right] d\eta,$$

where $\boldsymbol{k}$ is a unit vector normal to the two-dimensional plane and $\eta$ is a parameter along the line $L_\eta$ where the pressure $p$ is applied.

4. Prove that, by using a different cyclic permutation than that used to derive Equation (8.23), the following alternative form of Equation (8.24) can be found for the case of an enclosed fixed boundary with uniform surface pressure:

$$D\delta W^p_{\text{ext}}(\phi,\delta\boldsymbol{v})[\boldsymbol{u}] = \int_{A_\xi} p\boldsymbol{x}\cdot\left[\left(\frac{\partial\delta\boldsymbol{v}}{\partial\xi}\times\frac{\partial\boldsymbol{u}}{\partial\eta}\right) - \left(\frac{\partial\delta\boldsymbol{v}}{\partial\eta}\times\frac{\partial\boldsymbol{u}}{\partial\xi}\right)\right] d\xi d\eta.$$

5. Prove that, by assuming a constant pressure interpolation over the integration volume in Equation (8.44), a constant pressure technique equivalent to the mean dilatation method is obtained.

6. A six-field Hu–Washizu type of variational principle with independent volumetric and deviatoric variables is given as

$$\Pi_{HW}(\boldsymbol{\phi}, \bar{J}, \boldsymbol{F}, p, \boldsymbol{P}', \gamma) = \int_V \hat{\Psi}(\boldsymbol{C})\, dV + \int_V U(\bar{J})\, dV + \int_V p(J - \bar{J})\, dV$$
$$+ \int_V \boldsymbol{P}' : (\boldsymbol{\nabla}_0 \boldsymbol{\phi} - \boldsymbol{F})\, dV + \int_V \gamma\, \boldsymbol{P}' : \boldsymbol{F}\, dV,$$

where $\boldsymbol{C} = \boldsymbol{F}^T \boldsymbol{F}$, $J = \det(\boldsymbol{\nabla}_0 \boldsymbol{\phi})$, and $\boldsymbol{P}'$ denotes the deviatoric component of the first Piola–Kirchhoff stress tensor. Find the stationary conditions with respect to each variable. Explain the need to introduce the Lagrange multiplier $\gamma$. Derive the formulation that results from assuming that all the fields except for the motion are constant over the integration volume.

CHAPTER NINE

# DISCRETIZATION AND SOLUTION

## 9.1 INTRODUCTION

The equilibrium equations and their corresponding linearizations have been established in terms of a material or a spatial description. Either of these descriptions can be used to derive the discretized equilibrium equations and their corresponding tangent matrix. Irrespective of which configuration is used, the resulting quantities will be identical. It is, however, generally simpler to establish the discretized quantities in the spatial configuration.

Establishing the discretized equilibrium equations is relatively standard, with the only additional complication being the calculation of the stresses, which obviously depend upon nonlinear kinematic terms that are a function of the deformation gradient. Deriving the coefficients of the tangent matrix is somewhat more involved, requiring separate evaluation of constitutive, initial stress, and external force components. The latter deformation-dependent external force component is restricted to the case of enclosed normal pressure. In order to deal with near incompressibility the mean dilatation method is employed.

Having discretized the governing equations, the Newton–Raphson solution technique is reintroduced together with line search and arc length method enhancements.

## 9.2 DISCRETIZED KINEMATICS

The discretization is established in the initial configuration using isoparametric elements to interpolate the initial geometry in terms of the particles $\boldsymbol{X}_a$ defining the initial position of the element nodes as

$$\boldsymbol{X} = \sum_{a=1}^{n} N_a(\xi_1, \xi_2, \xi_3)\boldsymbol{X}_a, \tag{9.1}$$

where $N_a(\xi_1, \xi_2, \xi_3)$ are the standard shape functions and $n$ denotes the number of nodes per element. It should be emphasized that during the motion, nodes and elements are permanently attached to the material particles with which they were initially associated. Consequently, the subsequent motion is fully described in terms of the current position $\boldsymbol{x}_a(t)$ of the nodal particles as (Figure 9.1)

$$\boldsymbol{x} = \sum_{a=1}^{n} N_a \boldsymbol{x}_a(t). \tag{9.2}$$

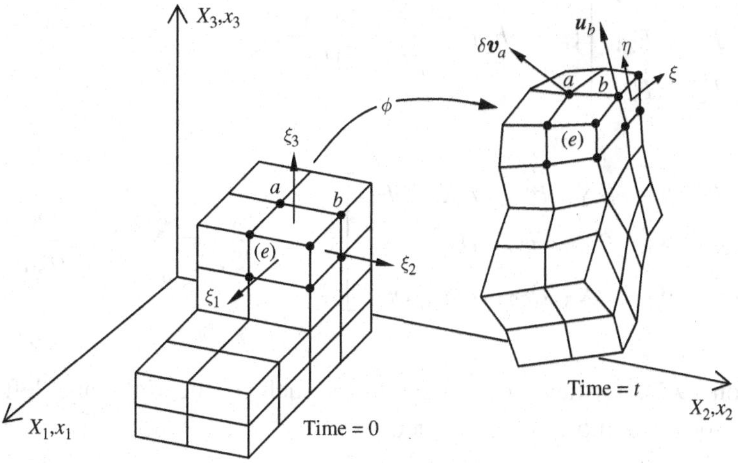


**FIGURE 9.1** Discretization.

Differentiating Equation (9.2) with respect to time gives the real or virtual velocity interpolation as

$$\boldsymbol{v} = \sum_{a=1}^{n} N_a \boldsymbol{v}_a; \qquad \delta\boldsymbol{v} = \sum_{a=1}^{n} N_a \delta\boldsymbol{v}_a. \tag{9.3}$$

Similarly, restricting the motion brought about by an arbitrary increment $\boldsymbol{u}$ to be consistent with Equation (9.2) implies that the displacement $\boldsymbol{u}$ is also interpolated as

$$\boldsymbol{u} = \sum_{a=1}^{n} N_a \boldsymbol{u}_a. \tag{9.4}$$

The fundamental deformation gradient tensor $\boldsymbol{F}$ is interpolated over an element by differentiating Equation (9.2) with respect to the initial coordinates to give, after using Equation (2.135a),

$$\boldsymbol{F} = \sum_{a=1}^{n} \boldsymbol{x}_a \otimes \boldsymbol{\nabla}_0 N_a, \tag{9.5}$$

where $\boldsymbol{\nabla}_0 N_a = \partial N_a/\partial \boldsymbol{X}$ can be related to $\boldsymbol{\nabla}_\xi N_a = \partial N_a/\partial \boldsymbol{\xi}$ in the standard manner by using the chain rule and Equation (9.1) to give

$$\frac{\partial N_a}{\partial \boldsymbol{X}} = \left(\frac{\partial \boldsymbol{X}}{\partial \boldsymbol{\xi}}\right)^{-T} \frac{\partial N_a}{\partial \boldsymbol{\xi}}; \qquad \frac{\partial \boldsymbol{X}}{\partial \boldsymbol{\xi}} = \sum_{a=1}^{n} \boldsymbol{X}_a \otimes \boldsymbol{\nabla}_\xi N_a. \tag{9.6a,b}$$

Equations (9.5) and (9.6a,b)$_b$ are sufficiently fundamental to justify expansion in detail in order to facilitate their eventual programming. To this effect, these equations are written in an explicit matrix form as

$$\boldsymbol{F} = \begin{bmatrix} F_{11} & F_{12} & F_{13} \\ F_{21} & F_{22} & F_{23} \\ F_{31} & F_{32} & F_{33} \end{bmatrix}; \qquad F_{iJ} = \sum_{a=1}^{n} x_{a,i} \frac{\partial N_a}{\partial X_J}; \tag{9.7}$$

and

$$\frac{\partial \boldsymbol{X}}{\partial \boldsymbol{\xi}} = \begin{bmatrix} \partial X_1/\partial \xi_1 & \partial X_1/\partial \xi_2 & \partial X_1/\partial \xi_3 \\ \partial X_2/\partial \xi_1 & \partial X_2/\partial \xi_2 & \partial X_2/\partial \xi_3 \\ \partial X_3/\partial \xi_1 & \partial X_3/\partial \xi_2 & \partial X_3/\partial \xi_3 \end{bmatrix}; \qquad \frac{\partial X_I}{\partial \xi_\alpha} = \sum_{a=1}^{n} X_{a,I} \frac{\partial N_a}{\partial \xi_\alpha}. \tag{9.8}$$

From Equation (9.5), further strain magnitudes such as the right and left Cauchy–Green tensors $\boldsymbol{C}$ and $\boldsymbol{b}$ can be obtained as

$$\boldsymbol{C} = \boldsymbol{F}^T \boldsymbol{F} = \sum_{a,b} (\boldsymbol{x}_a \cdot \boldsymbol{x}_b) \boldsymbol{\nabla}_0 N_a \otimes \boldsymbol{\nabla}_0 N_b; \quad C_{IJ} = \sum_{k=1}^{3} F_{kI} F_{kJ}; \tag{9.9a,b}$$

$$\boldsymbol{b} = \boldsymbol{F}\boldsymbol{F}^T = \sum_{a,b} (\boldsymbol{\nabla}_0 N_a \cdot \boldsymbol{\nabla}_0 N_b)\, \boldsymbol{x}_a \otimes \boldsymbol{x}_b; \quad b_{ij} = \sum_{K=1}^{3} F_{iK} F_{jK}. \tag{9.9c,d}$$

The discretization of the real (or virtual rate of deformation) tensor and the linear strain tensor can be obtained by introducing Equation (9.3) into the definition of $\boldsymbol{d}$ given in Equation (4.101), and Equation (9.4) into Equation (8.12a), to give

$$\boldsymbol{d} = \frac{1}{2} \sum_{a=1}^{n} (\boldsymbol{v}_a \otimes \boldsymbol{\nabla} N_a + \boldsymbol{\nabla} N_a \otimes \boldsymbol{v}_a); \tag{9.10a}$$

$$\delta \boldsymbol{d} = \frac{1}{2} \sum_{a=1}^{n} (\delta \boldsymbol{v}_a \otimes \boldsymbol{\nabla} N_a + \boldsymbol{\nabla} N_a \otimes \delta \boldsymbol{v}_a); \tag{9.10b}$$

$$\boldsymbol{\varepsilon} = \frac{1}{2} \sum_{a=1}^{n} (\boldsymbol{u}_a \otimes \boldsymbol{\nabla} N_a + \boldsymbol{\nabla} N_a \otimes \boldsymbol{u}_a); \tag{9.10c}$$

where, as in Equation (9.6a,b), $\boldsymbol{\nabla} N_a = \partial N_a/\partial \boldsymbol{x}$ can be obtained from the derivatives of the shape functions with respect to the isoparametric coordinates as

$$\frac{\partial N_a}{\partial \boldsymbol{x}} = \left(\frac{\partial \boldsymbol{x}}{\partial \boldsymbol{\xi}}\right)^{-T} \frac{\partial N_a}{\partial \boldsymbol{\xi}}; \quad \frac{\partial \boldsymbol{x}}{\partial \boldsymbol{\xi}} = \sum_{a=1}^{n} \boldsymbol{x}_a \otimes \boldsymbol{\nabla}_{\xi} N_a; \quad \frac{\partial x_i}{\partial \xi_\alpha} = \sum_{a=1}^{n} x_{a,i} \frac{\partial N_a}{\partial \xi_\alpha}. \tag{9.11a,b}$$

Although Equations (9.10) will eventually be expressed in a standard matrix form, if necessary the component tensor products can be expanded in a manner entirely analogous to Equations (9.6a,b) and (9.7).

### EXAMPLE 9.1: Discretization

This simple example illustrates the discretization and subsequent calculation of key shape function derivatives. Because the initial and current geometries comprise right-angled triangles, these are easily checked.

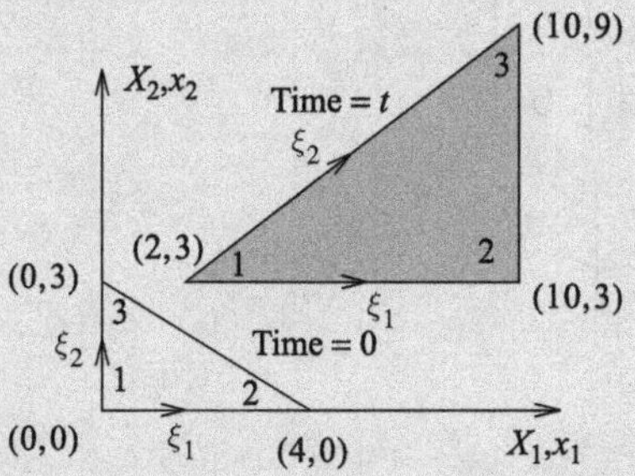


The initial $\boldsymbol{X}$ and current $\boldsymbol{x}$ nodal coordinates are

$$X_{1,1} = 0; \quad X_{2,1} = 4; \quad X_{3,1} = 0;$$
$$X_{1,2} = 0; \quad X_{2,2} = 0; \quad X_{3,2} = 3;$$
$$x_{1,1} = 2; \quad x_{2,1} = 10; \quad x_{3,1} = 10;$$
$$x_{1,2} = 3; \quad x_{2,2} = 3; \quad x_{3,2} = 9.$$

The shape functions and related derivatives are

$$N_1 = 1 - \xi_1 - \xi_2;$$
$$N_2 = \xi_1; \quad \frac{\partial N_1}{\partial \boldsymbol{\xi}} = \begin{bmatrix} -1 \\ -1 \end{bmatrix}; \quad \frac{\partial N_2}{\partial \boldsymbol{\xi}} = \begin{bmatrix} 1 \\ 0 \end{bmatrix}; \quad \frac{\partial N_3}{\partial \boldsymbol{\xi}} = \begin{bmatrix} 0 \\ 1 \end{bmatrix};$$
$$N_3 = \xi_2.$$

Equations (9.1) and (9.6a,b)$_b$ yield the initial position derivatives with respect to the nondimensional coordinates as

*(continued)*

**EXAMPLE 9.1:** *(cont.)*

$$\begin{matrix} X_1 = 4\xi_1 \\ X_2 = 3\xi_2 \end{matrix}\,; \quad \frac{\partial \boldsymbol{X}}{\partial \boldsymbol{\xi}} = \begin{bmatrix} 4 & 0 \\ 0 & 3 \end{bmatrix}; \quad \left(\frac{\partial \boldsymbol{X}}{\partial \boldsymbol{\xi}}\right)^{-T} = \frac{1}{12}\begin{bmatrix} 3 & 0 \\ 0 & 4 \end{bmatrix},$$

from which the derivatives of the shape functions with respect to the material coordinate system are found as

$$\frac{\partial N_1}{\partial \boldsymbol{X}} = \frac{1}{12}\begin{bmatrix} 3 & 0 \\ 0 & 4 \end{bmatrix}\begin{bmatrix} -1 \\ -1 \end{bmatrix} = -\frac{1}{12}\begin{bmatrix} 3 \\ 4 \end{bmatrix};$$

$$\frac{\partial N_2}{\partial \boldsymbol{X}} = \frac{1}{12}\begin{bmatrix} 3 \\ 0 \end{bmatrix}; \quad \frac{\partial N_3}{\partial \boldsymbol{X}} = \frac{1}{12}\begin{bmatrix} 0 \\ 4 \end{bmatrix}.$$

A similar set of manipulations using Equations (9.2) and (9.11a,b) yields the derivatives of the shape functions with respect to the spatial coordinate system as

$$\frac{\partial N_1}{\partial \boldsymbol{x}} = \frac{1}{24}\begin{bmatrix} 3 & 0 \\ -4 & 4 \end{bmatrix}\begin{bmatrix} -1 \\ -1 \end{bmatrix} = -\frac{1}{24}\begin{bmatrix} 3 \\ 0 \end{bmatrix};$$

$$\frac{\partial N_2}{\partial \boldsymbol{x}} = \frac{1}{24}\begin{bmatrix} 3 \\ -4 \end{bmatrix}; \quad \frac{\partial N_3}{\partial \boldsymbol{x}} = \frac{1}{24}\begin{bmatrix} 0 \\ 4 \end{bmatrix}.$$

**EXAMPLE 9.2: Discretized kinematics**

Following Example 9.1, the scene is set for the calculation of the deformation gradient $\boldsymbol{F}$ using Equation (9.7) to give

$$F_{iJ} = x_{1,i}\frac{\partial N_1}{\partial X_J} + x_{2,i}\frac{\partial N_2}{\partial X_J} + x_{3,i}\frac{\partial N_3}{\partial X_J}; i, J = 1, 2; \boldsymbol{F} = \frac{1}{3}\begin{bmatrix} 6 & 8 \\ 0 & 6 \end{bmatrix}.$$

Assuming plane strain deformation, the right and left Cauchy–Green tensors can be obtained from Equations (9.9a,b) and (9.9c,d) as

$$\boldsymbol{C} = \boldsymbol{F}^T\boldsymbol{F} = \frac{1}{9}\begin{bmatrix} 36 & 48 & 0 \\ 48 & 100 & 0 \\ 0 & 0 & 9 \end{bmatrix}; \quad \boldsymbol{b} = \boldsymbol{F}\boldsymbol{F}^T = \frac{1}{9}\begin{bmatrix} 100 & 48 & 0 \\ 48 & 36 & 0 \\ 0 & 0 & 9 \end{bmatrix}.$$

Finally, the Jacobian $J$ is found as

$$J = \det \boldsymbol{F} = \det\left(\frac{1}{3}\begin{bmatrix} 6 & 8 & 0 \\ 0 & 6 & 0 \\ 0 & 0 & 3 \end{bmatrix}\right) = 4.$$

## 9.3 DISCRETIZED EQUILIBRIUM EQUATIONS

### 9.3.1 General Derivation

In order to obtain the discretized spatial equilibrium equations, recall the spatial virtual work Equation (5.27) given as the total virtual work done by the residual force $\boldsymbol{r}$ as

$$\delta W(\boldsymbol{\phi}, \delta \boldsymbol{v}) = \int_v \boldsymbol{\sigma} : \delta \boldsymbol{d}\, dv - \int_v \boldsymbol{f} \cdot \delta \boldsymbol{v}\, dv - \int_{\partial v} \boldsymbol{t} \cdot \delta \boldsymbol{v}\, da. \tag{9.12}$$

At this stage, it is easier to consider the contribution to $\delta W(\boldsymbol{\phi}, \delta \boldsymbol{v})$ caused by a single virtual nodal velocity $\delta \boldsymbol{v}_a$ occurring at a typical node $a$ of element $(e)$. Introducing the interpolation for $\delta \boldsymbol{v}$ and $\delta \boldsymbol{d}$ given by Equations (9.3) and (9.10) gives

$$\delta W^{(e)}(\boldsymbol{\phi}, N_a \delta \boldsymbol{v}_a) = \int_{v^{(e)}} \boldsymbol{\sigma} : (\delta \boldsymbol{v}_a \otimes \boldsymbol{\nabla} N_a) dv - \int_{v^{(e)}} \boldsymbol{f} \cdot (N_a \delta \boldsymbol{v}_a)\, dv - \int_{\partial v^{(e)}} \boldsymbol{t} \cdot (N_a \delta \boldsymbol{v}_a)\, da, \tag{9.13}$$

where the symmetry of $\boldsymbol{\sigma}$ has been used to concatenate the internal energy term. Observing that the virtual nodal velocities are independent of the integration and Equation (2.52b), that is, $\boldsymbol{\sigma} : (\boldsymbol{u} \otimes \boldsymbol{v}) = \boldsymbol{u} \cdot \boldsymbol{\sigma} \boldsymbol{v}$ for any vectors $\boldsymbol{u}, \boldsymbol{v}$, enables the summation to be rearranged to give

$$\delta W^{(e)}(\boldsymbol{\phi}, N_a \delta \boldsymbol{v}_a) = \delta \boldsymbol{v}_a \cdot \left( \int_{v^{(e)}} \boldsymbol{\sigma} \boldsymbol{\nabla} N_a\, dv - \int_{v^{(e)}} N_a \boldsymbol{f}\, dv - \int_{\partial v^{(e)}} N_a \boldsymbol{t}\, da \right). \tag{9.14}$$

The virtual work per element $(e)$ per node $a$ can, alternatively, be expressed in terms of the internal and external equivalent nodal forces $\boldsymbol{T}_a^{(e)}$ and $\boldsymbol{F}_a^{(e)}$ as

$$\delta W^{(e)}(\boldsymbol{\phi}, N_a \delta \boldsymbol{v}_a) = \delta \boldsymbol{v}_a \cdot (\boldsymbol{T}_a^{(e)} - \boldsymbol{F}_a^{(e)}), \tag{9.15a}$$

where

$$\boldsymbol{T}_a^{(e)} = \int_{v^{(e)}} \boldsymbol{\sigma} \boldsymbol{\nabla} N_a\, dv; \qquad T_{a,i}^{(e)} = \sum_{j=1}^{3} \int_{v^{(e)}} \sigma_{ij} \frac{\partial N_a}{\partial x_j}\, dv; \tag{9.15b}$$

$$\boldsymbol{F}_a^{(e)} = \int_{v^{(e)}} N_a \boldsymbol{f}\, dv + \int_{\partial v^{(e)}} N_a \boldsymbol{t}\, da. \tag{9.15c}$$

In this equation the Cauchy stress $\boldsymbol{\sigma}$ is found from the appropriate constitutive equation given in Chapter 6, which will involve the calculation of the left

Cauchy–Green tensor $\boldsymbol{b} = \boldsymbol{F}\boldsymbol{F}^T$; for example, see Equations (6.29) or (6.55a,b) and Example 9.3.

**EXAMPLE 9.3: Equivalent nodal forces $\boldsymbol{T}_a$**

Building on the previous example, the calculation of the equivalent nodal forces is now demonstrated. A compressible neo-Hookean material will be considered for which $\mu = 3$ and $\lambda = 2$, which, using Equation (6.29), yields the Cauchy stresses (rounded for convenience) for this plane strain case as

$$\boldsymbol{\sigma} = \begin{bmatrix} \sigma_{11} & \sigma_{12} & 0 \\ \sigma_{21} & \sigma_{22} & 0 \\ 0 & 0 & \sigma_{33} \end{bmatrix} = \frac{\mu}{J}(\boldsymbol{b} - \boldsymbol{I}) + \frac{\lambda}{J}(\ln J)\boldsymbol{I} \approx \begin{bmatrix} 8 & 4 & 0 \\ 4 & 3 & 0 \\ 0 & 0 & 0.7 \end{bmatrix}.$$

From Equation (9.15b), the equivalent nodal internal forces are

$$T_{a,i} = \int_{v^{(e)}} \left( \sigma_{i1} \frac{\partial N_a}{\partial x_1} + \sigma_{i2} \frac{\partial N_a}{\partial x_2} \right) dv;$$

$$\begin{matrix} a = 1,2,3 \\ i = 1,2 \end{matrix}\; ; \quad \begin{matrix} T_{1,1} = -24t \\ T_{1,2} = -12t \end{matrix}\; ; \quad \begin{matrix} T_{2,1} = 8t \\ T_{2,2} = 0 \end{matrix}\; ; \quad \begin{matrix} T_{3,1} = 16t; \\ T_{3,2} = 12t \end{matrix}\; ;$$

where $t$ is the element thickness. Clearly, these forces are in equilibrium.

The contribution to $\delta W(\boldsymbol{\phi}, N_a \delta \boldsymbol{v}_a)$ from all elements $e$ (1 to $m_a$) containing node $a$ ($e \ni a$) is

$$\delta W(\boldsymbol{\phi}, N_a \delta \boldsymbol{v}_a) = \sum_{\substack{e=1 \\ e \ni a}}^{m_a} \delta W^{(e)}(\boldsymbol{\phi}, N_a \delta \boldsymbol{v}_a) = \delta \boldsymbol{v}_a \cdot (\boldsymbol{T}_a - \boldsymbol{F}_a), \tag{9.16a}$$

where the assembled equivalent nodal forces are

$$\boldsymbol{T}_a = \sum_{\substack{e=1 \\ e \ni a}}^{m_a} \boldsymbol{T}_a^{(e)}; \qquad \boldsymbol{F}_a = \sum_{\substack{e=1 \\ e \ni a}}^{m_a} \boldsymbol{F}_a^{(e)}. \tag{9.16b,c}$$

Finally, the contribution to $\delta W(\boldsymbol{\phi}, \delta \boldsymbol{v})$ from all nodes $N$ in the finite element mesh is

$$\delta W(\boldsymbol{\phi}, \delta \boldsymbol{v}) = \sum_{a=1}^{N} \delta W(\boldsymbol{\phi}, N_a \delta \boldsymbol{v}_a) = \sum_{a=1}^{N} \delta \boldsymbol{v}_a \cdot (\boldsymbol{T}_a - \boldsymbol{F}_a) = 0. \tag{9.17}$$

Because the virtual work equation must be satisfied for any arbitrary virtual nodal velocities, the discretized equilibrium equations, in terms of the nodal residual force $\boldsymbol{R}_a$, emerge as

$$\boldsymbol{R}_a = \boldsymbol{T}_a - \boldsymbol{F}_a = \boldsymbol{0}. \tag{9.18}$$

Consequently, the equivalent internal nodal forces are in equilibrium with the equivalent external forces at each node $a = 1, 2, \ldots, N$.

For convenience, all the nodal equivalent forces are assembled into single arrays to define the complete internal and external forces $\mathsf{T}$ and $\mathsf{F}$ respectively, as well as the complete residual force $\mathsf{R}$, as

$$\mathsf{T} = \begin{bmatrix} \boldsymbol{T}_1 \\ \boldsymbol{T}_2 \\ \vdots \\ \boldsymbol{T}_N \end{bmatrix}; \quad \mathsf{F} = \begin{bmatrix} \boldsymbol{F}_1 \\ \boldsymbol{F}_2 \\ \vdots \\ \boldsymbol{F}_N \end{bmatrix}; \quad \mathsf{R} = \begin{bmatrix} \boldsymbol{R}_1 \\ \boldsymbol{R}_2 \\ \vdots \\ \boldsymbol{R}_N \end{bmatrix}. \tag{9.19}$$

These definitions enable the discretized virtual work Equation (9.17) to be rewritten as

$$\delta W(\boldsymbol{\phi}, \delta \boldsymbol{v}) = \delta \mathsf{v}^T \mathsf{R} = \delta \mathsf{v}^T (\mathsf{T} - \mathsf{F}) = 0, \tag{9.20}$$

where the complete virtual velocity vector $\delta \mathsf{v}^T = [\delta \boldsymbol{v}_1^T, \delta \boldsymbol{v}_2^T, \ldots, \delta \boldsymbol{v}_N^T]$.

Finally, recalling that the internal equivalent forces are nonlinear functions of the current nodal positions $\boldsymbol{x}_a$ and defining a complete vector $\mathsf{x}$ of unknowns as the array containing all nodal positions as

$$\mathsf{x} = \begin{bmatrix} \boldsymbol{x}_1 \\ \boldsymbol{x}_2 \\ \vdots \\ \boldsymbol{x}_N \end{bmatrix} \tag{9.21}$$

enables the complete nonlinear equilibrium equations to be symbolically assembled as

$$\mathsf{R}(\mathsf{x}) = \mathsf{T}(\mathsf{x}) - \mathsf{F}(\mathsf{x}) = \mathbf{0}. \tag{9.22}$$

These equations represent the finite element discretization of the pointwise differential equilibrium Equation (5.16).

### 9.3.2 Derivation in Matrix Notation

The discretized equilibrium equations will now be recast in the more familiar matrix-vector notation.* To achieve this requires a reinterpretation of the symmetric stress tensor as a vector comprising six independent components as

$$\underline{\sigma} = [\sigma_{11}, \sigma_{22}, \sigma_{33}, \sigma_{12}, \sigma_{13}, \sigma_{23}]^T. \tag{9.23}$$

* Also known as *Voigt notation*.

Similarly, the symmetric rate of deformation tensor can be re-established in a corresponding manner as

$$\mathbf{d} = [d_{11}, d_{22}, d_{33}, 2d_{12}, 2d_{13}, 2d_{23}]^T, \tag{9.24}$$

where the off-diagonal terms have been doubled to ensure that the product $\mathbf{d}^T\underline{\boldsymbol{\sigma}}$ gives the correct internal energy as

$$\int_v \boldsymbol{\sigma} : \boldsymbol{d}\, dv = \int_v \mathbf{d}^T \underline{\boldsymbol{\sigma}}\, dv. \tag{9.25}$$

The rate of deformation vector $\mathbf{d}$ can be expressed in terms of the usual $\mathbf{B}$ matrix and the nodal velocities as

$$\mathbf{d} = \sum_{a=1}^{n} \mathbf{B}_a \boldsymbol{v}_a; \qquad \mathbf{B}_a = \begin{bmatrix} \frac{\partial N_a}{\partial x_1} & 0 & 0 \\ 0 & \frac{\partial N_a}{\partial x_2} & 0 \\ 0 & 0 & \frac{\partial N_a}{\partial x_3} \\ \frac{\partial N_a}{\partial x_2} & \frac{\partial N_a}{\partial x_1} & 0 \\ \frac{\partial N_a}{\partial x_3} & 0 & \frac{\partial N_a}{\partial x_1} \\ 0 & \frac{\partial N_a}{\partial x_3} & \frac{\partial N_a}{\partial x_2} \end{bmatrix}. \tag{9.26}$$

Introducing Equation (9.26) into Equation (9.25) for the internal energy enables the discretized virtual work Equation (9.13) to be rewritten as

$$\delta W(\boldsymbol{\phi}, N_a \delta \boldsymbol{v}_a) = \int_{v^{(e)}} (\mathbf{B}_a \delta \boldsymbol{v}_a)^T \underline{\boldsymbol{\sigma}}\, dv - \int_{v^{(e)}} \boldsymbol{f} \cdot (N_a \delta \boldsymbol{v}_a)\, dv$$
$$- \int_{\partial v^{(e)}} \boldsymbol{t} \cdot (N_a \delta \boldsymbol{v}_a)\, da. \tag{9.27}$$

Following the derivation given in the previous section leads to an alternative expression for the element equivalent nodal forces $\boldsymbol{T}_a^{(e)}$ for node $a$ as

$$\boldsymbol{T}_a^{(e)} = \int_{v^{(e)}} \mathbf{B}_a^T \underline{\boldsymbol{\sigma}}\, dv. \tag{9.28}$$

Observe that, because of the presence of zeros in the matrix $\mathbf{B}_a$, expression (9.15b) is computationally more efficient than Equation (9.28).

## 9.4 DISCRETIZATION OF THE LINEARIZED EQUILIBRIUM EQUATIONS

Equation (9.22) represents a set of nonlinear equilibrium equations with the current nodal positions as unknowns. The solution of these equations is achieved using a Newton–Raphson iterative procedure that involves the discretization of the linearized equilibrium equations given in Section 8.2. For notational convenience the virtual work Equation (9.12) is split into internal and external work components as

$$\delta W(\boldsymbol{\phi}, \delta \boldsymbol{v}) = \delta W_{\text{int}}(\boldsymbol{\phi}, \delta \boldsymbol{v}) - \delta W_{\text{ext}}(\boldsymbol{\phi}, \delta \boldsymbol{v}), \tag{9.29}$$

which can be linearized in the direction $\boldsymbol{u}$ to give

$$D\delta W(\boldsymbol{\phi}, \delta \boldsymbol{v})[\boldsymbol{u}] = D\delta W_{\text{int}}(\boldsymbol{\phi}, \delta \boldsymbol{v})[\boldsymbol{u}] - D\delta W_{\text{ext}}(\boldsymbol{\phi}, \delta \boldsymbol{v})[\boldsymbol{u}], \tag{9.30}$$

where the linearization of the internal virtual work can be further subdivided into constitutive and initial stress components as

$$\begin{aligned} D\delta W_{\text{int}}(\boldsymbol{\phi}, \delta \boldsymbol{v})[\boldsymbol{u}] &= D\delta W_c(\boldsymbol{\phi}, \delta \boldsymbol{v})[\boldsymbol{u}] + D\delta W_\sigma(\boldsymbol{\phi}, \delta \boldsymbol{v})[\boldsymbol{u}] \\ &= \int_v \delta \boldsymbol{d} : \mathcal{c} : \boldsymbol{\varepsilon}\, dv + \int_v \boldsymbol{\sigma} : [(\boldsymbol{\nabla} \boldsymbol{u})^T \boldsymbol{\nabla} \delta \boldsymbol{v}]\, dv. \end{aligned} \tag{9.31}$$

Before continuing with the discretization of the linearized equilibrium Equation (9.30), it is worth reiterating the general discussion of Section 8.2 to inquire in more detail why this is likely to yield a *tangent stiffness* matrix. Recall that Equation (9.15a), that is, $\delta W^{(e)}(\boldsymbol{\phi}, N_a \delta \boldsymbol{v}_a) = \delta \boldsymbol{v}_a \cdot (\boldsymbol{T}_a^{(e)} - \boldsymbol{F}_a^{(e)})$, essentially expresses the contribution of the nodal equivalent forces $\boldsymbol{T}_a^{(e)}$ and $\boldsymbol{F}_a^{(e)}$ to the overall equilibrium of node $a$. Observing that $\boldsymbol{F}_a^{(e)}$ may be position-dependent, linearization of Equation (9.15c) in the direction $N_b \boldsymbol{u}_b$, with $N_a \delta \boldsymbol{v}_a$ remaining constant, expresses the change in the nodal equivalent forces $\boldsymbol{T}_a^{(e)}$ and $\boldsymbol{F}_a^{(e)}$, at node $a$, due to a change $\boldsymbol{u}_b$ in the current position of node $b$ as

$$\begin{aligned} D\delta W^{(e)}(\boldsymbol{\phi}, N_a \delta \boldsymbol{v}_a)[N_b \boldsymbol{u}_b] &= D(\delta \boldsymbol{v}_a \cdot (\boldsymbol{T}_a^{(e)} - \boldsymbol{F}_a^{(e)}))[N_b \boldsymbol{u}_b] \\ &= \delta \boldsymbol{v}_a \cdot D(\boldsymbol{T}_a^{(e)} - \boldsymbol{F}_a^{(e)})[N_b \boldsymbol{u}_b] \\ &= \delta \boldsymbol{v}_a \cdot \boldsymbol{K}_{ab}^{(e)}\, \boldsymbol{u}_b. \end{aligned} \tag{9.32}$$

The relationship between changes in forces at node $a$ due to changes in the current position of node $b$ is furnished by the tangent stiffness matrix $\boldsymbol{K}_{ab}^{(e)}$, which is clearly seen to derive from the linearization of the virtual work equation. In physical terms the tangent stiffness provides the Newton–Raphson procedure with the operator that adjusts current nodal positions so that the deformation-dependent equivalent nodal forces tend toward being in equilibrium with the external equivalent nodal forces.

### 9.4.1 Constitutive Component: Indicial Form

The constitutive contribution to the linearized virtual work Equation (9.31) for element $(e)$ linking nodes $a$ and $b$ is

$$D\delta W_c^{(e)}(\boldsymbol{\phi}, N_a\delta\boldsymbol{v}_a)[N_b\boldsymbol{u}_b]$$
$$= \int_{v^{(e)}} \frac{1}{2}(\delta\boldsymbol{v}_a \otimes \boldsymbol{\nabla}N_a + \boldsymbol{\nabla}N_a \otimes \delta\boldsymbol{v}_a) : \boldsymbol{\mathcal{c}} : \frac{1}{2}(\boldsymbol{u}_b \otimes \boldsymbol{\nabla}N_b + \boldsymbol{\nabla}N_b \otimes \boldsymbol{u}_b)\, dv. \tag{9.33}$$

In order to make progress it is necessary to temporarily resort to indicial notation, which enables the above equation to be rewritten as

$$D\delta W_c^{(e)}(\boldsymbol{\phi}, N_a\delta\boldsymbol{v}_a)[N_b\boldsymbol{u}_b]$$
$$= \sum_{i,j,k,l=1}^{3} \int_{v^{(e)}} \frac{1}{2}\left(\delta v_{a,i}\frac{\partial N_a}{\partial x_j} + \delta v_{a,j}\frac{\partial N_a}{\partial x_i}\right)\mathcal{c}_{ijkl}\,\frac{1}{2}\left(u_{b,k}\frac{\partial N_b}{\partial x_l} + u_{b,l}\frac{\partial N_b}{\partial x_k}\right) dv$$
$$= \sum_{i,j,k,l=1}^{3} \delta v_{a,i}\left(\int_{v^{(e)}} \frac{\partial N_a}{\partial x_j}\,\mathcal{c}_{ijkl}\,\frac{\partial N_b}{\partial x_l}\, dv\right) u_{b,k}$$
$$= \delta\boldsymbol{v}_a \cdot \boldsymbol{K}_{c,ab}^{(e)}\boldsymbol{u}_b, \tag{9.34}$$

where the constitutive component of the tangent matrix relating node $a$ to node $b$ in element $(e)$ is

$$[\boldsymbol{K}_{c,ab}]_{ij} = \int_{v^{(e)}} \sum_{k,l=1}^{3} \frac{\partial N_a}{\partial x_k}\,\mathcal{c}_{ikjl}\,\frac{\partial N_b}{\partial x_l}\, dv; \qquad i,j = 1,2,3. \tag{9.35}$$

**EXAMPLE 9.4: Constitutive component of tangent matrix $[\boldsymbol{K}_{c,ab}]$**

The previous example is revisited in order to illustrate the calculation of the tangent matrix component connecting nodes 2 and 3. Omitting zero derivative terms, the summation given by Equation (9.35) yields

$$[\boldsymbol{K}_{c,23}]_{11} = \left[\left(\frac{1}{8}\right)(\mathcal{c}_{1112})\left(\frac{1}{6}\right) - \left(\frac{1}{6}\right)(\mathcal{c}_{1212})\left(\frac{1}{6}\right)\right](24t);$$
$$[\boldsymbol{K}_{c,23}]_{12} = \left[\left(\frac{1}{8}\right)(\mathcal{c}_{1122})\left(\frac{1}{6}\right) - \left(\frac{1}{6}\right)(\mathcal{c}_{1222})\left(\frac{1}{6}\right)\right](24t);$$
$$[\boldsymbol{K}_{c,23}]_{21} = \left[\left(\frac{1}{8}\right)(\mathcal{c}_{2112})\left(\frac{1}{6}\right) - \left(\frac{1}{6}\right)(\mathcal{c}_{2212})\left(\frac{1}{6}\right)\right](24t);$$
$$[\boldsymbol{K}_{c,23}]_{22} = \left[\left(\frac{1}{8}\right)(\mathcal{c}_{2122})\left(\frac{1}{6}\right) - \left(\frac{1}{6}\right)(\mathcal{c}_{2222})\left(\frac{1}{6}\right)\right](24t);$$

*(continued)*

**EXAMPLE 9.4:** *(cont.)*

where $t$ is the thickness of the element. Substituting for $\mathcal{c}_{ijkl}$ from Equations (6.40) and (6.41) yields the stiffness coefficients as

$$[\boldsymbol{K}_{c,23}]_{11} = -\frac{2}{3}\mu' t; \quad [\boldsymbol{K}_{c,23}]_{12} = \frac{1}{2}\lambda' t; \quad [\boldsymbol{K}_{c,23}]_{21} = \frac{1}{2}\mu' t;$$

$$[\boldsymbol{K}_{c,23}]_{22} = -\frac{2}{3}(\lambda' + 2\mu')t;$$

where $\lambda' = \lambda/J$ and $\mu' = (\mu - \lambda \ln J)/J$.

### 9.4.2 Constitutive Component: Matrix Form

The constitutive contribution to the linearized virtual work Equation (9.31) for element $(e)$ can alternatively be expressed in matrix notation by defining the small strain vector $\underline{\varepsilon}$ in a similar manner to Equation (9.26) for $\mathbf{d}$ as

$$\underline{\varepsilon} = [\varepsilon_{11}, \varepsilon_{22}, \varepsilon_{33}, 2\varepsilon_{12}, 2\varepsilon_{13}, 2\varepsilon_{23}]^T; \qquad \underline{\varepsilon} = \sum_{a=1}^{n} \mathbf{B}_a \boldsymbol{u}_a. \tag{9.36a,b}$$

The constitutive component of the linearized internal virtual work (see Equation (9.31)) can now be rewritten in matrix-vector notation as

$$D\delta W_c(\boldsymbol{\phi}, \delta\boldsymbol{v})[\boldsymbol{u}] = \int_V \delta\boldsymbol{d} : \mathcal{c} : \boldsymbol{\varepsilon}\, dv = \int_v \delta\mathbf{d}^T \mathbf{D}\underline{\varepsilon}\, dv, \tag{9.37}$$

where the spatial constitutive matrix $\mathbf{D}$ is constructed from the components of the fourth-order tensor $\mathcal{c}$ by equating the tensor product $\delta\boldsymbol{d} : \mathcal{c} : \boldsymbol{\varepsilon}$ to the matrix product $\delta\mathbf{d}^T\mathbf{D}\underline{\varepsilon}$ to give, after some algebra,

$$\mathbf{D} = \begin{bmatrix} \mathcal{c}_{1111} & \mathcal{c}_{1122} & \mathcal{c}_{1133} & \mathcal{c}_{1112} & \mathcal{c}_{1113} & \mathcal{c}_{1123} \\ & \mathcal{c}_{2222} & \mathcal{c}_{2233} & \mathcal{c}_{2212} & \mathcal{c}_{2213} & \mathcal{c}_{2223} \\ & & \mathcal{c}_{3333} & \mathcal{c}_{3312} & \mathcal{c}_{3313} & \mathcal{c}_{3323} \\ & & & \mathcal{c}_{1212} & \mathcal{c}_{1213} & \mathcal{c}_{1223} \\ & \text{sym.} & & & \mathcal{c}_{1313} & \mathcal{c}_{1323} \\ & & & & & \mathcal{c}_{2323} \end{bmatrix}. \tag{9.38}$$

In the particular case of a neo-Hookean material (see Equation (6.29)), $\mathbf{D}$ becomes

$$\mathbf{D} = \begin{bmatrix} \lambda' + 2\mu' & \lambda' & \lambda' & 0 & 0 & 0 \\ \lambda' & \lambda' + 2\mu' & \lambda' & 0 & 0 & 0 \\ \lambda' & \lambda' & \lambda' + 2\mu' & 0 & 0 & 0 \\ 0 & 0 & 0 & \mu' & 0 & 0 \\ 0 & 0 & 0 & 0 & \mu' & 0 \\ 0 & 0 & 0 & 0 & 0 & \mu' \end{bmatrix};$$

$$\lambda' = \frac{\lambda}{J}; \qquad \mu' = \frac{\mu - \lambda \ln J}{J}. \tag{9.39}$$

Substituting for $\delta\mathbf{d}$ and $\underline{\varepsilon}$ from Equations (9.26) and (9.36a,b) respectively into the right-hand side of Equation (9.37) enables the contribution from element $(e)$ associated with nodes $a$ and $b$ to emerge as

$$\begin{aligned} D\delta W_c^{(e)}(\boldsymbol{\phi}, N_a \delta\boldsymbol{v}_a)[N_b \boldsymbol{u}_b] &= \int_{v^{(e)}} (\mathbf{B}_a \delta\boldsymbol{v}_a)^T \mathbf{D} (\mathbf{B}_b \boldsymbol{u}_b)\, dv \\ &= \delta\boldsymbol{v}_a \cdot \left( \int_{v^{(e)}} \mathbf{B}_a^T \mathbf{D} \mathbf{B}_b \, dv \right) \boldsymbol{u}_b. \end{aligned} \tag{9.40}$$

The term in brackets defines the constitutive component of the tangent matrix relating node $a$ to node $b$ in element $(e)$ as

$$\boldsymbol{K}_{c,ab}^{(e)} = \int_{v^{(e)}} \mathbf{B}_a^T \mathbf{D} \mathbf{B}_b \, dv. \tag{9.41}$$

### 9.4.3 Initial Stress Component

Concentrating attention on the second term in the linearized equilibrium Equation (9.31), note first that the gradients of $\delta\boldsymbol{v}$ and $\boldsymbol{u}$ can be interpolated from Equations (9.3)–(9.4) as

$$\boldsymbol{\nabla} \delta\boldsymbol{v} = \sum_{a=1}^{n} \delta\boldsymbol{v}_a \otimes \boldsymbol{\nabla} N_a; \tag{9.42a}$$

$$\boldsymbol{\nabla} \boldsymbol{u} = \sum_{b=1}^{n} \boldsymbol{u}_b \otimes \boldsymbol{\nabla} N_b. \tag{9.42b}$$

Introducing these two equations into the second term of Equation (9.31) and noting Equation (2.52b), that is, $\boldsymbol{\sigma} : (\boldsymbol{u} \otimes \boldsymbol{v}) = \boldsymbol{u} \cdot \boldsymbol{\sigma} \boldsymbol{v}$ for any vectors $\boldsymbol{u}, \boldsymbol{v}$, enables the initial stress contribution to the linearized virtual work Equation (9.31) for element $(e)$ linking nodes $a$ and $b$ to be found as

$$
\begin{aligned}
D\delta W_\sigma(\boldsymbol{\phi}, N_a\delta\boldsymbol{v}_a)[N_b\boldsymbol{u}_b] &= \int_v \boldsymbol{\sigma} : [(\boldsymbol{\nabla}\boldsymbol{u}_b)^T \boldsymbol{\nabla}\delta\boldsymbol{v}_a]\, dv \\
&= \int_{v^{(e)}} \boldsymbol{\sigma} : [(\delta\boldsymbol{v}_a \cdot \boldsymbol{u}_b)\boldsymbol{\nabla}N_b \otimes \boldsymbol{\nabla}N_a]\, dv \\
&= (\delta\boldsymbol{v}_a \cdot \boldsymbol{u}_b) \int_{v^{(e)}} \boldsymbol{\nabla}N_a \cdot \boldsymbol{\sigma}\boldsymbol{\nabla}N_b\, dv. \qquad (9.43)
\end{aligned}
$$

Observing that the integral in Equation (9.43) is a scalar, and noting that $\delta\boldsymbol{v}_a \cdot \boldsymbol{u}_b = \delta\boldsymbol{v}_a \cdot \boldsymbol{I}\boldsymbol{u}_b$, the expression can be rewritten in matrix form as

$$D\delta W_\sigma(\boldsymbol{\phi}, N_a\delta\boldsymbol{v}_a)[N_b\boldsymbol{u}_b] = \delta\boldsymbol{v}_a \cdot \boldsymbol{K}^{(e)}_{\sigma,ab}\boldsymbol{u}_b, \qquad (9.44a)$$

where the components of the so-called *initial stress matrix* $\boldsymbol{K}^{(e)}_{\sigma,ab}$ are

$$\boldsymbol{K}^{(e)}_{\sigma,ab} = \int_{v^{(e)}} (\boldsymbol{\nabla}N_a \cdot \boldsymbol{\sigma}\boldsymbol{\nabla}N_b)\boldsymbol{I}\, dv; \qquad (9.44b)$$

$$[\boldsymbol{K}^{(e)}_{\sigma,ab}]_{ij} = \int_{v^{(e)}} \sum_{k,l=1}^{3} \frac{\partial N_a}{\partial x_k}\sigma_{kl}\frac{\partial N_b}{\partial x_l}\delta_{ij}\, dv; \qquad i,j = 1,2,3. \qquad (9.44c)$$

**EXAMPLE 9.5: Initial stress component of tangent matrix $[\boldsymbol{K}_{\sigma,ab}]$**

Using the same configuration as in Examples 9.1–9.4, a typical initial stress tangent matrix component connecting nodes 1 and 2 can be found from Equation (9.44c) as

$$
\begin{aligned}
[\boldsymbol{K}_{\sigma,12}] &= \int_{v^{(e)}} \begin{bmatrix} \frac{\partial N_1}{\partial x_1} & \frac{\partial N_1}{\partial x_2} \end{bmatrix} \begin{bmatrix} \sigma_{11} & \sigma_{12} \\ \sigma_{21} & \sigma_{22} \end{bmatrix} \begin{bmatrix} \frac{\partial N_2}{\partial x_1} \\ \frac{\partial N_2}{\partial x_2} \end{bmatrix} \begin{bmatrix} 1 & 0 \\ 0 & 1 \end{bmatrix} dv \\
&= \left(\left(-\frac{1}{8}\right)8\left(\frac{1}{8}\right) + \left(-\frac{1}{8}\right)4\left(-\frac{1}{6}\right)\right) \begin{bmatrix} 1 & 0 \\ 0 & 1 \end{bmatrix} 24t \\
&= \begin{bmatrix} -1t & 0 \\ 0 & -1t \end{bmatrix}.
\end{aligned}
$$

### 9.4.4 External Force Component

As explained in Section 8.5, the body forces are invariably independent of the motion and consequently do not contribute to the linearized virtual work. However, for the particular case of enclosed normal pressure discussed in Section 8.5.2, the linearization of the associated virtual work term is given by Equation (8.24) as

$$D\delta W^{p}_{\text{ext}}(\boldsymbol{\phi}, \delta\boldsymbol{v})[\boldsymbol{u}] = \frac{1}{2}\int_{A_\xi} p\frac{\partial \boldsymbol{x}}{\partial \xi}\cdot\left[\left(\frac{\partial \delta\boldsymbol{v}}{\partial \eta}\times\boldsymbol{u}\right) - \left(\delta\boldsymbol{v}\times\frac{\partial \boldsymbol{u}}{\partial \eta}\right)\right] d\xi d\eta$$
$$-\frac{1}{2}\int_{A_\xi} p\frac{\partial \boldsymbol{x}}{\partial \eta}\cdot\left[\left(\frac{\partial \delta\boldsymbol{v}}{\partial \xi}\times\boldsymbol{u}\right) - \left(\delta\boldsymbol{v}\times\frac{\partial \boldsymbol{u}}{\partial \xi}\right)\right] d\xi d\eta. \tag{9.45}$$

Implicit in the isoparametric volume interpolation is a corresponding surface representation in terms of $\xi$ and $\eta$ as (see Figure 9.1)

$$\boldsymbol{x}(\xi, \eta) = \sum_{a=1}^{n} N_a \boldsymbol{x}_a, \tag{9.46}$$

where $n$ is the number of nodes per surface element. Similar surface interpolations apply to both $\delta\boldsymbol{v}$ and $\boldsymbol{u}$ in Equation (9.45). Considering, as before, the contribution to the linearized external virtual work term in Equation (9.30) from surface element $(e)$ associated with nodes $a$ and $b$ gives

$$\begin{aligned} D\delta W^{p(e)}{}_{\text{ext}}&(\boldsymbol{\phi}, N_a\delta\boldsymbol{v}_a)[N_b\boldsymbol{u}_b] \\ &= (\delta\boldsymbol{v}_a\times\boldsymbol{u}_b)\cdot\frac{1}{2}\int_{A_\xi} p\frac{\partial \boldsymbol{x}}{\partial \xi}\left(\frac{\partial N_a}{\partial \eta}N_b - \frac{\partial N_b}{\partial \eta}N_a\right) d\xi d\eta \\ &\quad - (\delta\boldsymbol{v}_a\times\boldsymbol{u}_b)\cdot\frac{1}{2}\int_{A_\xi} p\frac{\partial \boldsymbol{x}}{\partial \eta}\left(\frac{\partial N_a}{\partial \xi}N_b - \frac{\partial N_b}{\partial \xi}N_a\right) d\xi d\eta \\ &= (\delta\boldsymbol{v}_a\times\boldsymbol{u}_b)\cdot\boldsymbol{k}_{p,ab}, \end{aligned} \tag{9.47}$$

where the vector of stiffness coefficients $\boldsymbol{k}_{p,ab}$ is

$$\begin{aligned} \boldsymbol{k}_{p,ab} &= \frac{1}{2}\int_{A_\xi} p\frac{\partial \boldsymbol{x}}{\partial \xi}\left(\frac{\partial N_a}{\partial \eta}N_b - \frac{\partial N_b}{\partial \eta}N_a\right) d\xi d\eta \\ &\quad -\frac{1}{2}\int_{A_\xi} p\frac{\partial \boldsymbol{x}}{\partial \eta}\left(\frac{\partial N_a}{\partial \xi}N_b - \frac{\partial N_b}{\partial \xi}N_a\right) d\xi d\eta. \end{aligned} \tag{9.48}$$

Equation (9.47) can now be reinterpreted in terms of tangent matrix components as

$$D\delta W^{p(e)}_{\text{ext}}(\boldsymbol{\phi}, N_a\delta\boldsymbol{v}_a)[N_b\boldsymbol{u}_b] = \delta\boldsymbol{v}_a\cdot\boldsymbol{K}^{(e)}_{p,ab}\,\boldsymbol{u}_b, \tag{9.49a}$$

where the external pressure component of the tangent matrix is

$$\boldsymbol{K}^{(e)}_{p,ab} = \boldsymbol{\mathcal{E}}\boldsymbol{k}^{(e)}_{p,ab}; \quad [\boldsymbol{K}^{(e)}_{p,ab}]_{ij} = \sum_{k=1}^{3}\mathcal{E}_{ijk}[\boldsymbol{k}^{(e)}_{p,ab}]_k; \quad i,j = 1,2,3; \tag{9.49b}$$

where $\boldsymbol{\mathcal{E}}$ is the third-order alternating tensor ($\mathcal{E}_{ijk} = \pm 1$ or 0, depending on the parity of the $ijk$ permutation).

**EXAMPLE 9.6: External pressure component of tangent matrix $[\boldsymbol{K}_{p,ab}]$**

Consider the same triangle of Example 9.1 now representing a face on which pressure $p$ is applied. If the isoparametric coordinates are renamed $\xi, \eta$, the vectors $\partial \boldsymbol{x}/\partial \xi$ and $\partial \boldsymbol{x}/\partial \eta$ in the vector of stiffness coefficients given in Equation (9.48) are constant and depend upon the geometry of the particular surface element, whereas the terms in parentheses depend only on the element type. Noting that if $a = b$ the terms in parentheses are zero, the resulting simple integration yields

$$N_1 = 1 - \xi - \eta; N_2 = \xi; N_3 = \eta;$$

$$\frac{\partial \boldsymbol{x}}{\partial \xi} = \begin{bmatrix} 8 \\ 0 \\ 0 \end{bmatrix} \quad \frac{\partial \boldsymbol{x}}{\partial \eta} = \begin{bmatrix} 8 \\ 6 \\ 0 \end{bmatrix};$$

$$\boldsymbol{k}_{p,12} = \frac{1}{2} p \frac{\partial \boldsymbol{x}}{\partial \xi}\left(-\frac{1}{6}\right) - \frac{1}{2} p \frac{\partial \boldsymbol{x}}{\partial \eta}\left(-\frac{1}{3}\right);$$

$$\boldsymbol{k}_{p,13} = \frac{1}{2} p \frac{\partial \boldsymbol{x}}{\partial \xi}\left(-\frac{1}{3}\right) - \frac{1}{2} p \frac{\partial \boldsymbol{x}}{\partial \eta}\left(-\frac{1}{6}\right);$$

$$\boldsymbol{k}_{p,23} = \frac{1}{2} p \frac{\partial \boldsymbol{x}}{\partial \xi}\left(-\frac{1}{6}\right) - \frac{1}{2} p \frac{\partial \boldsymbol{x}}{\partial \eta}\left(+\frac{1}{6}\right).$$

Nonzero pressure stiffness submatrices are now found from Equation (9.49b) as

$$[\boldsymbol{K}_{p,12}] = \frac{p}{2}\begin{bmatrix} 0 & 0 & -2 \\ 0 & 0 & \frac{8}{6} \\ 2 & -\frac{8}{6} & 0 \end{bmatrix}; \quad [\boldsymbol{K}_{p,13}] = \frac{p}{2}\begin{bmatrix} 0 & 0 & -1 \\ 0 & 0 & -\frac{8}{6} \\ 1 & \frac{8}{6} & 0 \end{bmatrix};$$

$$[\boldsymbol{K}_{p,23}] = \frac{p}{2}\begin{bmatrix} 0 & 0 & 1 \\ 0 & 0 & -\frac{8}{3} \\ -1 & \frac{8}{3} & 0 \end{bmatrix}.$$

## 9.4.5 Tangent Matrix

The linearized virtual work Equation (9.30) can now be discretized for element $(e)$ linking nodes $a$ and $b$ (see Equation (9.32) and Figure 9.2(a)), in terms of the total substiffness matrix $\boldsymbol{K}_{ab}$ obtained by combining Equations (9.35), (9.44a,b,c), and (9.49a,b) to give

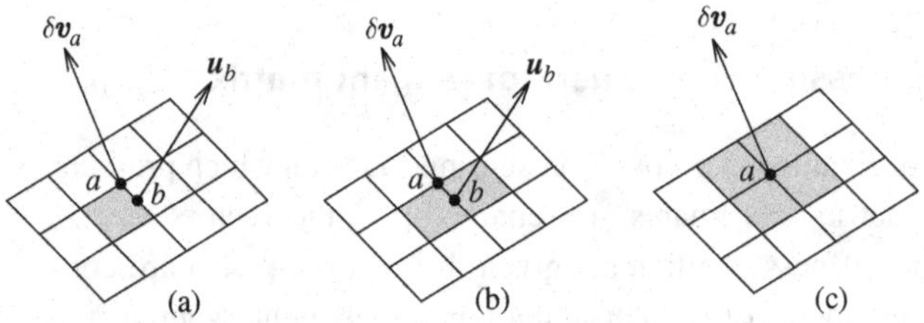


**FIGURE 9.2** Assembly of linearized virtual work.

$$D\delta W^{(e)}(\phi, N_a\delta\boldsymbol{v}_a)[N_b\boldsymbol{u}_b] = \delta\boldsymbol{v}_a \cdot \boldsymbol{K}^{(e)}_{ab}\,\boldsymbol{u}_b; \tag{9.50a}$$

$$\boldsymbol{K}^{(e)}_{ab} = \boldsymbol{K}^{(e)}_{c,ab} + \boldsymbol{K}^{(e)}_{\sigma,ab} - \boldsymbol{K}^{(e)}_{p,ab}. \tag{9.50b}$$

The assembly of the total linearized virtual work can now be accomplished by establishing (i) the contribution to node $a$ from node $b$ associated with all elements $(e)$ (1 to $m_{a,b}$) containing nodes $a$ and $b$ (Figure 9.2(b)); (ii) summing these contributions to node $a$ from all nodes $b = 1, n_a$, where $n_a$ is the number of nodes connected to node $a$ (Figure 9.2(c)); (iii) summing contributions from all nodes $a = 1, N$. This assembly process is summarized as

$$\text{(i)}\ D\delta W(\phi, N_a\delta\boldsymbol{v}_a)[N_b\boldsymbol{u}_b] = \sum_{\substack{e=1\\ e\ni a,b}}^{m_{a,b}} D\delta W^{(e)}(\phi, N_a\delta\boldsymbol{v}_a)[N_b\boldsymbol{u}_b]; \tag{9.51a}$$

$$\text{(ii)}\ D\delta W(\phi, N_a\delta\boldsymbol{v}_a)[\boldsymbol{u}] = \sum_{b=1}^{n_a} D\delta W(\phi, N_a\delta\boldsymbol{v}_a)[N_b\boldsymbol{u}_b]; \tag{9.51b}$$

$$\text{(iii)}\ D\delta W(\phi, \delta\boldsymbol{v})[\boldsymbol{u}] = \sum_{a=1}^{N} D\delta W(\phi, N_a\delta\boldsymbol{v}_a)[\boldsymbol{u}]. \tag{9.51c}$$

This standard finite element assembly procedure can alternatively be expressed using the complete virtual velocity vector given in Equation (9.20) together with the corresponding nodal displacements $\mathbf{u}^T = [\boldsymbol{u}_1^T, \boldsymbol{u}_2^T, \ldots, \boldsymbol{u}_N^T]$ and the assembled tangent stiffness matrix $\mathbf{K}$ to yield

$$D\delta W(\phi, \delta\boldsymbol{v})[\boldsymbol{u}] = \delta\mathbf{v}^T\mathbf{K}\mathbf{u}, \tag{9.52a}$$

where the tangent stiffness matrix $\mathbf{K}$ is defined by assembling the nodal components as

$$\mathbf{K} = \begin{bmatrix} \boldsymbol{K}_{11} & \boldsymbol{K}_{12} & \cdots & \boldsymbol{K}_{1N} \\ \boldsymbol{K}_{21} & \boldsymbol{K}_{22} & \cdots & \boldsymbol{K}_{2N} \\ \vdots & \vdots & \ddots & \vdots \\ \boldsymbol{K}_{N1} & \boldsymbol{K}_{N2} & \cdots & \boldsymbol{K}_{NN}. \end{bmatrix}. \tag{9.52b}$$

## 9.5 MEAN DILATATION METHOD FOR INCOMPRESSIBILITY

The standard discretization presented above is unfortunately not applicable to simulations involving incompressible or nearly incompressible three-dimensional or plain strain behavior. It is well known that without further development such a formulation is kinematically overconstrained, resulting in the overstiff phenomenon known as *volumetric locking*. These deficiencies in the standard formulation can be overcome using the three-field Hu–Washizu variational approach together with an appropriate distinction being made between the discretization of distortional and volumetric components. The resulting independent volumetric variables $p$ and $\bar{J}$ can now be interpolated either continuously or discontinuously across element boundaries. In the former case, new nodal unknowns are introduced into the final solution process, which leads to a cumbersome formulation that will not be pursued herein. In the latter case, the volumetric variables $p$ and $\bar{J}$ pertain only to an element and can be eliminated at the element level. In such a situation, the simplest discontinuous interpolation is to make $p$ and $\bar{J}$ constant throughout the element. This is the so-called *mean dilatation technique* discussed in Section 8.6.5. Observe, however, that for simple constant stress elements such as the linear triangle and tetrahedron, the mean dilatation method coincides with the standard formulation and therefore suffers the detrimental locking phenomenon.

### 9.5.1 Implementation of the Mean Dilatation Method

Recall from Section 8.6.5 that the mean dilatation approach for a given volume $v$ leads to a constant pressure over the volume, as indicated by Equations (8.53a–b). When this formulation is applied to each element $(e)$ in a finite element mesh the pressure becomes constant over the element volume. In particular, assuming for instance that the potential shown in Equation (8.54) is used, the uniform element pressure is given as

$$p^{(e)} = \kappa \frac{v^{(e)} - V^{(e)}}{V^{(e)}}, \tag{9.53}$$

where $V^{(e)}$ and $v^{(e)}$ are the initial and current element volumes.

The internal equivalent nodal forces for a typical element $(e)$ are given by Equation (9.15b), where now the Cauchy stress is evaluated from

$$\boldsymbol{\sigma} = \boldsymbol{\sigma}' + p^{(e)} \boldsymbol{I} \tag{9.54}$$

and the deviatoric stress $\boldsymbol{\sigma}'$ is evaluated using the appropriate constitutive equation given by Equation (6.55a,b) or (6.107).

Continuing with the discretization, recall the modified linearized virtual work Equation (8.63), which for an element $(e)$ is

$$D\delta W_{\text{int}}^{(e)}(\phi,\delta v)[u] = \int_{v^{(e)}} \delta d : \mathcal{c} : \varepsilon \, dv + \int_{v^{(e)}} \sigma : [(\nabla u)^T \nabla \delta v] \, dv$$
$$+ \bar{\kappa} v^{(e)} (\overline{\operatorname{div} u})(\overline{\operatorname{div} \delta v}); \qquad \bar{\kappa} = \frac{\kappa v^{(e)}}{V^{(e)}}, \tag{9.55}$$

where the elasticity tensor is $\mathcal{c} = \hat{\mathcal{c}} + \mathcal{c}_p$ and $\hat{\mathcal{c}}$ is the distortional component that depends upon the material used, and $\mathcal{c}_p$ is given by Equation (6.59b) as

$$\mathcal{c}_p = p(I \otimes I - 2\mathcal{i}). \tag{9.56}$$

The average divergences are now redefined for an element $(e)$ as

$$\overline{\operatorname{div} u} = \frac{1}{v^{(e)}} \int_{v^{(e)}} \operatorname{div} u \, dv = \frac{1}{v^{(e)}} \int_{v^{(e)}} \left( \sum_{a=1}^{n} u_a \cdot \nabla N_a \right) dv; \tag{9.57a,b}$$

$$\overline{\operatorname{div} \delta v} = \frac{1}{v^{(e)}} \int_{v^{(e)}} \operatorname{div} \delta v \, dv = \frac{1}{v^{(e)}} \int_{v^{(e)}} \left( \sum_{a=1}^{n} \delta v_a \cdot \nabla N_a \right) dv. \tag{9.57c,d}$$

Discretization of the first two terms in Equation (9.55) is precisely as given in the previous section, but the final dilatation term needs further attention. For element $(e)$ the contribution to the linearized internal virtual work related to the dilatation and associated, as before, with nodes $a$ and $b$ is

$$\begin{aligned} D\delta W_{\kappa}^{(e)}&(\phi, N_a \delta v_a)[N_b u_b] \\ &= \frac{\kappa}{V^{(e)}} \left( \int_{v^{(e)}} \delta v_a \cdot \nabla N_a \, dv \right) \left( \int_{v^{(e)}} u_b \cdot \nabla N_b \, dv \right) \\ &= \delta v_a \cdot \left[ \frac{\kappa}{V^{(e)}} \left( \int_{v^{(e)}} \nabla N_a \, dv \right) \otimes \left( \int_{v^{(e)}} \nabla N_b \, dv \right) \right] u_b \\ &= \delta v_a \cdot K_{\kappa,ab}^{(e)} \, u_b, \end{aligned} \tag{9.58}$$

where the dilatational tangent stiffness component is obtained in terms of the average Cartesian derivatives of the shape functions as

$$K_{\kappa,ab}^{(e)} = \bar{\kappa} v^{(e)} \overline{\nabla N_a} \otimes \overline{\nabla N_b}\,; \quad \overline{\nabla N_a} = \frac{1}{v^{(e)}} \int_{v^{(e)}} \nabla N_a \, dv. \tag{9.59a,b}$$

The complete discretization of Equation (9.55) can now be written in terms of the total element tangent substiffness matrix as

$$K_{ab}^{(e)} = K_{c,ab}^{(e)} + K_{\sigma,ab}^{(e)} + K_{\kappa,ab}^{(e)}, \tag{9.60}$$

in which the surface pressure component $\boldsymbol{K}_{p,ab}^{(e)}$ (see Equation (9.50)) may, if appropriate, be included. Assembly of the complete linearized virtual work and hence the tangent matrix follows the procedure given in Equations (9.51)–(9.53).

## 9.6 NEWTON–RAPHSON ITERATION AND SOLUTION PROCEDURE

### 9.6.1 Newton–Raphson Solution Algorithm

In the previous sections it was shown that the equilibrium equation was discretized as $\delta W(\boldsymbol{\phi}, \delta\boldsymbol{v}) = \delta\mathbf{v}^T\mathbf{R}$, whereas the linearized virtual work term is expressed in terms of the tangent matrix as $D\delta W(\boldsymbol{\phi}, \delta\boldsymbol{v})[\boldsymbol{u}] = \delta\mathbf{v}^T\mathbf{K}\mathbf{u}$. Consequently, the Newton–Raphson equation $\delta W(\delta\boldsymbol{v}, \boldsymbol{\phi}_k) + D\delta W(\boldsymbol{\phi}_k, \delta\boldsymbol{v})[\boldsymbol{u}] = 0$ given in Equation (8.3) is expressed in a discretized form as

$$\delta\mathbf{v}^T\mathbf{K}\mathbf{u} = -\delta\mathbf{v}^T\mathbf{R}. \tag{9.61}$$

Because the nodal virtual velocities are arbitrary, a discretized Newton–Raphson scheme is formulated as

$$\mathbf{K}\mathbf{u} = -\mathbf{R}(\mathbf{x}_k); \mathbf{x}_{k+1} = \mathbf{x}_k + \mathbf{u}. \tag{9.62}$$

Although it is theoretically possible to achieve a direct solution for a given load case, it is more practical to consider the external load $\mathbf{F}$ as being applied in a series of increments as

$$\mathbf{F} = \sum_{i=1}^{l} \Delta\mathbf{F}_i, \tag{9.63}$$

where $l$ is the total number of load increments. Clearly, the more increments taken, the easier it becomes to find a converged solution for each individual load step. Observe that in the case of a hyperelastic material the final solution is independent of the manner in which the load increments are applied. If, however, the material is not hyperelastic, this conclusion may not hold.

An outline of the complete solution algorithm is shown in Box 9.1.

**BOX 9.1: Solution Algorithm**

- INPUT geometry, material properties and solution parameters
- INITIALIZE $\mathbf{F} = 0, \mathbf{x} = \mathbf{X}, \mathbf{R} = 0$
- LOOP over load increments

*(continued)*

**BOX 9.1:** *(cont.)*

- FIND $\Delta\mathbf{F}$ using (9.15c)
- SET $\mathbf{F} = \mathbf{F} + \Delta\mathbf{F}$
- SET $\mathbf{R} = \mathbf{R} - \Delta\mathbf{F}$
- DO WHILE ($\|\mathbf{R}\|/\|\mathbf{F}\| >$ tolerance)
  - FIND $\mathbf{K}$ using (9.50b)
  - SOLVE $\mathbf{Ku} = -\mathbf{R}$
  - UPDATE $\mathbf{x} = \mathbf{x} + \mathbf{u}$
  - FIND $\boldsymbol{F}^{(e)}$, $b^{(e)}$ and $\boldsymbol{\sigma}^{(e)}$ using (9.5), (9.9a,b,d) and typically (6.29)
  - FIND $\mathbf{T}$ using (6.15b)
  - FIND $\mathbf{R} = \mathbf{T} - \mathbf{F}$
- ENDDO
- ENDLOOP

**Remark 9.1**: By comparing Equation (9.62) with the example given in Section 2.3.4 relating to the linearization of a system of algebraic equations, it is evident that the tangent stiffness matrix can be found directly as

$$\mathbf{K} = \frac{\partial \mathbf{R}}{\partial \mathbf{x}}; \quad K_{ij} = \frac{\partial R_i}{\partial x_j}; \quad i,j = 1, ndof, \tag{9.64}$$

where $ndof$ is the number of degrees of freedom in the problem. For some situations, such as finite deformation thin shell analysis, where the discretization of the kinematic quantities is very much algorithm-dependent, such a direct approach, though tedious, may be the only method of obtaining the tangent matrix coefficients.

### 9.6.2 Line Search Method

The Newton–Raphson process is generally capable of reaching the convergence of the equilibrium equations in a small number of iterations. Nevertheless, during the course of complex deformation processes, situations may occur where the straight application of the Newton–Raphson method becomes insufficient. A powerful technique often used to improve the convergence rate is the *line search method*. This technique consists of interpreting the displacement vector $\mathbf{u}$ obtained from Equation (9.62) as an optimal direction of advance toward the solution but allowing the magnitude of the step to be controlled by a parameter $\eta$ as

$$\mathbf{x}_{k+1} = \mathbf{x}_k + \eta\mathbf{u}. \tag{9.65}$$

The value of $\eta$ is normally chosen so that the total potential energy, $\Pi(\eta) = \Pi(\mathbf{x}_k + \eta\mathbf{u})$, at the end of the iteration is minimized in the direction of $\mathbf{u}$. This is equivalent to the requirement that the residual force at the end of each iteration, that is, $\mathbf{R}(\mathbf{x}_k + \eta\mathbf{u})$, is orthogonal to the direction of advance $\mathbf{u}$. This yields a scalar equation for $\eta$ as (Figure 9.3)

$$R(\eta) = \mathbf{u}^T\mathbf{R}(\mathbf{x}_k + \eta\mathbf{u}) = 0. \tag{9.66}$$

Because of the extreme nonlinearity of the function $R(\eta)$, the condition (9.66) is too stringent and in practice it is sufficient to obtain a value of $\eta$ such that

$$|R(\eta)| < \rho|R(0)|, \tag{9.67}$$

where, typically, a value of $\rho = 0.5$ is used. Under normal conditions, the value $\eta = 1$, which corresponds to the standard Newton–Raphson method, automatically satisfies Equation (9.67), and therefore few extra operations are involved. Occasionally, this is not the case, and a more suitable value of $\eta$ must be obtained. For this purpose it is convenient to approximate $R(\eta)$ as a quadratic in $\eta$. To achieve this requires knowledge of the value $R(0)$, together with the derivative $dR/d\eta$ at $\eta = 0$, which, recalling Remark 9.1, is obtained from Equation (9.66) as

$$\frac{dR}{d\eta} = \mathbf{u}^T\frac{\partial\mathbf{R}}{\partial\mathbf{x}}\bigg|_{\mathbf{x}=\mathbf{x}_k}\mathbf{u} = \mathbf{u}^T\mathbf{K}(\mathbf{x}_k)\,\mathbf{u} = -\mathbf{u}^T\mathbf{R}(\mathbf{x}_k) = -R(0); \tag{9.68}$$

and, finally, a third value, which typically is the standard value of the residual force for which $\eta = 1$, as

$$R(1) = \mathbf{u}^T\mathbf{R}(\mathbf{x}_k + \mathbf{u}). \tag{9.69}$$

The quadratic approximation thus obtained with these coefficients gives

$$R(\eta) \approx (1 - \eta)R(0) + R(1)\eta^2 = 0, \tag{9.70}$$

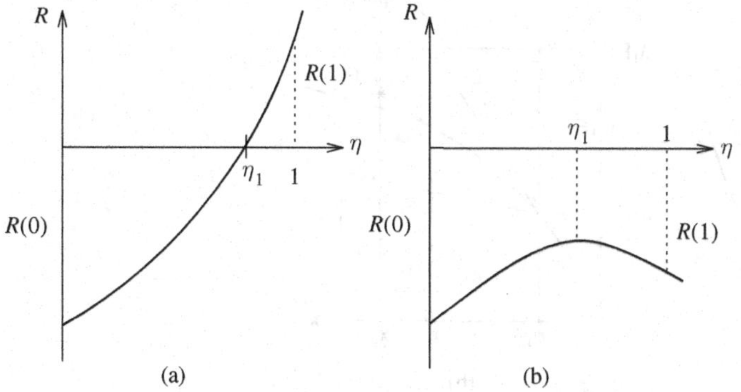


**FIGURE 9.3** Quadratic line search.

which yields a value for $\eta$ as

$$\eta = \frac{\alpha}{2} \pm \sqrt{\left(\frac{\alpha}{2}\right)^2 - \alpha}; \quad \alpha = \frac{R(0)}{R(1)}. \tag{9.71}$$

If $\alpha < 0$, the square root is real and the current value for $\eta$ emerges as

$$\eta_1 = \frac{\alpha}{2} + \sqrt{\left(\frac{\alpha}{2}\right)^2 - \alpha}\,. \tag{9.72}$$

Alternatively, if $\alpha > 0$ (see Figure 9.3(b)), then $\eta$ can be simply obtained by using the value that minimizes the quadratic function, that is, $\eta_1 = \alpha/2$. Within each (DO WHILE) iteration, the quadratic approximation procedure is repeated with the three values $R(0)$, $R'(0)$, and $R(\eta_1)$, until Equation (9.67) is satisfied.

### 9.6.3 Arc Length Method

Although the line search method will improve the convergence rate of the Newton–Raphson method, it will not enable the solution to traverse the so-called *limit points* on the equilibrium path. Figure 9.4(a) shows two such limit points A and B in an example of *snap-back behavior*. If the load is incremented, the solution will experience convergence problems in the neighborhood of A and may jump to the alternative equilibrium position A′. In many cases the equilibrium path can be followed beyond A by prescribing a characteristic displacement and calculating the load as the corresponding reaction. In Figure 9.4(a) this technique would enable the solution to progress to the neighborhood of the limit point B, but, again, the solution is likely to jump to point B′.

Various ad hoc schemes combining load and displacement incrementation have been devised to deal with limit point problems, but these have been superseded by arc length methods that constrain the iterative solution to follow a certain route

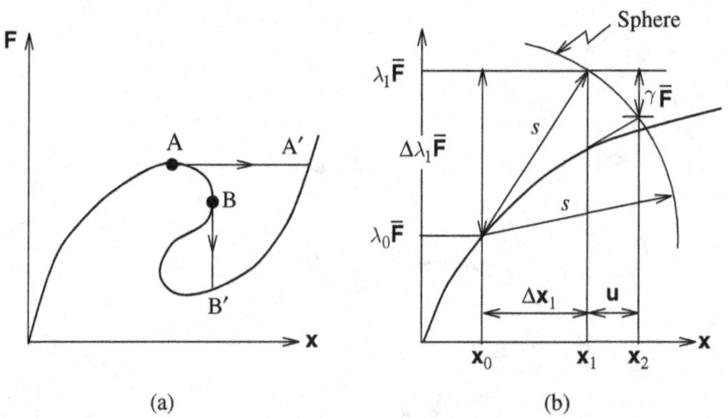


**FIGURE 9.4** (a) Snap-back behavior; (b) Spherical arc length method.

toward the equilibrium path. This is achieved by controlling the magnitude of the external loads by a parameter $\lambda$ so that the equilibrium equations become

$$\mathbf{R}(\mathbf{x}, \lambda) = \mathbf{T}(\mathbf{x}) - \lambda\overline{\mathbf{F}} = 0, \tag{9.73}$$

where $\overline{\mathbf{F}}$ is a representative equivalent nodal load. The value of $\lambda$ is now allowed to vary during the Newton–Raphson iteration process by imposing an additional constraint equation. As a consequence of the proportional loading, load increment $i$ is defined by an increment in the value of $\lambda$ over the value at the end of the previous load increment $i-1$ as

$$\Delta\mathbf{F}_i = \Delta\lambda\overline{\mathbf{F}}; \quad \Delta\lambda = \lambda - \lambda_{i-1}. \tag{9.74a,b}$$

Similarly, the total change in position over the load increment is denoted as $\Delta\mathbf{x}$, that is,

$$\Delta\mathbf{x} = \mathbf{x} - \mathbf{x}_{i-1}. \tag{9.75}$$

A number of arc length methods have been proposed by imposing different constraint equations to evaluate the additional unknown $\lambda$. A robust candidate is the spherical arc length method, in which the additional constraint equation is

$$\Delta\mathbf{x}^T\Delta\mathbf{x} + \Delta\lambda^2\psi^2\overline{\mathbf{F}}^T\overline{\mathbf{F}} = s^2. \tag{9.76}$$

Figure 9.4(b) illustrates this equation, where the constant $s$ can be thought of as the magnitude of the generalized vector

$$\mathbf{s} = \begin{bmatrix} \Delta\mathbf{x} \\ \Delta\lambda\psi F \end{bmatrix} \tag{9.77}$$

which defines a generalized sphere, which is the surface upon which the iterative solution is constrained to move as convergence toward the equilibrium path progresses. The term $\psi^2$ is a scaling factor that, in principle at least, renders (9.76) dimensionally consistent.

The Newton–Raphson process is now established by linearizing Equation (9.73), taking into account the possible change in $\lambda$, to give a set of linear equations at iteration $k$ as

$$\mathbf{R}(\mathbf{x}_k, \lambda_k) + \mathbf{K}(\mathbf{x}_k)\mathbf{u} - \gamma\overline{\mathbf{F}} = 0, \tag{9.78a}$$

where $\mathbf{u}$ represents the iterative change in position and $\gamma$ denotes the iterative change in $\lambda$ as

$$\mathbf{x}_{k+1} = \mathbf{x}_k + \mathbf{u}; \quad \Delta\mathbf{x}_{k+1} = \Delta\mathbf{x}_k + \mathbf{u}; \tag{9.78b}$$

$$\lambda_{k+1} = \lambda_k + \gamma; \quad \Delta\lambda_{k+1} = \Delta\lambda_k + \gamma. \tag{9.78c}$$

Solving the above equations gives the iterative displacements $\mathbf{u}$ in terms of the, as yet, unknown parameter change $\gamma$ and the auxiliary displacements $\mathbf{u}_R$ and $\mathbf{u}_F$ as

$$\mathbf{u} = \mathbf{u}_R + \gamma \mathbf{u}_F; \quad \mathbf{u}_R = -\mathbf{K}(\mathbf{x}_k)^{-1}\mathbf{R}(\mathbf{x}_k, \lambda_k); \quad \mathbf{u}_F = \mathbf{K}(\mathbf{x}_k)^{-1}\overline{\mathbf{F}}. \tag{9.79a,b,c}$$

In (9.78a), $\mathbf{u}$ and $\gamma$ must be such that the constraint Equation (9.76) remains satisfied; hence

$$(\Delta\mathbf{x}_k + \mathbf{u})^T(\Delta\mathbf{x}_k + \mathbf{u}) + (\Delta\lambda_k + \gamma)^2\psi^2\overline{\mathbf{F}}^T\overline{\mathbf{F}} = s^2. \tag{9.80}$$

Expanding this equation using (9.76) and (9.79a,b,c) gives a quadratic equation for the unknown iterative load parameter change $\gamma$ as

$$a_1\gamma^2 + a_2\gamma + a_3 = 0, \tag{9.81a}$$

where

$$a_1 = \mathbf{u}_F^T\mathbf{u}_F + \psi^2\overline{\mathbf{F}}^T\overline{\mathbf{F}}; \tag{9.81b}$$

$$a_2 = 2\mathbf{u}_F^T(\Delta\mathbf{x}_k + \mathbf{u}_R) + 2\Delta\lambda_k\psi^2\overline{\mathbf{F}}^T\overline{\mathbf{F}}; \tag{9.81c}$$

$$a_3 = \mathbf{u}_R^T(2\Delta\mathbf{x}_k + \mathbf{u}_R) + \Delta\mathbf{x}_k^T\Delta\mathbf{x}_k + \Delta\lambda_k^2\psi^2\overline{\mathbf{F}}^T\overline{\mathbf{F}} - s^2. \tag{9.81d}$$

There are two solutions $\gamma^{(1)}$ and $\gamma^{(2)}$ to (9.81a), which when substituted into (9.79a,b,c), (9.78b,c), and (9.76) give two revised generalized vectors $\mathbf{s}_{k+1}^{(1)}$ and $\mathbf{s}_{k+1}^{(2)}$. The correct parameter, $\gamma^{(1)}$ or $\gamma^{(2)}$, is that which gives the minimum "angle" $\theta$ between $\mathbf{s}_k$ and $\mathbf{s}_{k+1}^{(j)}$, where $\theta$ is obtained from

$$\cos\theta^{(j)} = \frac{\mathbf{s}_k^T\mathbf{s}_{k+1}^{(j)}}{s^2}; \quad \mathbf{s}_k = \begin{bmatrix} \Delta\mathbf{x}_k \\ \Delta\lambda_k\psi\overline{\mathbf{F}} \end{bmatrix}; \quad \mathbf{s}_{k+1}^{(j)} = \begin{bmatrix} \Delta\mathbf{x}_k + \mathbf{u}^{(j)} \\ (\Delta\lambda_k + \gamma^{(j)})\psi\overline{\mathbf{F}} \end{bmatrix}. \tag{9.82}$$

In practice, where many degrees of freedom exist, the scaling factor $\psi$ is taken as zero; alternatively, if the representative load $\overline{\mathbf{F}}$ is normalized and $\psi = 1$, the second term in the constraint Equation (9.76) reduces to $\Delta\lambda^2$.

## Exercises

1. A three-noded plane strain linear triangle finite element of unit thickness is deformed as shown in Example 9.1. The material is defined by a compressible neo-Hookean material with $\lambda = 2$ and $\mu = 3$; see Equation (6.29). Calculate the following items:
   (a) deformation gradient tensor $\boldsymbol{F}$;
   (b) Cauchy–Green tensors $\boldsymbol{C}$ and $\boldsymbol{b}$;

(c) second Piola–Kirchhoff and Cauchy stress tensors, $\boldsymbol{S}$ and $\boldsymbol{\sigma}$ respectively;
(d) vector of internal nodal $\boldsymbol{T}_a$ forces for each node $a$;
(e) component of the tangent stiffness $\boldsymbol{K}_{23}$ connecting nodes $2-3$.

2. Prove that the equivalent internal nodal forces can be expressed with respect to the initial configuration as

$$\boldsymbol{T}_a = \int_{V^{(e)}} \boldsymbol{F}\boldsymbol{S}\boldsymbol{\nabla}_0 N_a \, dV,$$

and then validate this equation by recalculating the equivalent nodal forces found in Example 9.3.

3. Prove that the initial stress matrix can be expressed with respect to the initial configuration as

$$\boldsymbol{K}^{(e)}_{\sigma,ab} = \int_{V^{(e)}} (\boldsymbol{\nabla}_0 N_a \cdot \boldsymbol{S}\boldsymbol{\nabla}_0 N_b)\boldsymbol{I} \, dV,$$

and then validate this equation by recalculating the initial stress matrix $\boldsymbol{K}_{\sigma,12}$ found in Example 9.5.

4. Show that the constitutive component of the tangent matrix can be expressed at the initial configuration as

$$\left[\boldsymbol{K}^{(e)}_{c,ab}\right]_{ij} = \sum_{I,J,K,L=1}^{3} \int_{V^{(e)}} F_{iI}\frac{\partial N_a}{\partial X_J}\mathcal{C}_{IJKL}\frac{\partial N_b}{\partial X_K}F_{jL} \, dV.$$

5. With the help of Exercise 2 of Chapter 8, derive a two-dimensional equation for the external pressure component of the tangent matrix $\boldsymbol{K}_p$ for a line element along an enclosed boundary of a two-dimensional body under uniform pressure $p$.

6. Recalling the line search method discussed in Section 9.6.2, show that minimizing $\Pi(\eta) = \Pi(\mathbf{x}_k + \eta\mathbf{u})$ with respect to $\eta$ gives the orthogonality condition

$$R(\eta) = \mathbf{u}^T\mathbf{R}(\mathbf{x}_k + \eta\mathbf{u}) = 0.$$

CHAPTER TEN

# COMPUTER IMPLEMENTATION

## 10.1 INTRODUCTION

We have seen in the previous chapters that the solution to the nonlinear equilibrium equations is basically achieved using the Newton–Raphson iterative method. In addition, in a finite element context it is advisable to apply the external forces in a series of increments. This has the advantage of enhancing the converging properties of the solution and, where appropriate, provides possible intermediate equilibrium states. Moreover, for path-dependent materials such as those exhibiting plasticity, these intermediate states represent the loading path which needs to be accurately followed. Furthermore, it is clear that the two fundamental quantities that facilitate the Newton–Raphson solution are the residual force and the tangent matrix. In this chapter we shall describe the MATLAB implementation of the solution procedure in the teaching program FLagSHyP (**F**inite element **La**rge **S**train **Hy**perelasto-plastic **P**rogram).

It is expected that the reader already has some familiarity with the computer implementation of the finite element method in the linear context. Consequently, this chapter will emphasize only those aspects of the implementation that are of particular relevance in the nonlinear finite deformation context.

The program description includes user instructions, a dictionary of variables and functions, and sample input and output for a few typical examples. The program can deal with three-dimensional truss elements and a number of two-dimensional and three-dimensional elements, together with a variety of compressible and nearly incompressible hyperelastic constitutive equations including simple Von Mises hyperelastic–plastic behavior. It can be obtained, together with sample data, as a download from the website `www.flagshyp.com`. Alternatively, it can be obtained by email request to any of the authors: `a.j.gil@swansea.ac.uk`, `r.d.wood@swansea.ac.uk` or `j.bonet@swansea.ac.uk`.

The master m-file `FLagSHyP.m`, which controls the overall organization of the program, is divided into three sections. The first section includes a series of statements designed to add the necessary directories to the path of the program. The second section includes the function `input_data_and_initialisation.m`, which is devoted to the reading of input/control data and the initialization of critical variables including equivalent nodal forces and the initial tangent stiffness matrix. The third section calls one of the three fundamental algorithms incorporated into FLagSHyP, depending on the problem-dependent control variables chosen by the user:

`ALG1` The function `Newton_Raphson_algorithm.m` is the basic Newton–Raphson algorithm, to be explained thoroughly in the remainder of this chapter.

`ALG2` The function `Line_Search_Newton_Raphson_algorithm.m` combines the basic Newton–Raphson algorithm with the line search procedure described in Section 9.6.2.

`ALG3` The function `Arc_Length_Newton_Raphson_algorithm.m` combines the basic Newton–Raphson algorithm with the arc length procedure described in Section 9.6.3.

The basic Newton–Raphson algorithm is described in Box 9.1 and amplified in Box 10.1 to include the line search and arc length options. The items in parentheses in Box 10.1 refer to program segments, each presented in Section 10.7 in a separate box in which comments have been removed and only MATLAB instructions are shown, along with a short description.

**BOX 10.1: Solution Algorithm (`ALG1`)**

- INPUT geometry, material properties and solution parameters (*FLagSHyP segment 2*)
- INITIALIZE $\mathbf{F} = 0$, $\mathbf{x} = \mathbf{X}$ (initial geometry), $\mathbf{R} = 0$ (*FLagSHyP segment 2*)
- FIND initial $\mathbf{K}$ (*FLagSHyP segment 2*)
- LOOP over load increments (*FLagSHyP segment 5*)
  - SET $\lambda = \lambda + \Delta\lambda$, $\mathbf{F} = \lambda\bar{\mathbf{F}}$, $\mathbf{R} = \mathbf{R} - \Delta\lambda\bar{\mathbf{F}}$ (*FLagSHyP segment 5*)
  - IF PRESCRIBED DISPLACEMENTS: (*FLagSHyP segment 5*)
    - UPDATE GEOMETRY (*FLagSHyP segment 5*)
    - FIND $\mathbf{F}, \mathbf{T}, \mathbf{K}, \mathbf{R} = \mathbf{T} - \mathbf{F}$ (*FLagSHyP segment 5*)
  - END IF (*FLagSHyP segment 5*)
  - DO WHILE ($\|\mathbf{R}\|/\|\mathbf{F}\| >$ tolerance ) (*FLagSHyP segment 6*)
    - SOLVE $\mathbf{Ku} = -\mathbf{R}$ (*FLagSHyP segment 6*)
    - IF LINE SEARCH FIND $\eta$ (`ALG2`)
    - IF ARC-LENGTH FIND $\lambda$ (`ALG3`)
    - UPDATE GEOMETRY $\mathbf{x} = \mathbf{x} + \mathbf{u}$ (*FLagSHyP segment 6*)
    - FIND $\mathbf{F}, \mathbf{T}, \mathbf{K}, \mathbf{R} = \mathbf{T} - \lambda\bar{\mathbf{F}}$ (*FLagSHyP segment 6*)
  - END DO (*FLagSHyP segment 6*)
  - OUTPUT INCREMENT RESULTS (*FLagSHyP segment 7*)
- ENDLOOP (*FLagSHyP segment 8*)

A feature of the FLagSHyP program is the arrangement of variables (and the grouping of those sharing a similar nature) into data structures. The use of data structures facilitates the understanding of all variables in the program, although their use can introduce some computational overheads, noticeable in a large-scale problem. Fundamental data structures are named in capital letters and their "children" in lower-case letters. The following table presents the parent data structures with their general description.

| Name | Comments |
|---|---|
| `PRO.` | Program run mode, input/output files, problem title |
| `GEOM.` | Nodes, initial and spatial coordinates, elemental volumes |
| `FEM.` | Finite element information (shape functions) and connectivity |
| `QUADRATURE.` | Gauss point information: location and weights |
| `KINEMATICS.` | Kinematics quantities: $\boldsymbol{F}$, $\boldsymbol{b}$, $J, \ldots$ |
| `MAT.` | Material properties: $\rho$, $\lambda$, $\mu$, $\kappa, \ldots$ |
| `BC.` | Boundary conditions: fixed and free |
| `LOAD.` | Loads: point loads, gravity, pressure load |
| `GLOBAL.` | Global vectors and matrices: residual and stiffness |
| `PLAST.` | Plasticity variables |
| `CON.` | Program control parameters: line search, arc length, ... |

As an example, the table below describes some of the children of parent data structures. A glossary of all the variables in the program is presented in Appendix 10.13 at the end of this chapter.

| Parent | Child | Comments |
|---|---|---|
| `PRO.` | `outputfile_name` | Output file name |
| `GEOM.` | `ndime` | Number of dimensions |
| `FEM.` | `mesh.` | Data structure with mesh information |
| `QUADRATURE.` | `element.` | Gauss point element information |
| `KINEMATICS.` | `F` | Deformation gradient at Gauss points |
| `MAT.` | `props` | Material properties |
| `BC.` | `freedof` | Free degree-of-freedom vector |
| `LOAD.` | `n_pressure_loads` | Number of pressure surface elements |
| `GLOBAL.` | `Residual` | Global residual force vector |
| `PLAST.` | `yield.` | Data structure for the yield function |
| `CON.` | `dlamb` | Incremental load parameter |

## 10.2 USER INSTRUCTIONS

In the execution of the master m-file FLagSHyP.m, the directories (and subdirectories) code and job_folder are added to the directory path of the program. For simplicity, all the necessary MATLAB functions comprising the FLagSHyP program are located within subdirectories of the directory code, whereas input/output data files are located within subdirectories of the directory job_folder. Moreover, the program expects the input file (e.g. FLagSHyP_input_file.dat) to be located in the directory job_folder/FLagSHyP_input_file. Notice that while the name (e.g. FLagSHyP_input_file) or extension (e.g. .dat) of the input file is arbitrary, it must coincide precisely with the name of the subdirectory which contains it (e.g. job_folder/FLagSHyP_input_file); otherwise an error will occur and the program will not progress. The input file required by FLagSHyP is described in the following table. The file is free-formatted, so items within a line are simply separated by commas or spaces.

| | |
|---|---|
| **Item 1**: Title | **Number of lines:** 1 |
| PRO.title | Problem title |
| **Item 2**: Type of element | **Number of lines:** 1 |
| FEM.mesh.element_type | Element type (see Note 1) |
| **Item 3**: Number of nodes | **Number of lines:** 1 |
| GEOM.npoin | Number of mesh nodes |
| **Item 4**: Nodal information | **Number of lines:** GEOM.npoin |
| ip, | Node number |
| BC.icode(ip), | Boundary code (see Note 2) |
| GEOM.x(i,ip) | $x, y, z$ coordinates |
| [i=1:GEOM.ndime] | [Number of dimensions] |
| **Item 5**: Number of elements | **Number of lines:** 1 |
| FEM.mesh.nelem | Number of elements |
| **Item 6**: Element information | **Number of lines:** FEM.mesh.nelem |
| ie, | Element number |
| MAT.matno(ie), | Material number |
| FEM.mesh.connectivity(ie,i) | Connectivities |
| [i=1:FEM.mesh.n_nodes_elem] | [Number of nodes per element] |
| **Item 7**: Number of materials | **Number of lines:** 1 |
| MAT.nmats | Number of different materials |

| **Item 8**: Material properties | **Number of lines:** MAT.nmats |
|---|---|
| im, | Material number |
| MAT.matyp(im), | Constitutive equation type |
| MAT.props(ipr,im) | Properties (see Note 3) |
| [ipr=1:npr(dependent upon material)] | [Number of properties] |
| **Item 9**: Load information | **Number of lines:** 1 |
| n_point_loads, | Number of loaded nodes |
| BC.n_prescribed_displacements, | Number of nonzero prescribed displacements |
| LOAD.n_pressure_loads, | Number of surface (line) elements with pressure |
| LOAD.gravt(i) | Gravity vector |
| [i=1:GEOM.ndime] | [Number of dimensions] |
| **Item 10**: Nodal loads | **Number of lines:** n_point_loads |
| ip, | Node number |
| force_value(i), | Force vector |
| [i=1:GEOM.ndime] | [Number of dimensions] |
| **Item 11**: Prescribed displacements | **Number of lines:** BC.n_prescribed_displacements |
| ip, | Node number |
| id, | Spatial direction |
| prescribed_value | Nominal prescribed displacement |
| **Item 12**: Pressure loads | **Number of lines:** LOAD.n_pressure_loads |
| ie, | Surface element number |
| FEM.mesh.connectivity_faces(ie,in), | Force vector |
| LOAD.pressure(ie) | Nominal pressure (see Note 4) |
| [in=1:FEM.mesh.connectivity_faces] | [Number of nodes per element] |
| **Item 13**: Control information | **Number of lines:** 1 |
| CON.nincrm | Number of load/displacement increments |

| | |
|---|---|
| CON.xlmax | Maximum value of load-scaling parameter |
| CON.dlamb | Load parameter increment |
| CON.miter | Maximum number of iterations per increment |
| CON.cnorm | Convergence tolerance |
| CON.searc | Line search parameter (if 0.0 not in use) |
| CON.ARCLEN.arcln | Arc length parameter (if 0.0 not in use)* |
| CON.OUTPUT.incout | Output counter (e.g. 5 for every five increments) |
| CON.ARCLEN.itarget | Target iterations per increment (see Note 5 below) |
| CON.OUTPUT.nwant | Single output node (0 if not used) |
| CON.OUTPUT.iwant | Output degree of freedom (0 if not used) |

**Note 1:** The following element types (FEM.mesh.element_type) are recognized (see also Section 10.4 for more details about these elements):

truss2: 2-noded truss
tria3: 3-noded linear triangle
tria6: 6-noded quadratic triangle
quad4: 4-noded bilinear quadrilateral
tetr4: 4-noded linear tetrahedron
tetr10: 10-noded quadratic tetrahedron
hexa8: 8-noded trilinear hexahedron.

Different element types cannot be mixed in a mesh. Given the element name, the program automatically sets the number of nodes per element (FEM.mesh.n_nodes_elem), the number of dimensions (GEOM.ndime), and the number of Gauss points (QUADRATURE.element.ngauss). It also identifies the associated type of surface or line element for pressure loads and sets the corresponding number of nodes per element (FEM.mesh.n_face_nodes_elem) and Gauss points (QUADRATURE.boundary.ngauss).

**Note 2:** The boundary codes (BC.icode) are as follows:

0: free
1: $x$ prescribed
2: $y$ prescribed
3: $x, y$ prescribed

* Note that the arc length option in data item 13 cannot be used if nonzero prescribed displacements are employed.

4: $z$ prescribed
5: $x, z$ prescribed
6: $y, z$ prescribed
7: $x, y, z$ prescribed.

Prescribed degrees of freedom are assumed to be fixed (that is, no displacement) unless otherwise prescribed to be different from zero in input item 9 or 11. If a displacement is imposed in input item 9 or 11, the corresponding degree of freedom must have already been indicated as prescribed in input item 4.

**Note 3:** The following constitutive equations have been implemented (see also Section 10.6):

1: plane strain or three-dimensional compressible neo-Hookean
2: one-dimensional stretch-based hyperelastic plastic (`truss2` only)
3: plane strain or three-dimensional hyperelastic in principal directions
4: plane stress hyperelastic in principal directions
5: plane strain or three-dimensional nearly incompressible neo-Hookean
6: plane stress incompressible neo-Hookean
7: plane strain or three-dimensional nearly incompressible hyperelas ticity in principal directions
8: plane stress incompressible hyperelasticity in principal directions
17: plane strain or three-dimensional nearly incompressible hyperelastic plastic in principal directions.

The corresponding list of properties `MAT.props(i,im)` to be read in Item 8† are shown in the following table. For simplicity, the columns of the table have been named as `props(i)`, which refer to the specific variable name `MAT.props(i,im)`. In this table, $\rho$ represents the density in the reference configuration, $\lambda$ and $\mu$ are the Lamé coefficients, $\kappa = \lambda + 2\mu/3$ is the bulk modulus,* $h$ is the thickness for plane stress cases, $E$ is the Young's modulus, $\nu$ is Poisson's ratio, and $area$ is the initial cross-sectional area. Finally, $\tau_y$ and $H$ are the yield stress and hardening parameter, respectively.

**Note 4:** For surface elements the direction of positive pressure is given by the right-hand screw rule following the surface element numbering. For line elements the positive pressure is at 90° turning counterclockwise to the direction of increasing node numbering.

† For material 17, only $\rho$, $\mu$, $\lambda$, $\tau_y$, and $H$ are entered; $\kappa$ is calculated by the program and stored in `MAT.props(4,im)`.

| Type | props(1) | props(2) | props(3) | props(4) | props(5) | props(6) |
|---|---|---|---|---|---|---|
| 1 | $\rho$ | $\mu$ | $\lambda$ | – | – | – |
| 2 | $\rho$ | $E$ | $\nu$ | *area* | $\tau_y$ | $H$ |
| 3 | $\rho$ | $\mu$ | $\lambda$ | – | – | – |
| 4 | $\rho$ | $\mu$ | $\lambda$ | $h$ | – | – |
| 5 | $\rho$ | $\mu$ | $\kappa$ | – | – | – |
| 6 | $\rho$ | $\mu$ | $h$ | – | – | – |
| 7 | $\rho$ | $\mu$ | $\kappa$ | – | – | – |
| 8 | $\rho$ | $\mu$ | $h$ | – | – | – |
| 17 | $\rho$ | $\mu$ | $\lambda$ | – | $\tau_y$ | $H$ |

**Note 5:** Typically, a nonlinear structural analysis solution is carried out in a number of incremental steps, `CON.incrm=1:CON.nincr`, where `CON.nincr` is the chosen maximum number of steps.

The applied *loads* can be point forces, pressure forces, or even prescribed displacements. Any input value of these items is nominal and will be multiplied by the load parameter increment `CON.dlamb`. The nominal load multiplied by `CON.dlamb` is called an *increment in load*, $\Delta\mathbf{F}$ (or $\Delta\mathbf{u}$ for prescribed displacements).

Typically, at each load step `CON.incrm`, the applied load is increased by an amount equal to $\Delta\mathbf{F}$. This means that the nominal load has been multiplied by `lambda=CON.incrm×CON.dlamb` to give the actual load $\mathbf{F}$, where `lambda` is called the *load-scaling parameter*.

If arc length control is employed then `lambda` is controlled indirectly; see Section 9.6.3. A positive value of `CON.ARCLEN.arcln` will produce a variable arc length, the value of which is determined by the desired number of iterations per increment `itarget`. Irrespective of the magnitude input by the user (i.e. `CON.ARCLEN.arcln=` 1), the program will work out the most appropriate value, discarding that entered by the user. A negative value (simply as an indicator) for `CON.ARCLEN.arcln` will provide a constant arc length. In this latter case, some experimentation with values will be necessary. If the arc length option is not to be used, input `CON.ARCLEN.arcln` $=0$ and `CON.ARCLEN.itarget` $=0$.

There is an additional control item, `CON.xlmax`, which is the maximum value of the load-scaling parameter, and the program will stop when (CON.incrm $\times$ CON.dlamb) $>$ CON.xlmax, even if CON.incrm $<$ CON.nincr. Alternatively, the program will stop when CON.incrm $=$ CON.nincr even if (CON.nincr $\times$ CON.dlamb) $<$ CON.xlmax.

At each load step the Newton–Raphson iteration attempts to achieve equilibrium within a maximum allowed number of iterations, input as CON.miter. Equilibrium is achieved when the residual force $||\mathbf{R}|| <$ CON.cnorm, where CON.cnorm is the convergence tolerance, usually about $10^{-6}$.

Convergence toward equilibrium can be improved by using the line search option, which minimizes the residual force in the direction of the Newton–Raphson iterative change in the position of the structure. Line search is activated and controlled by the parameter CON.searc, which should have a value of about 0.5. The line search algorithm cannot be used in conjunction with the arc length method (a warning message is displayed in that case and the program stops).

To facilitate easy plotting of load displacement graphs, output from a single node and single degree of freedom (at that node) can be specified. Parameter CON.OUTPUT.nwant specifies the node and CON.OUTPUT.iwant the degree of freedom. The output is in a file always called flag.out which contains the increment number, the coordinate relating to the degree of freedom, the force relating to the degree of freedom, the current value of CON.xlamb, and (if used) the current arc length value CON.ARCLEN.arcln.

The somewhat contrived example shown in Figure 10.1 has been chosen to illustrate as many diverse features of these input instructions as possible. The required input file is listed in Box 10.2. Note that point loads, gravity loads,

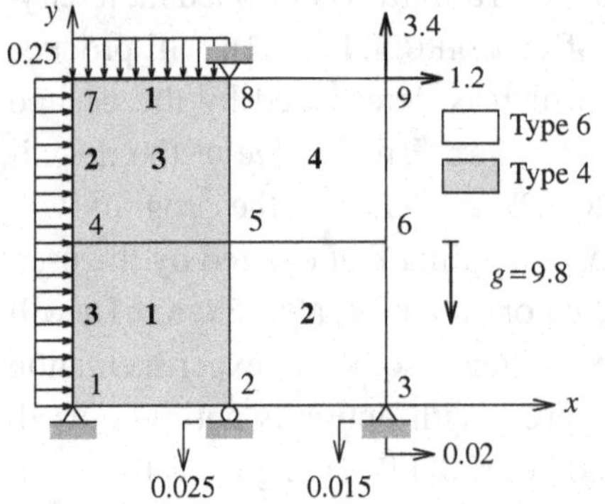


**FIGURE 10.1** Simple two-dimensional example.

pressure loads, and prescribed displacements are all subject to the same incrementation in the solution procedure.

**BOX 10.2: Input File for Example in Figure 10.1**

```
2-D Example quad4
9
1 3 0.0 0.0
2 2 1.0 0.0
3 3 2.0 0.0
4 0 0.0 1.0
5 0 1.0 1.0
6 0 2.0 1.0
7 0 0.0 2.0
8 3 1.0 2.0
9 0 2.0 2.0
4
1 1 1 2 5 4
2 2 6 5 2 3
3 1 5 8 7 4
4 2 5 6 9 8
2
1 4
1.0 100. 100. 0.1
2 6
1.0 100. 0.1
1 3 3 0.0 −9.8
9 1.2 3.4
3 1 0.02
2 2 −0.025
3 2 −0.015
1 8 7 0.25
2 7 4 0.25
3 1 4 −0.25
2 10.0 5.0 25 1.e−10 0.0 0.0 1 5 7 1
```

## 10.3 OUTPUT FILE DESCRIPTION

The program FLagSHyP produces a largely unannotated output file that is intended as an input file for a postprocessor to be supplied by the user. The output file is only produced every `CON.OUTPUT.incout` increments. The contents and structure of this file are shown in the table below.

| | |
|---|---|
| **Item 1**: Title | **Number of lines:** 1 |
| `PRO.title,` | Problem title |
| `' at increment: ', num2str (CON.incrm),` | Increment number |
| `' load: ', num2str(CON.xlamb)` | Current load parameter |
| **Item 2**: Type of element | **Number of lines:** 1 |
| `FEM.mesh.element_type` | Element type |
| **Item 3**: Number of nodes | **Number of lines:** 1 |
| `GEOM.npoin` | Number of mesh nodes |
| **Item 4**: Nodal information | **Number of lines:** `GEOM.npoin` |
| `ip,` | Node number |
| `BC.icode(ip),` | Boundary code |
| `GEOM.x(i,ip),` | $x, y, z$ coordinates |
| `force(i)` | External force or reaction |
| `[i=1:GEOM.ndime]` | [Number of dimensions] |
| **Item 5**: Number of elements | **Number of lines:** 1 |
| `FEM.mesh.nelem` | Number of elements |
| **Item 6**: Element information | **Number of lines:** `FEM.mesh.nelem` |
| `ie,` | Element number |
| `MAT.matno(ie),` | Material number |
| `FEM.mesh.connectivity(ie,i)` | Connectivities |
| `[i=1:FEM.mesh.n_nodes_elem]` | [Number of nodes per element] |
| **Item 7**: Stress information | **Number of lines:** `QUADRATURE.element.ngauss` × `FEM.mesh.nelem` |
| `Stress(i,ig,ie)` | Cauchy stress for each Gauss point |
| `[i=1:stress_components]` | (2-D) stress: $\sigma_{xx}, \sigma_{xy}, \sigma_{yy}$ (plane strain) |
| `[ig=1:QUADRATURE.element.ngauss]` | (2-D) stress: $\sigma_{xx}, \sigma_{xy}, \sigma_{yy}, h$ (plane stress) |
| `[ie=1:FEM.mesh.nelem]` | (3-D) stress: $N_x$ (truss) |
| | (3-D) stress: $\sigma_{xx}, \sigma_{xy}, \sigma_{xz}, \sigma_{yy}, \sigma_{yz}, \sigma_{zz}$ |

The output file produced by FLagSHyP for the simple example in Figure 10.1 is listed in Box 10.3.

**BOX 10.3: Output File for Example in Figure 10.1**

```
2-D Example        at increment:        1,         load:  5
quad4
9
1 3  0.0000E+00  0.0000E+00 -3.3614E+00  9.5002E-01
2 2  1.1889E+00 -1.2500E-01  0.0000E+00 -2.1952E+00
3 3  2.1000E+00 -7.5000E-02 -1.2617E+00 -2.2113E+00
4 0  2.9056E-01  7.8088E-01  0.0000E+00 -2.4500E+00
5 0  1.2833E+00  1.0620E+00  0.0000E+00 -4.9000E+00
6 0  2.0531E+00  1.2262E+00  0.0000E+00 -2.4500E+00
7 0  5.0207E-02  1.6092E+00  0.0000E+00 -1.2250E+00
8 3  1.0000E+00  2.0000E+00 -3.8769E+00 -4.3497E-02
9 0  2.3964E+00  3.8249E+00  6.0000E+00  1.5775E+01
4
1 1 1 2 5 4
2 2 6 5 2 3
3 1 5 8 7 4
4 2 5 6 9 8
 3.1165E+01  1.6636E+01 -2.9752E+01  9.9858E-02
 3.7922E+01  7.0235E+00  2.9804E+01  9.2369E-02
 9.8170E+00  2.8948E+01  2.3227E+01  9.6515E-02
-9.1664E+00  5.2723E+01 -5.2341E+01  1.0566E-01
-3.1460E+01  9.0191E+00  6.9610E+01  9.7692E-02
-4.4255E+01  1.9009E+01  4.0029E+01  1.0422E-01
-1.0503E+01  1.4344E+01  5.8661E+01  9.4115E-02
-1.0937E+00  4.3534E+00  8.4855E+01  8.8759E-02
 2.9733E+00  4.9849E+00 -8.6633E+00  1.0056E-01
-2.5993E+00  1.0535E+01 -4.9380E+00  1.0075E-01
-1.0028E+01  1.6380E+01 -2.4223E+01  1.0326E-01
-3.7416E+00  1.0076E+01 -2.8318E+01  1.0306E-01
 1.8711E+01  2.7033E+01  1.2770E+02  8.0604E-02
 5.8710E+01  9.3889E+01  5.0464E+02  5.2100E-02
 1.4861E+02  2.3372E+02  7.0689E+02  3.9520E-02
 1.3288E+02  1.6687E+02  3.5422E+02  5.4008E-02

2-D Example        at increment:        2,         load:  10
quad4
9
1 3  0.0000E+00  0.0000E+00 -6.0853E+00  2.5627E+00
2 2  1.3519E+00 -2.5000E-01  0.0000E+00 -3.9191E+00
3 3  2.2000E+00 -1.5000E-01 -2.4435E+00 -2.9205E+00
4 0  5.4010E-01  6.6991E-01  0.0000E+00 -4.9000E+00
5 0  1.5590E+00  1.1437E+00  0.0000E+00 -9.8000E+00
6 0  2.2245E+00  1.2882E+00  0.0000E+00 -4.9000E+00
```

*(continued)*

**BOX 10.3:** *(cont.)*

```
7 0  1.9116E-01  1.3055E+00  0.0000E+00 -2.4500E+00
8 3  1.0000E+00  2.0000E+00 -8.4712E+00 -2.7232E+00
9 0  3.3987E+00  6.1513E+00  1.2000E+01  3.1550E+01
4
1 1 1 2 5 4
2 2 6 5 2 3
3 1 5 8 7 4
4 2 5 6 9 8
 6.2596E+01  2.1249E+01 -3.2758E+01  9.6870E-02
 6.1948E+01  9.8381E+00  5.4321E+01  8.5200E-02
 2.1019E+01  4.4812E+01  4.5486E+01  9.2526E-02
-1.5069E+01  1.0427E+02 -1.0393E+02  1.1028E-01
-5.0536E+01  1.8529E+01  1.0546E+02  1.0025E-01
-5.4947E+01  3.3161E+01  9.1325E+01  1.0362E-01
-1.1718E+01  3.2504E+01  1.1797E+02  8.9494E-02
-9.7154E+00  1.7872E+01  1.2969E+02  8.6976E-02
 2.1962E+01  8.2142E+00 -4.1974E+00  9.8174E-02
-2.0453E-02  1.3036E+01  7.8808E+00  9.9204E-02
-3.3571E+01  3.7568E+01 -3.3248E+01  1.0611E-01
-2.7830E+00  2.9571E+01 -4.8372E+01  1.0477E-01
 8.3822E+01  6.9453E+01  3.6196E+02  5.1329E-02
 1.6278E+02  4.2678E+02  1.7024E+03  2.9913E-02
 4.7504E+02  9.9613E+02  2.7018E+03  1.8205E-02
 4.1084E+02  6.3881E+02  1.3761E+03  2.4400E-02
```

## 10.4 ELEMENT TYPES

Nodes and Gauss points in a given finite element can be numbered in a variety of ways. The numbering scheme chosen in FLagSHyP is shown in Figures 10.2 and 10.3.

In order to avoid the common repetitious use of shape functions for each mesh element, FLagSHyP (with the exception of `truss2`) stores in

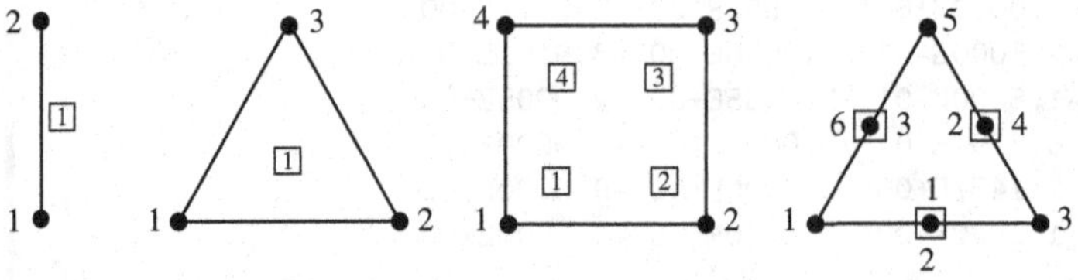

**FIGURE 10.2** Numbering of two-dimensional elements.

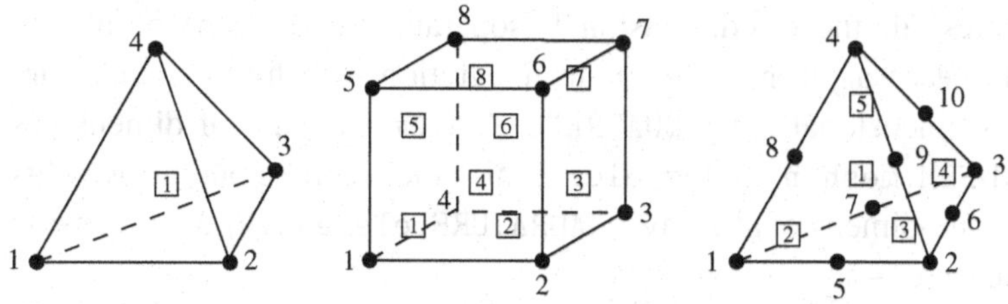


**FIGURE 10.3** Numbering of three-dimensional elements.

memory the shape functions and their nondimensional derivatives for each Gauss point of the chosen element type. This information is stored in the two-dimensional array `FEM.interpolation.element.N` and the three-dimensional array `FEM.interpolation.element.DN_chi` used to store shape functions and derivatives, respectively. Specifically, the dimensions of `FEM.interpolation.element.N` are the number of nodes per element (`FEM.mesh.n_nodes_elem`) and the number of Gauss points per element (`QUADRATURE.element.ngauss`), which can be depicted as follows:

$$\begin{bmatrix} N_1(\xi_1,\eta_1,\zeta_1) & \cdots & N_1(\xi_m,\eta_m,\zeta_m) \\ \cdots & \cdots & \cdots \\ N_n(\xi_1,\eta_1,\zeta_1) & \cdots & N_n(\xi_m,\eta_m,\zeta_m) \end{bmatrix} \quad \begin{array}{l} m = \texttt{QUADRATURE.element} \\ \texttt{.ngauss;} \\ n = \texttt{FEM.mesh.n\_nodes\_elem.} \end{array}$$

Analogously, the dimensions of `FEM.interpolation.element.DN_chi` are the number of dimensions (`GEOM.ndime`), the number of nodes per element (`FEM.mesh.n_nodes_elem`), and the number of Gauss points per element (`QUADRATURE.element.ngauss`):

$$\begin{bmatrix} \frac{\partial N_1}{\partial \xi}(\xi_i,\eta_i,\zeta_i) & \cdots & \frac{\partial N_n}{\partial \xi}(\xi_i,\eta_i,\zeta_i) \\ \frac{\partial N_1}{\partial \eta}(\xi_i,\eta_i,\zeta_i) & \cdots & \frac{\partial N_n}{\partial \eta}(\xi_i,\eta_i,\zeta_i) \\ \frac{\partial N_1}{\partial \zeta}(\xi_i,\eta_i,\zeta_i) & \cdots & \frac{\partial N_n}{\partial \zeta}(\xi_i,\eta_i,\zeta_i) \end{bmatrix} \quad \begin{array}{l} i = 1 : \texttt{QUADRATURE.element} \\ \texttt{.ngauss;} \\ n = \texttt{FEM.mesh.n\_nodes\_elem.} \end{array}$$

The coordinates (in the nondimensional isoparametric domain) of all the quadrature points necessary for the accurate integration of a finite element are stored in the two-dimensional array `QUADRATURE.element.Chi`, of dimensions `GEOM.ndime` and `FEM.mesh.n_nodes_elem`, while their corresponding weights are stored in the one-dimensional array `QUADRATURE.element.W`, of dimension `FEM.mesh.n_nodes_elem`.

Exactly the same type of arrays are constructed for the line or surface elements and stored in `FEM.interpolation.boundary.N` (for shape functions), `FEM.interpolation.boundary.DN_chi` (for shape functions derivatives), and `QUADRATURE.boundary.W` and `QUADRATURE.boundary.Chi` (for Gauss point information).

## 10.5 SOLVER DETAILS

For the solution of the resulting system of linear algebraic equations $\mathbf{Ku} = -\mathbf{R}$, FLagSHyP uses the MATLAB command $\mathbf{u} = -\mathbf{K}\backslash\mathbf{R}$, where $\mathbf{K}$ denotes the assembled global stiffness matrix (`GLOBAL.K`) and $\mathbf{R}$ is the assembled global residual vector (`GLOBAL.RESIDUAL`). MATLAB will aim to take advantage of possible symmetries in the problem, redirecting to the most appropriate linear algebra solver, with the final objective of minimizing the computational time while maintaining the accuracy of the final solution. Specifically, tailor-made solvers can be utilized for dealing with a sparse matrix $\mathbf{K}$ as in the present case.

In order to take advantage of the sparse solver capabilities, the final assembled stiffness matrix $\mathbf{K}_c + \mathbf{K}_\sigma + \mathbf{K}_\kappa - \mathbf{K}_p$ is stored in a temporary one-dimensional vector array (`global_stiffness`), along with two other temporary one-dimensional vector arrays (of the same length as `global_stiffness`), namely (`indexi`) and (`indexj`), where the rows (`indexi`) and columns (`indexj`) corresponding to the entries of the stiffness matrix are saved. By making use of the MATLAB intrinsic function `sparse.m`, these vectors are assembled into `GLOBAL.K`.

Finally, it is important to remark that the linear algebra solver is not restricted to the typical case of a symmetric stiffness matrix. In other words, it allows for the most general case of pressure loads which can lead to an unsymmetric stiffness matrix. This results in the need to store both lower and upper triangular entries of the assembled stiffness matrix $\mathbf{K}$. However, as stated above, if the symmetry of $\mathbf{K}$ is detected when the MATLAB command $\mathbf{Ku} = -\mathbf{R}$ is executed, a specific symmetric solver will then be employed.

Algorithm 1 shown below depicts in a simplified diagram the way in which the element tangent stiffness matrix is formed as a series of loops beginning with element and Gauss quadrature and cascading down to nodes and dimensions loops. Upon completion of these calculations, the sparse assembly process is carried out. Notice that this process is repeated for the assembly of each of the various components of the assembled global stiffness matrix, namely the constitutive matrix component $\mathbf{K}_c$ (refer to the function constitutive_matrix.m), the geometric or initial stress matrix component $\mathbf{K}_\sigma$ (refer to the function geometric_matrix.m), the volumetric mean dilatation stiffness component $\mathbf{K}_\kappa$ (refer to the function volumetric_mean_dilatation_matrix.m), and the external force pressure stiffness component $\mathbf{K}_p$ (refer to the function pressure_load_matrix.m).

**Algorithm 1:** Calculation and storage of assembled constitutive matrix entries

```
Input  : counter=0; indexi(:)=0; indexj(:)=0;
         global_stiffness(:)=0
Output: GLOBAL.K

for ielement=1:FEM.mesh.nelem do
    for igauss=1:QUADRATURE.element.ngauss do
        for anode=1:FEM.mesh.n_nodes_elem do
            for bnode=1:FEM.mesh.n_nodes_elem do
                for i=1:GEOM.ndime do
                    for j=1:GEOM.ndime do
                        Evaluate indexi(counter);
                        Evaluate indexj(counter);
                        Evaluate stiffness coefficient k;
                        Assign global_stiffness(counter)← k;
                        counter ← counter + 1;
                    end
                end
            end
        end
    end
end
GLOBAL.K=sparse(indexi,indexj,global_stiffness)
```

To illustrate the assembly of the global stiffness matrix, consider the somewhat artificial example in Figure 10.4, comprising three triangular elements with one degree of freedom per node and two Gauss points per element. The figure shows the process by which a stiffness coefficient $k$ per Gauss point, shown in Algorithm 1, is assigned to the appropriate location in the `global_stiffness` vector in a sequence determined by `counter`. Also observe in Figure 10.4 how the MATLAB function `sparse.m` assembles a typical global stiffness entry from its Gauss point contributions.

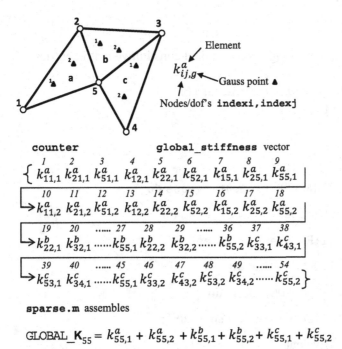

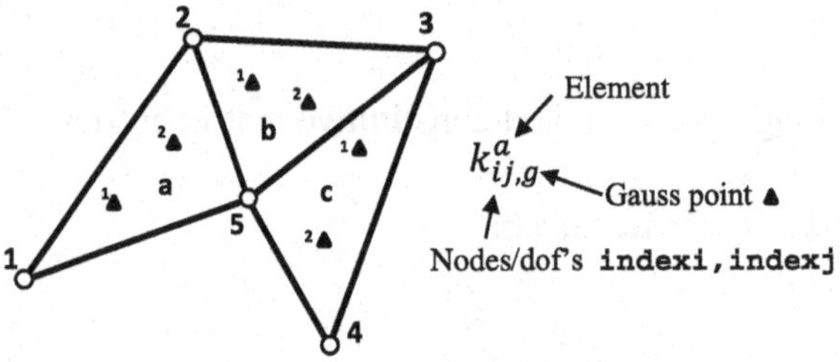


**counter** **global_stiffness** vector

$$\begin{aligned}
&\overset{1}{k^a_{11,1}}\;\overset{2}{k^a_{21,1}}\;\overset{3}{k^a_{51,1}}\;\overset{4}{k^a_{12,1}}\;\overset{5}{k^a_{22,1}}\;\overset{6}{k^a_{52,1}}\;\overset{7}{k^a_{15,1}}\;\overset{8}{k^a_{25,1}}\;\overset{9}{k^a_{55,1}}\\
&\overset{10}{k^a_{11,2}}\;\overset{11}{k^a_{21,2}}\;\overset{12}{k^a_{51,2}}\;\overset{13}{k^a_{12,2}}\;\overset{14}{k^a_{22,2}}\;\overset{15}{k^a_{52,2}}\;\overset{16}{k^a_{15,2}}\;\overset{17}{k^a_{25,2}}\;\overset{18}{k^a_{55,2}}\\
&\overset{19}{k^b_{22,1}}\;\overset{20}{k^b_{32,1}}\;\ldots\;\overset{27}{k^b_{55,1}}\;\overset{28}{k^b_{22,2}}\;\overset{29}{k^b_{32,2}}\;\ldots\;\overset{36}{k^b_{55,2}}\;\overset{37}{k^c_{33,1}}\;\overset{38}{k^c_{43,1}}\\
&\overset{39}{k^c_{53,1}}\;\overset{40}{k^c_{34,1}}\;\ldots\;\overset{45}{k^c_{55,1}}\;\overset{46}{k^c_{33,2}}\;\overset{47}{k^c_{43,2}}\;\overset{48}{k^c_{53,2}}\;\overset{49}{k^c_{34,2}}\;\ldots\;\overset{54}{k^c_{55,2}}
\end{aligned}$$

**sparse.m** assembles

$$\text{GLOBAL\_}\mathbf{K}_{55} = k^a_{55,1} + k^a_{55,2} + k^b_{55,1} + k^b_{55,2} + k^c_{55,1} + k^c_{55,2}$$

**FIGURE 10.4** Stiffness matrix assembly; `counter` refers to the location of the stiffness coefficient $k$ in the `global stiffness vector`.

## 10.6 PROGRAM STRUCTURE

In order to give an overview of the structure of the program FLagSHyP, Box 10.4 lists the main functions comprising the program. The remainder are either functions common to standard linear elasticity finite element codes or subsidiary functions that are not crucial to an understanding of the flow of the program and can be examined via a program download (see `www.flagshyp.com`). The functions in italic typeface are described in detail in the following sections.

## BOX 10.4: FLagSHyP Structure

```
FLagSHyP ............................................................m-file
 └─code ............................................Main code directory
    ├─solution_equations ......................Solution algorithms directory
    │  ├─Newton_Raphson_algorithm ..............Newton–Raphson master function
    │  ├─Arc_Length_Newton_Raphson_algorithm .........Arc length master function
    │  ├─Line_Search_Newton_Raphson_algorithm .......Line search master function
    │  ├─arclen ...........................................Arc length algorithm
    │  ├─linear_solver ......................Solver of linear system of equations
    │  └─search ...........................................Line search algorithm
    ├─input_reading ..............................Reading of input data directory
    │  ├─boundary_codes ..........................Reading of boundary conditions
    │  ├─elinfo ..................................Pre-allocate element information
    │  ├─incontr .................................Reading input control parameters
    │  ├─inelems .......................................Reading element information
    │  ├─inloads .......................................Reading loading information
    │  ├─innodes ..........................................Reading nodal information
    │  ├─input_data_and_initialisation ...........Initialization of some variables
    │  ├─matprop ........................................Reading material properties
    │  ├─reading_input_file ...................................Reading input function
    │  └─welcome ...........................................Read input file name
    ├─initialisation ..............................Initialization functions directory
    ├─numerical_integration ...............Numerical integration functions directory
    ├─FEM_shape_functions ............................Shape functions directory
    ├─kinematics ..................................Kinematics calculations directory
    │  ├─gradients .........................Computation of fundamental kinematics
    │  ├─kinematics_gauss_point .............Extraction of quantities at Gauss point
    │  ├─normal_vector_boundary .............................Normal vector at surface
    │  └─thickness_plane_stress ................Plane stress thickness computation
    ├─constitutive laws ..............................Constitutive behavior directory
    │  ├─Cauchy_type_selection ........... Cauchy stress tensor (material selection)
    │  ├─elasticity_modulus_selection ....... Elasticity tensor (material selection)
    │  ├─muab_choice ........... Checking function for materials in principal directions
    │  ├─elasticity_tensor ..................................ElasitICY tensor directory
    │  │  ├─ctens1 ................................... Elasticity tensor material 1
    │  │  ├─ctens3 ................................... Elasticity tensor material 3
    │  │  ├─ctens4 ................................... Elasticity tensor material 4
    │  │  ├─ctens5 ................................... Elasticity tensor material 5
    │  │  ├─ctens6 ................................... Elasticity tensor material 6
    │  │  ├─ctens7 ................................... Elasticity tensor material 7
    │  │  ├─ctens8 ................................... Elasticity tensor material 8
    │  │  └─ctens17 .................................. Elasticity tensor material 17
    │  └─stress ...............................................Stress tensor directory
    │     ├─stress1 ..................................... Stress tensor material 1
    │     ├─stress3 ..................................... Stress tensor material 3
    │     ├─stress4 ..................................... Stress tensor material 4
    │     ├─stress5 ..................................... Stress tensor material 5
    │     ├─stress6 ..................................... Stress tensor material 6
    │     ├─stress7 ..................................... Stress tensor material 7
    │     ├─stress8 ..................................... Stress tensor material 8
    │     └─stress17 ....................................Stress tensor material 17
    ├─plasticity ......................................Plasticity functions directory
    ├─element_calculations ...........................Element calculations directory
    │  ├─element_gravity_vector ...........................Load vector due to gravity
    │  ├─element_gravity_vector_truss ...........Load vector due to gravity (truss)
    │  ├─element_force_and_stiffness ...........Force vector and stiffness matrix
    │  ├─element_force_and_stiffness_truss ......Truss element force and stiffness
    │  ├─constitutive_matrix ..............................Element constitutive matrix
    │  ├─geometric_matrix ..............................Element initial stress matrix
    │  ├─mean_dilatation_pressure ...........Mean dilatation pressure computation
    │  ├─mean_dilatation_pressure_addition ........ Adds pressure to stress tensor
    │  ├─mean_dilatation_volumetric_matrix .............Mean dilatation matrix
    │  ├─pressure_element_load_and_stiffness ........ Pressure load and stiffness
    │  └─pressure_load_matrix .............................Element pressure load matrix
    ├─global_assembly ..............................Assembly functions directory
    │  ├─external_force_update ..........................Update of the external forces
    │  ├─force_vectors_assembly ..............Assembly of the external force vector
    │  ├─gravity_vector_assembly ..................Assembly of the gravity vector
    │  ├─pressure_load_and_stiffness_assembly ... Assembly pressure residual and
    │  │                                                                 stiffness
    │  └─residual_and_stiffness_assembly .........Assembly residual and stiffness
    ├─solution_update ...................................................Update geometry
    ├─convergence_check ........................ Newton–Raphson convergence check
    ├─solution_write ..................................................Output solution
    └─support ....................................................Auxiliary functions
```

## 10.7 MASTER m-FILE FLagSHyP

As stated at the beginning of this chapter, the master m-file FLagSHyP.m is divided into three sections, presented below.

**FLagSHyP segment 1 – Master m-file**

```
clear all;close all;clc;

%%SECTION 1

if isunix()
   dirsep =  '/';
else
   dirsep =  '\';
end
basedir_fem = mfilename('fullpath');
basedir_fem = fileparts(basedir_fem);
basedir_fem = strrep(basedir_fem,['code' dirsep 'support'],'');
addpath(fullfile(basedir_fem));
addpath(genpath(fullfile(basedir_fem,'code')));
addpath((fullfile(basedir_fem,'job_folder')));

%%SECTION 2

[PRO,FEM,GEOM,QUADRATURE,BC,MAT,LOAD,CON,CONSTANT,...
 GLOBAL,PLAST,KINEMATICS] = input_data_and_initialisation(basedir_fem);

%%SECTION 3

if abs(CON.ARCLEN.arcln)==0
   if ~CON.searc
       Newton_Raphson_algorithm(PRO,FEM,GEOM,QUADRATURE,BC,MAT,LOAD,...
       CON,CONSTANT,GLOBAL,PLAST,KINEMATICS);
   else
       Line_Search_Newton_Raphson_algorithm(PRO,FEM,GEOM,QUADRATURE,...
       BC,MAT,LOAD,CON,CONSTANT,GLOBAL,PLAST,KINEMATICS);
   end
else
       Arc_Length_Newton_Raphson_algorithm(PRO,FEM,GEOM,QUADRATURE,...
       BC,MAT,LOAD,CON,CONSTANT,GLOBAL,PLAST,KINEMATICS);
end
```

The first section sets up the directory path, which has been previously explained in the user instructions. The second section is the function input_data_and_initialisation.m, for the reading of input/control data and the initialization of variables including the initial assembled tangent stiffness and equivalent nodal forces. At this stage, it is appropriate to get an overview of the third section before dealing in detail with the input data and initialization function. Depending on the values chosen by the user for some of the problem-dependent control variables, this redirects the program to one of the three fundamental algorithms incorporated into FLagSHyP, namely `ALG1`, `ALG2`, and `ALG3`.

The three `ALG` MATLAB functions share many sub-functions, but `ALG1` constitutes the majority of the program. As a result, we will focus on the description of this basic algorithm in order to give a general overview of a standard large strain nonlinear finite element computer program. Once this function has been fully understood, it is expected that readers can then proceed to study the other two functions (with the help of Sections 9.6.2 and 9.6.3).

Within the function `Newton_Raphson_algorithm.m`, it is essential to understand one crucial function, namely, `residual_and_stiffness_assembly.m`, which assembles the residual force (`GLOBAL.Residual`) and the global tangent stiffness matrix (`GLOBAL.K`). This requires the computation of the equivalent nodal forces due to internal stress and the main components of the stiffness matrix for each element, which is carried out in `element_force_and_stiffness.m`. Consequently, it provides a vehicle for examining those aspects of the computation that are particular to finite deformation analysis. In addition, the function `pressure_load_and_stiffness_assembly.m` evaluates the equivalent nodal forces due to surface pressure and the corresponding tangent matrix components. Additional functions included are `linear_solver.m`, which calls the linear solver for the computation of the unknown displacements $\mathbf{u}$, `update_geometry.m`, which updates the geometry of the deformed solid, and `check_residual_norm.m`, which verifies whether convergence has been achieved.

Within the function `residual_and_stiffness_assembly.m` mentioned above, it is important to distinguish between truss and continuum elements. Specifically, for truss elements the function `element_force_and_stiffness_truss.m` evaluates the equivalent nodal forces and the components of the tangent stiffness matrix. This function is not described as it is transparent in its replication of the equations in the text. For any other kind of finite element, the counterpart function `element_force_and_stiffness.m` evaluates the same quantities. This latter function is more complex, hence lengthier, as it is not restricted to simplified truss kinematics and it allows for the consideration of a large variety of kinematics hypotheses, for example, near incompressibility, analysis along principal directions, or plane stress.

The function `element_force_and_stiffness.m` calls the utility functions `Cauchy_type_selection.m` and `elasticity_modulus_selection.m` to compute the stress tensor and tangent modulus, respectively, depending upon the material type selection, either a pure hyperelastic or an elasto-plastic material. The remaining functions in the program are either relatively similar to standard finite element elasticity codes or are a direct implementation of equations contained in the book.

Returning to a detailed consideration of the second section of FLagSHyP, this is presented below.

**FLagSHyP segment 2 – Input data and initialization**

```
function [PRO,FEM,GEOM,QUADRATURE,BC,MAT,LOAD,CON,CONSTANT,GLOBAL,...
          PLAST,KINEMATICS] = input_data_and_initialisation(basedir_fem)

PRO = welcome(basedir_fem);
if ~PRO.rest
   fid = PRO.fid_input;
   [FEM,GEOM,QUADRATURE,BC,MAT,LOAD,CON,PRO,...
    GLOBAL]= reading_input_file(PRO,fid);

   CONSTANT = constant_entities(GEOM.ndime);
   CON.xlamb = 0;
   CON.incrm = 0;
   [GEOM,LOAD,GLOBAL,PLAST,KINEMATICS] = initialisation(FEM,GEOM,...
    QUADRATURE,MAT,LOAD,CONSTANT,CON,GLOBAL);
   cd(PRO.job_folder)
   save_restart_file(PRO,FEM,GEOM,QUADRATURE,BC,MAT,LOAD,CON,CONSTANT,...
                     GLOBAL,PLAST,KINEMATICS,'internal')
```

The program can start either from an input data file or, when convenient, using data from a restart file written during a previous incomplete analysis. The mode of operation is recorded in the variable `PRO.rest` read by the `welcome.m` function. Function `initialisation.m` sets global residual vector `GLOBAL.Residual` to zero and defines the initial geometry `GEOM.x0` as the current geometry `GEOM.x`. In addition, the equivalent nodal forces due to gravity are evaluated and the function `residual_and_stiffness_assembly.m` is called to compute the global tangent stiffness matrix at the unstressed configuration (this coincides with the small strain linear elastic stiffness matrix). The function `initial_volume.m` evaluates the total mesh volume and prints this on the screen. This is particularly useful in order to check the validity of the geometry given by the input file but also fundamental in the case of dealing with a nearly incompressible material, where the initial elemental volume is necessary for the mean dilatation algorithm. All the relevant information is dumped to a restart file.

**FLagSHyP segment 3 – Restart**

```
else
  cd(PRO.job_folder);
  load(PRO.restartfile_name);
  [GLOBAL,PLAST] = residual_and_stiffness_assembly(CON.xlamb,GEOM,...
    MAT,FEM,GLOBAL,CONSTANT,QUADRATURE.element,PLAST,KINEMATICS);
end
```

The program can restart from a previously converged incremental step in order to progress to subsequent increments. This can be useful when wishing to continue the analysis to obtain further incremental solutions. In the following it is assumed that the data are read for the first time.

**FLagSHyP segment 4 – Reading Input data**

```
function [FEM,GEOM,QUADRATURE,BC,MAT,LOAD,CON,PRO,GLOBAL] =...
          reading_input_file(PRO,fid)

print_statement('beginning_reading')
PRO.title = strtrim(fgets(fid));
[FEM,GEOM,QUADRATURE] = elinfo(fid);

switch FEM.mesh.element_type
 case 'truss2'
   FEM.interpolation.element = [];
   FEM.interpolation.boundary = [];
 otherwise
   QUADRATURE.element = element_quadrature_rules(FEM.mesh.element_type);
   QUADRATURE.boundary = edge_quadrature_rules(FEM.mesh.element_type);
   FEM = shape_functions_iso_derivs(QUADRATURE,FEM,GEOM.ndime);
end

[GEOM,BC,FEM] = innodes(GEOM,fid,FEM);
[FEM,MAT] = inelems(FEM,fid);
BC = find_fixed_free_dofs(GEOM,FEM,BC);
MAT = matprop(MAT,FEM,fid);
[LOAD,BC,FEM,GLOBAL] = inloads(GEOM,FEM,BC,fid);
CON = incontr(BC,fid);

fclose('all');
print_statement('end_reading')
```

Function `elinfo.m` reads the element type `FEM.mesh.element_type` and establishes the basic information regarding the specific type of finite element to be used. Function `innodes.m` reads nodal input data, namely initial coordinates (`GEOM.x`) and boundary conditions (`BC.icode`). Function `inelems.m` reads relevant element information, namely connectivity (`FEM.mesh.connectivity`) and material type (`MAT.matno`). Function `find_fixed_free_dofs.m` classifies the degrees of freedom in fixed and free, based on the nodal boundary condition information. Function `matprop.m` reads in material property data (`MAT.props`). Function `inloads.m` inputs loading and prescribed displacement items. Finally, function `incontr.m` reads the solution control parameters (`CON`).

**FLagSHyP segment 5 – Increment loop**

```
function Newton_Raphson_algorithm(PRO,FEM,GEOM,QUADRATURE,BC,MAT,...
LOAD,CON,CONSTANT,GLOBAL,PLAST,KINEMATICS)

while ((CON.xlamb < CON.xlmax) && (CON.incrm < CON.nincr))
  CON.incrm = CON.incrm + 1;
  CON.xlamb = CON.xlamb + CON.dlamb;

  [GLOBAL.Residual,GLOBAL.external_load] = ...
  external_force_update(GLOBAL.nominal_external_load,...
  GLOBAL.Residual,GLOBAL.external_load,CON.dlamb);

  if LOAD.n_pressure_loads
     GLOBAL = pressure_load_and_stiffness_assembly(GEOM,MAT,...
     FEM,GLOBAL,LOAD,QUADRATURE.boundary,CON.dlamb);
  end

  if  BC.n_prescribed_displacements > 0
      GEOM.x = update_prescribed_displacements(BC.dofprescribed,...
      GEOM.x0,GEOM.x,CON.xlamb,BC.presc_displacement);
      [GLOBAL,updated_PLAST] = ...
      residual_and_stiffness_assembly(CON.xlamb,...
      GEOM,MAT,FEM,GLOBAL,CONSTANT,QUADRATURE.element,PLAST,KINEMATICS);

      if LOAD.n_pressure_loads
          GLOBAL = pressure_load_and_stiffness_assembly(GEOM,...
          MAT,FEM,GLOBAL,LOAD,QUADRATURE.boundary,CON.xlamb);
      end

  end
```

The `while` controls the load or prescribed displacement incrementation. This remains active while the load-scaling parameter `CON.xlamb` is less than the maximum `CON.xlmax` and the increment number is smaller than the total number of increments `CON.nincr`.

The imposition of an increment of point or gravity loads immediately creates a residual force. This is added to any small residual carried over from the previous increment by the function `external_force_update.m`. This prevents errors in the converged solution (for instance, due to a large tolerance value `CON.cnorm`) from propagating throughout the solution process. The further addition to the residual forces due to an increment in applied surface pressure is evaluated by calling `pressure_load_and_stiffness_assembly.m` with the last argument set to the load parameter increment `CON.dlamb`. At the same time `pressure_load_and_stiffness_assembly.m` evaluates the addition to the tangent matrix caused by possible geometrical changes to the pressure surface and the increment in the magnitude of the pressure.

If prescribed displacements have been imposed, then the current increment in their value will immediately change the current geometry and necessitate a complete re-evaluation of equivalent internal and surface pressure forces and the tangent matrix (this is effectively a geometry update). In this case, function `update_prescribed_displacements.m` will reset the current geometry for those nodes with prescribed displacements based on their initial coordinates, the nominal value of the prescribed displacement, and the current load-scaling parameter `CON.xlamb`. Function `residual_and_stiffness_assembly.m` recomputes the residual vector and the global tangent stiffness matrix due to the new updated configuration. Subsequently, function `pressure_load_and_stiffness_assembly.m` is now called with the last argument set to `CON.xlamb`, to add the total value of the nodal forces due to surface pressure and obtain the corresponding initial surface pressure tangent matrix component.

**FLagSHyP segment 6 – Newton–Raphson loop**

```
CON.niter = 0;
rnorm = 2*CON.cnorm;
while((rnorm > CON.cnorm) && (CON.niter < CON.miter))
    CON.niter = CON.niter + 1;
    displ = linear_solver(GLOBAL.K,-GLOBAL.Residual,BC.fixdof);
```

The `while` loop controls the Newton–Raphson iteration process. This continues until the residual norm `rnorm` is smaller than the tolerance `CON.cnorm` and, of course, while the iteration number `CON.niter` is smaller than the maximum allowed `CON.miter`. Note that `rnorm` is initialized (`rnorm = 2*CON.cnorm`) in such a way that this loop is completed at least once. The function `linear_solver.m` solves the linear system of equations to obtain the incremental displacements **u**.

**FLagSHyP segment 7 – Solution update and equilibrium check loop**

```
        GEOM.x = update_geometry(GEOM.x,1,displ,BC.freedof);

        [GLOBAL,updated_PLAST] = ...
        residual_and_stiffness_assembly(CON.xlamb,GEOM,MAT,...
        FEM,GLOBAL,CONSTANT,QUADRATURE.element,PLAST,KINEMATICS);

        if LOAD.n_pressure_loads
           GLOBAL = pressure_load_and_stiffness_assembly(GEOM,MAT,...
           FEM,GLOBAL,LOAD,QUADRATURE.boundary,CON.xlamb);
        end

        [rnorm,GLOBAL] = check_residual_norm(CON,BC,GLOBAL,BC.freedof);
        if (abs(rnorm)>1e7 || isnan(rnorm))
           CON.niter=CON.miter;
           break;
        end
    end
```

The function update_geometry.m adds the displacements evaluated in segment 6 to the current nodal positions GEOM.x. Given this new geometry, the two functions residual_and_stiffness_assembly.m and pressure_load_and_stiffness_assembly.m serve the same purpose as described in segment 5. Finally, checkr_residual_norm.m computes the residual norm rnorm given the residual force vector GLOBAL.Residual and the total external force vector (including pressure loading).

**FLagSHyP segment 8 – Retry and output results**

```
      if( CON.niter >= CON.miter)
        load(PRO.internal_restartfile_name)
        fprintf('Solution not converged. Restart from previous...
        step by decreasing the load parameter increment to ...
        half its initally fixed value. {\mibf n}');
        CON.dlamb = CON.dlamb/2;
      else
        PLAST = save_output(updated_PLAST,PRO,FEM,GEOM,QUADRATURE,...
        BC,MAT,LOAD,CON,CONSTANT,GLOBAL,PLAST,KINEMATICS);
      end
  end
```

Function save_output.m writes out the output file described in Section 10.3. This occurs at every CON.OUTPUT.incout increment and is stored in the relevant example folder within job_folder with a unique identifier. In addition, the function utilizes the MATLAB function save.m to dump all variables within the existing workspace. This is continually refreshed to be available for an automatic restart in the event that the Newton–Raphson iteration does not converge, in which case the program will reduce the load increment by half (CON.dlamb←CON.dlamb/2).

## 10.8 FUNCTION residual_and_stiffness_assembly

This function assembles the residual force vector and the main components of the global tangent stiffness matrix for all elements other than those surface (line) elements subjected to pressure load, which are considered in Section 10.11.

residual_and_stiffness_assembly **part 1 – Initialization**

```
function [GLOBAL,updated_PLAST] = residual_and_stiffness_assembly(xlamb,...
GEOM,MAT,FEM,GLOBAL,CONSTANT,QUADRATURE,PLAST,KINEMATICS)

updated_PLAST = PLAST;
GLOBAL.external_load = xlamb*GLOBAL.nominal_external_load;

switch FEM.mesh.element_type
  case 'truss2'
     n_components = FEM.mesh.nelem*FEM.mesh.n_dofs_elem^2;
  otherwise
    n_components_mean_dilatation =...
    MAT.n_nearly_incompressible*(FEM.mesh.n_dofs_elem^2);
    n_components_displacement_based = ...
    FEM.mesh.nelem*(FEM.mesh.n_dofs_elem^2+...
    (FEM.mesh.n_nodes_elem^2*GEOM.ndime))*QUADRATURE.ngauss;
    n_components = n_components_mean_dilatation +...
    n_components_displacement_based;
end

counter = 1;
indexi = zeros(n_components,1);
indexj = zeros(n_components,1);
global_stiffness = zeros(n_components,1);
GLOBAL.T_int = zeros(FEM.mesh.n_dofs,1);
```

The segment above initializes the dimension of the vectors indexi and indexj required for the allocation of the stiffness terms in the vector global_stiffness. In addition, the size of the vector GLOBAL.T_int is initialized based upon the total number of degrees of freedom FEM.mesh.n_dofs.

`residual_and_stiffness_assembly` **part 2 – Element loop**

```
for ielement=1:FEM.mesh.nelem
    global_nodes = FEM.mesh.connectivity(:,ielement);
    material_number = MAT.matno(ielement);
    matyp = MAT.matyp(material_number);
    properties = MAT.props(:,material_number);
    xlocal = GEOM.x(:,global_nodes);
    x0local = GEOM.x0(:,global_nodes);
    Ve = GEOM.Ve(ielement);
    PLAST_element = selecting_internal_variables_element(PLAST,...
    matyp,ielement);

    switch FEM.mesh.element_type
       case 'truss2'
            [T_internal,indexi,indexj,global_stiffness,counter,...
             PLAST_element] = ...
             element_force_and_stiffness_truss(properties,...
             xlocal,x0local,global_nodes,FEM,PLAST_element,...
             counter,indexi,indexj,global_stiffness,GEOM);
       otherwise
            [T_internal,indexi,indexj,global_stiffness,counter,...
            PLAST_element] = ...
            element_force_and_stiffness(FEM,xlocal,x0local,...
            global_nodes,Ve,QUADRATURE,properties,CONSTANT,...
            GEOM.ndime,matyp,PLAST_element,counter,KINEMATICS,...
            indexi,indexj,global_stiffness);
   end

   GLOBAL.T_int = force_vectors_assembly(T_internal,global_nodes,...
   GLOBAL.T_int,FEM.mesh.dof_nodes);

   updated_PLAST = plasticity_storage(PLAST_element,...
   updated_PLAST,matyp,ielement);
end

GLOBAL.K = sparse(indexi,indexj,global_stiffness);
GLOBAL.Residual = GLOBAL.T_int - GLOBAL.external_load;
```

For all element types the function `residual_and_stiffness_assembly.m` loops over all the elements. For each element other than `truss2`, the function `element_force_and_stiffness.m` is called to obtain the element equivalent force vector `T_internal` and the element contribution to the global stiffness matrix `global_stiffness`.

Within the element loop, `force_vectors_assembly.m` assembles the element equivalent force `T_internal` into the global equivalent force vector

GLOBAL.T_int. Once the loop over elements ends, the element stiffness contributions in global_stiffness are assembled into the global stiffness matrix GLOBAL.K, employing the MATLAB intrinsic function sparse.m. In addition, the global residual vector GLOBAL.Residual is obtained after subtracting the equivalent external force vector (GLOBAL.external_load) from the internal force vector (GLOBAL.T_int). When element type is truss2 the function element_force_and_stiffness_truss.m is used for the calculation of the equivalent nodal forces and tangent matrix. The coding is not discussed herein as it is self-explanatory.

The element_force_and_stiffness.m function called in part 2 above will now be considered in detail.

element_force_and_stiffness **part 1 – Prior to Gauss point loop**

```
function [T_internal,indexi,indexj,global_stiffness,counter,...
          PLAST_element] = element_force_and_stiffness(FEM,xlocal,...
          x0local,element_connectivity,Ve,QUADRATURE,properties,...
          CONSTANT,dim,matyp,PLAST,counter,KINEMATICS,indexi,...
          indexj,global_stiffness)

T_internal = zeros(FEM.mesh.n_dofs_elem,1);
KINEMATICS = gradients(xlocal,x0local,FEM.interpolation.element.DN_chi,...
QUADRATURE,KINEMATICS);

switch matyp
    case {5,7,17}
         [pressure,kappa_bar,DN_x_mean,ve] = ...
         mean_dilatation_pressure(FEM,dim,matyp,properties,...
         Ve,QUADRATURE,KINEMATICS);
    otherwise
         pressure = 0;
end
```

Here the element internal equivalent force vector T_internal is initialized based upon the number of element degrees of freedom FEM.mesh.n_dofs_elem. The function gradients.m computes all the kinematics quantities associated with a finite element, such as $\boldsymbol{F}$ and $\boldsymbol{b} = \boldsymbol{F}\boldsymbol{F}^T$ and the pointwise volume ratio $J = \det \boldsymbol{F}$ (or area ratio in the case of plane stress). For convenience, these quantities are computed at all Gauss points of the element and stored in the variables KINEMATICS.F, KINEMATICS.b, and KINEMATICS.J. Other elemental quantities of interest computed in gradients.m include the current Cartesian derivatives $\partial N_a/\partial \boldsymbol{x}$ of the shape functions (refer to Equation (9.11a,b)), which are stored in KINEMATICS.DN_x, and the weighted Jacobian per Gauss point, which is stored in

KINEMATICS.Jx_Chi. Using the left Cauchy–Green tensor $\boldsymbol{b}$, it is then straightforward to obtain the Cauchy stress and elasticity tensor for each of the seven material types implemented.

For nearly incompressible materials 5 and 7 and the elasto-plastic material 17, the mean dilatation technique described in Chapters 8 and 9 is used to obtain the internal element pressure $p$ (pressure) from the volume ratio $\bar{J}$ (Jbar in the program) using Equation (8.53a). This is implemented in function mean_dilatation_pressure.m.

Referring to Algorithm 1, we now consider the Gauss integration loop.

**element_force_and_stiffness part 2 – Gauss point loop**

```
for igauss=1:QUADRATURE.ngauss
   kinematics_gauss = kinematics_gauss_point(KINEMATICS,igauss);

   [Cauchy,PLAST,plast_gauss] = Cauchy_type_selection(kinematics_gauss,...
   properties,CONSTANT,dim,matyp,PLAST,igauss);

   c = elasticity_modulus_selection(kinematics_gauss,...
   properties,CONSTANT,dim,matyp,PLAST,plast_gauss,igauss);

   [Cauchy,c] = pressure_addition(Cauchy,c,CONSTANT,pressure,matyp);

   th = thickness_plane_stress(properties,kinematics_gauss.J,matyp);

   JW = kinematics_gauss.Jx_chi*QUADRATURE.W(igauss)*th;

   T = Cauchy*kinematics_gauss.DN_x;

   T_internal = T_internal + T(:)*JW;

   [indexi,indexj,global_stiffness,counter] = constitutive_matrix(FEM,...
   dim,element_connectivity,kinematics_gauss,c,JW,counter,indexi,...
   indexj,global_stiffness);

   DN_sigma_DN = kinematics_gauss.DN_x'*(Cauchy*kinematics_gauss.DN_x);
   [indexi,indexj,global_stiffness,counter] = geometric_matrix(FEM,...
   dim,element_connectivity,DN_sigma_DN,JW,counter,indexi,...
   indexj,global_stiffness);
end
switch matyp
   case {5,7,17}
        [indexi,indexj,global_stiffness,counter] = ...
        volumetric_mean_dilatation_matrix(FEM,dim,...
        element_connectivity,DN_x_mean,counter,...
        indexi,indexj,global_stiffness,kappa_bar,ve);
end
PLAST_element = PLAST;
```

For each Gauss point, and once the relevant kinematics quantities have been extracted, the function Cauchy_type_selection.m evaluates the Cauchy stress tensor as a function of the left Cauchy–Green tensor and the properties of the material of the element ielem under consideration. Within this function, depending on the value of the variable MAT.matyp(MAT.matno(ielem)), a specific stress function is called. Analogously, the function elasticity_modulus_selection.m evaluates the elasticity tensor as a function of the left Cauchy–Green tensor and the properties of the material of the element ielem.

For materials 5, 7, and 17, the mean dilatation procedure applies and the pressure evaluated in function mean_dilatation_pressure.m is now added to the deviatoric Cauchy stress tensor in the function pressure_addition.m. Recalling that we remain within the Gauss loop, the Cauchy stress components and the Cartesian derivatives of the shape functions are used to compute and assemble the equivalent nodal forces T employing Equations (9.15a–c). Function constitutive_matrix.m then evaluates the constitutive component of the tangent matrix according to the indicial Equation (9.35). The initial stress matrix is obtained and assembled in geometric_matrix.m using Equation (9.44c). This function will be described in detail in Section 10.10.

In addition, and once the Gauss integration loop is completed, the dilatational component $\mathbf{K}_\kappa$ of the tangent matrix is computed at the element level and assembled in function mean_dilatation_volumetric_matrix.m using Equation (9.59a,b).

Cauchy_type_selection – **Stress tensor evaluation**

```
function [Cauchy,PLAST,PLAST_gauss] = Cauchy_type_selection(kinematics,...
          properties,cons,dim,matyp,PLAST,igauss)

PLAST_gauss = [];
switch matyp
   case 1
        Cauchy = stress1(kinematics,properties,cons);
   case 3
        Cauchy = stress3(kinematics,properties,dim);
   case 4
        Cauchy = stress4(kinematics,properties,dim);
   case 5
        Cauchy = stress5(kinematics,properties,cons,dim);
   case 6
        Cauchy = stress6(kinematics,properties,cons);
   case 7
        Cauchy = stress7(kinematics,properties,dim);
```

*(continued)*

*(cont.)*

```
    case 8
         Cauchy = stress8(kinematics,properties,dim);
    case 17
         PLAST_gauss.OLD.invCp = PLAST.invCp(:,:,igauss);
         PLAST_gauss.OLD.epbar = PLAST.epbar(igauss);
         [Cauchy,PLAST_gauss] = stress17(kinematics,...
         properties,dim,PLAST_gauss);
         PLAST = plasticity_update(PLAST_gauss.UPDATED,...
         PLAST,igauss,matyp);
end
```

Depending on the value of the variable `MAT.matyp(MAT.matno(ielem))`, the function `Cauchy_type_selection.m` selects a specific stress tensor function.

`elasticity_modulus_selection` – **Elasticity tensor evaluation**

```
function c_tensor = elasticity_modulus_selection(kinematics,...
         properties,cons,dimension,matyp,PLAST,plast_gauss,igauss)

c_tensor = [];
switch matyp
   case 1
        c_tensor = ctens1(kinematics,properties,cons);
   case 3
        c_tensor = ctens3(kinematics,properties,dimension);
   case 4
        c_tensor = ctens4(kinematics,properties,dimension);
   case 5
        c_tensor = ctens5(kinematics,properties,cons,dimension);
   case 6
        c_tensor = ctens6(kinematics,properties,cons);
   case 7
        c_tensor = ctens7(kinematics,properties,dimension);
   case 8
        c_tensor = ctens8(kinematics,properties,dimension);
   case 17
        plast_gauss.OLD.invCp = PLAST.invCp(:,:,igauss);
        plast_gauss.OLD.epbar = PLAST.epbar(igauss);
        c_tensor = ctens17(kinematics,properties,dimension,plast_gauss);
end
```

Similarly, depending on the value of `MAT.matyp(MAT.matno(ielem))`, the function `elasticity_modulus_selection.m` selects a specific elasticity tensor function.

It is now worthwhile discussing some aspects of the implementation of the stress tensor and elasticity tensor for the various materials. In particular, the stress calculations for the elasto-plastic material 17 will be considered in more detail. Recall that these calculations are particular to each Gauss point; for example, within an element, one Gauss point could be elastic while another could have become plastic.

**Material 1**. For a compressible neo-Hookean material, the functions `stress1.m` and `ctens1.m` are described in Section 6.4.3. The Cauchy stresses are evaluated in `stress1.m` in terms of the $\boldsymbol{b}$ tensor from Equation (6.29), whereas the coefficients of the elasticity tensor are obtained in `ctens1.m` using Equation (6.40).

**Material 3**. For a hyperelastic material in principal directions, the principal directions $\boldsymbol{n}_\alpha$ and principal stretches $\lambda_\alpha$ are computed by the function `gradients.m` and stored in `KINEMATICS.n` and `KINEMATICS.lambda`, respectively. Given these stretches and directions, the function `stress3.m` evaluates the Cartesian components of the Cauchy stress tensor with the help of Equations (6.81) and (6.94). Finally, the function `ctens3.m` uses Equations (6.90), (6.91), and (6.95) to compute the corresponding elasticity tensor.

**Material 4**. A plane stress hyperelastic material in principal directions is described in Section 6.6.7 and is implemented in a similar way to the previous material, the main difference being in the parameter $\gamma$, which is obtained in `stress4.m` and `ctens4.m`. For this material, the current thickness, computed in the function `thickness_plane_stress.m`, is updated in terms of the volume ratio $J$, the area ratio $j$, and the initial thickness stored in `MAT.props(4,im)` (see Exercise 4 of Chapter 4). The current thickness is stored as the fourth stress for output purposes.

**Material 5**. A nearly incompressible neo-Hookean material is discussed in Sections 6.5.2 and 6.5.3. The deviatoric Cauchy stresses are evaluated in `stress5.m` using Equation (6.55a,b). Note that in the mean dilatation method the pressure is constant over the element and therefore evaluated in segment 2 outside the Gauss point loop. This pressure value is added to the deviatoric components in `pressure_addition.m`. Analogously, function `ctens5.m` obtains the deviatoric component of the elasticity tensor and `pressure_addition.m` adds the volumetric component (refer to Equations (6.59a,b)).

**Material 6**. The plane stress incompressible neo-Hookean material is discussed in Exercise 2 of Chapter 6 and implemented in functions `stress6.m` and `ctens6.m`. Initially, the effective shear coefficient $\mu'$ is evaluated, dividing $\mu$ by $j^2 = \det_{2\times 2} \boldsymbol{b}$. As shown in Exercise 2 of Chapter 6, an effective lambda coefficient (see Equation (6.40)) emerges as twice the effective shear coefficient. The thickness is finally computed as in material 4, with the exception that due to incompressibility $J$ is equal to 1.

**Material 7**. The nearly incompressible material in principal directions is discussed in Section 6.6.6. Stretches and principal directions are first evaluated in `gradients.m`. Function `stress7.m` implements Equation (6.94) with $\lambda$ being set to $-2\mu/3$ to give the deviatoric Cauchy stresses in accordance with Equation (6.107). The internal hydrostatic pressure is subsequently added in `pressure_addition.m`. The deviatoric components of the elasticity tensor are obtained using the function `ctens7.m`. Finally, the volumetric component is obtained as per material 5.

**Material 8**. This material, implemented in functions `stress8.m` and `ctens8.m`, is identical to material 4, but because of incompressibility $\lambda \to \infty$ and hence $\gamma = 0$, $J = 1$, and $\bar{\lambda} = 2\mu$.

**Material 17**. This material is presented below in two parts. Part 1 evaluates the yield function and part 2 considers the consequences of this calculation, namely elastic or plastic behavior at the Gauss point.

**`stress17` part 1 – Cauchy stress and elasticity tensors**

```
function [Cauchy,PLAST] = stress17(kinematics,properties,dim,PLAST)

invCp = PLAST.OLD.invCp;
ep = PLAST.OLD.epbar;
mu = properties(2);
DIM = dim;
J = kinematics.J;
F(1:dim,1:dim) = kinematics.F;

be_trial(1:dim,1:dim) = F*(invCp*F');
[V,D] = eig(be_trial);
lambdae_trial = sqrt(diag(D));
na_trial = V;

tauaa_trial = (2*mu)*log(lambdae_trial)- (2*mu/3)*log(J);
switch dim
   case 2
        DIM = 3;
        lambdae_trial(3) = 1/det(invCp);
        tauaa_trial = (2*mu)*log(lambdae_trial)- (2*mu/3)*log(J);
        na_trial = [[na_trial(:,1);0] [na_trial(:,2);0] [0;0;1]];
        be_trial(3,3) = lambdae_trial(3);
end
tau_trial = zeros(DIM);
for idim=1:DIM
   tau_trial = tau_trial + ...
   tauaa_trial(idim)*(na_trial(:,idim)*na_trial(:,idim)');
end

f = Von_Mises_yield_function(tau_trial,ep,properties(5),properties(6));
```

In part 1 above, the trial left Cauchy–Green tensor, given by Equation (7.41), is first evaluated in order to calculate the associated principal directions and stretches. Subsequently, the trial principal (see Equation (7.43a,b)) Cartesian deviatoric Kirchhoff stresses are computed to enable the yield function to be evaluated in accordance with Equation (7.20a,b).

stress17 **part 2 – Cauchy stress and elasticity tensors**

```
if f>0
   [Dgamma,nu_a] = radial_return_algorithm(f,tau_trial,...
   tauaa_trial,mu,properties(6));
   lambdae = exp(log(lambdae_trial) - Dgamma*nu_a);
   norm_tauaa_trial = norm(tauaa_trial,'fro');
   tauaa = (1-2*mu*Dgamma/(sqrt(2/3)*norm_tauaa_trial))*tauaa_trial;
   tau = zeros(dim);
   be = zeros(dim);
   for idim=1:DIM
       tau = tau +...
       tauaa(idim)*(na_trial(1:dim,idim)*na_trial(1:dim,idim)');
       be = be + ...
       lambdae(idim)^2*(na_trial(1:dim,idim)*na_trial(1:dim,idim)');
   end
else
   tau = tau_trial(1:dim,1:dim);
   tauaa = tauaa_trial(1:dim);
   Dgamma = 0;
   nu_a = zeros(dim,1);
   be = be_trial(1:dim,1:dim);
end

Cauchy = tau/J;
Cauchyaa = tauaa/J;
PLAST.stress.Cauchy = Cauchy;
PLAST.stress.Cauchyaa = Cauchyaa;

invF = inv(F);
PLAST.UPDATED.invCp = invF*(be*invF');
PLAST.UPDATED.epbar = ep + Dgamma;

PLAST.trial.lambdae = lambdae_trial(1:dim);
PLAST.trial.tau = tau_trial(1:dim,1:dim);
PLAST.trial.n = na_trial(1:dim,1:dim);

PLAST.yield.f = f;
PLAST.yield.Dgamma = Dgamma;
PLAST.yield.nu_a = nu_a(1:dim);
```

For the elastic case, the same implementation as with material 3 is used to calculate the deviatoric Cauchy principal stresses. Note that the pressure found in `mean_dilatation_pressure.m` is added to the deviatoric Cartesian stress components in `pressure_addition.m`. Analogously, the deviatoric and volumetric components of the elasticity tensor are identical to those for materials 3 and 5 and computed in `ctens17.m` and `pressure_addition.m`, respectively.

For the plastic case, the radial return algorithm given by Equations (7.59) and (7.60) is implemented to find the incremental plastic multiplier and the deviatoric Kirchhoff principal stresses. In addition, the elastic left Cauchy–Green tensor is evaluated from Equation (7.49). The Cartesian deviatoric Cauchy stresses are found using Equation (6.81) and added to the pressure in `pressure_addition.m`. The deviatoric component of the tangent modulus, given by Equation (7.62a), is evaluated in `ctens17.m` and added to the volumetric component in `pressure_addition.m`. Finally, the state variables are updated, namely, the Von Mises equivalent strain (Equation (7.61)) and the inverse of plastic right Cauchy–Green tensor given by Equation $(7.51\text{a,b})_b$.

The radial return algorithm is described in Section 7.6.1 and summarized under Material 10.17 in Appendix 10.14 at the end of this chapter. The vector `nu_a`, normal to the yield surface, is given by Equation (7.54a,b) and the incremental plastic multiplier `Dgamma` by Equation (7.59). Trial stretches `lambdae` are evaluated using Equation (7.45a,b). The radial return operation gives the deviatoric Kirchhoff principal stresses `tauaa` using Equation (7.60). Finally, the updated left Cauchy–Green tensor be is evaluated using Equation (7.49).

Additionally, Appendix 10.14 contains the constitutive equations for all the material types presented in indicial form with appropriate references to equations within the text.

## 10.9 FUNCTION `constitutive_matrix`

This function implements the constitutive component of the tangent stiffness matrix. Recognizing the continual iterative updating of the geometric and constitutive terms, the function is identical to that found in a standard linear elasticity program.

```
constitutive_matrix

function [element_indexi,element_indexj,element_stiffness,counter] =...
 constitutive_matrix(FEM,dim,element_connectivity,kinematics_gauss,...
 c,JW,counter,element_indexi,element_indexj,element_stiffness)
```

*(continued)*

*(cont.)*

```
for bnode=1:FEM.mesh.n_nodes_elem
   for anode=1:FEM.mesh.n_nodes_elem
       for j=1:dim
           indexj = FEM.mesh.dof_nodes(j,element_connectivity(bnode));
           for i=1:dim
               indexi = FEM.mesh.dof_nodes(i,element_connectivity(anode));
               sum = 0;
               for k=1:dim
                   for l=1:dim
                       sum = sum +...
                       kinematics_gauss.DN_x(k,anode)*c(i,k,j,l)*...
                       kinematics_gauss.DN_x(l,bnode)*JW;
                   end
               end
               element_indexi(counter) = indexi;
               element_indexj(counter) = indexj;
               element_stiffness(counter) = sum;
               counter = counter + 1;
           end
       end
   end
end
```

The constitutive matrix calculation shown above is within the Gauss integration loop shown in Algorithm 1. The innmermost quantity sum is a direct implementation of Equation (9.35).

## 10.10 FUNCTION geometric_matrix

Although all stiffness matrix calculations are a direct implementation of the relevant equations given in the text, it is instructive to consider the computation and assembly of the initial stress matrix $\mathbf{K}_{\sigma}$, which is of particular significance in finite deformation analysis.

geometric_matrix

```
function [element_indexi,element_indexj,element_stiffness,...
          counter] = geometric_matrix(FEM,dim,element_connectivity,...
          DN_sigma_DN,JW,counter,element_indexi,element_indexj,...
          element_stiffness)
```

*(continued)*

*(cont.)*

```
for bnode=1:FEM.mesh.n_nodes_elem
    for anode=1:FEM.mesh.n_nodes_elem
        DNa_sigma_DNb = DN_sigma_DN(anode,bnode);

        element_indexi(counter:counter+dim-1) = ...
        FEM.mesh.dof_nodes(:,element_connectivity(anode));

        element_indexj(counter:counter+dim-1) = ...
        FEM.mesh.dof_nodes(:,element_connectivity(bnode));

        element_stiffness(counter:counter+dim-1) = ...
        DNa_sigma_DNb*JW;

        counter = counter + dim;
    end
end
```

This function is called from `element_force_and_stiffness.m` within a Gauss and element loop. The variable `DNa_sigma_DNb` contains the quantity $\boldsymbol{\nabla} N_a \cdot \boldsymbol{\sigma} \boldsymbol{\nabla} N_b$ computed according to Equation (9.44c), which is saved in the vector `element_stiffness`. The vectors `element_indexi` and `element_indexj` store the degree-of-freedom numbers corresponding to spatial directions of the nodes `anode` and `bnode`. Given the diagonal structure of every elemental initial stress matrix contribution $\boldsymbol{K}^{(e)}_{\sigma,ab}$, only diagonal entries are stored.

## 10.11 FUNCTION `pressure_load_and_stiffness_assembly`

For cases where there are surface or line elements with pressure applied, this function assembles the equivalent nodal forces and tangent stiffness matrix contributions.

`pressure_load_and_stiffness_assembly` – **Surface element loop**

```
function GLOBAL = pressure_load_and_stiffness_assembly(GEOM,...
MAT,FEM,GLOBAL,LOAD,QUADRATURE,xlamb)

GLOBAL.nominal_pressure = zeros(FEM.mesh.n_dofs,1);
GLOBAL.R_pressure = zeros(FEM.mesh.n_dofs,1);
GLOBAL.K_pressure = sparse([],[],[],FEM.mesh.n_dofs,FEM.mesh.n_dofs);
```

*(continued)*

*(cont.)*

```
n_components = ...
(FEM.mesh.n_face_dofs_elem^2*QUADRATURE.ngauss)*LOAD.n_pressure_loads;
indexi = zeros(n_components,1);
indexj = zeros(n_components,1);
global_stiffness = zeros(n_components,1);
counter = 1;

for ipressure = 1:LOAD.n_pressure_loads
   element = LOAD.pressure_element(ipressure);
   global_nodes = FEM.mesh.connectivity_faces(:,element);
   material_number = MAT.matno(element);
   matyp = MAT.matyp(material_number);
   properties = MAT.props(:,material_number);
   xlocal_boundary = GEOM.x(:,global_nodes);

   counter0 = counter;
   [R_pressure_0,indexi,indexj,global_stiffness,counter] = ...
    pressure_load_element_Residual_and_Stiffness(properties,...
    matyp,xlocal_boundary,global_nodes,GEOM.ndime,...
    QUADRATURE,FEM,counter,indexi,indexj,global_stiffness);

   nominal_pressure = LOAD.pressure(ipressure)*R_pressure_0;
   GLOBAL.nominal_pressure = force_vectors_assembly(nominal_pressure,...
   global_nodes,GLOBAL.nominal_pressure,FEM.mesh.dof_nodes);
   global_stiffness(counter0:counter-1) = ...
   LOAD.pressure(ipressure)*xlamb*global_stiffness(counter0:counter-1);
end

GLOBAL.K_pressure = sparse(indexi,indexj,global_stiffness,...
FEM.mesh.n_dofs,FEM.mesh.n_dofs);
GLOBAL.Residual = GLOBAL.Residual + xlamb*GLOBAL.nominal_pressure;
GLOBAL.K = GLOBAL.K - GLOBAL.K_pressure;
```

This function loops over the elements and, for each element, calls the function pressure_element_load_and_stiffness.m to evaluate the current unit pressure load vector, which is stored in the variable R_pressure_0. Within the latter function, a loop over the Gauss points of the surface or line elements is carried out, where the tangent vectors $\partial \boldsymbol{x}/\partial \xi$ and $\partial \boldsymbol{x}/\partial \eta$ are evaluated. This enables the equivalent nodal forces and the tangent stiffness matrix contributions due to surface pressure to be calculated. For two-dimensional cases, the vector $\partial \boldsymbol{x}/\partial \eta$ is set to $[0, 0, -1]^T$, so that when the cross product with $\partial \boldsymbol{x}/\partial \xi$ is taken the correct normal vector (normal_vector) is obtained. Using the current unit pressure load vector R_pressure_0, the nominal pressure value LOAD.pressure(ipressure), and the current load parameter CON.xlamb, the equivalent nodal force is obtained and used to assemble the global residual vector `GLOBAL.Residual` and to calculate and assemble the global stiffness matrix `GLOBAL.K_pressure`.

## 10.12 EXAMPLES

A limited number of examples are described in this section in order to show the capabilities of FLagSHyP and illustrate some of the difficulties that may arise in the analysis of highly nonlinear problems. Where appropriate, the yield stress is chosen to give obvious finite deformation elastic behavior prior to the onset of plasticity.

### 10.12.1 Simple Patch Test

As a simple check on the code, the nonlinear plane strain patch test example shown in Figure 10.5 is studied. Two irregular six-noded elements making up a square are employed and boundary displacements are prescribed as shown in order to obtain a uniform deformation gradient tensor given by

$$\boldsymbol{F} = \begin{bmatrix} 2 & 0 \\ 0 & 3/4 \end{bmatrix}; \quad J = 3/2.$$

Assuming material 1 with values $\lambda = 100$ and $\mu = 100$, the program gives the correct state of uniform stress that can be evaluated from Equation (6.29) as

$$\boldsymbol{\sigma} = \frac{\mu}{J}(\boldsymbol{b} - \boldsymbol{I}) + \frac{\lambda}{J}(\ln J)\boldsymbol{I} = \begin{bmatrix} 227.03 & 0 \\ 0 & -2.136 \end{bmatrix};$$

$$\boldsymbol{b} = \boldsymbol{F}\boldsymbol{F}^T = \begin{bmatrix} 4 & 0 \\ 0 & 9/16 \end{bmatrix}.$$

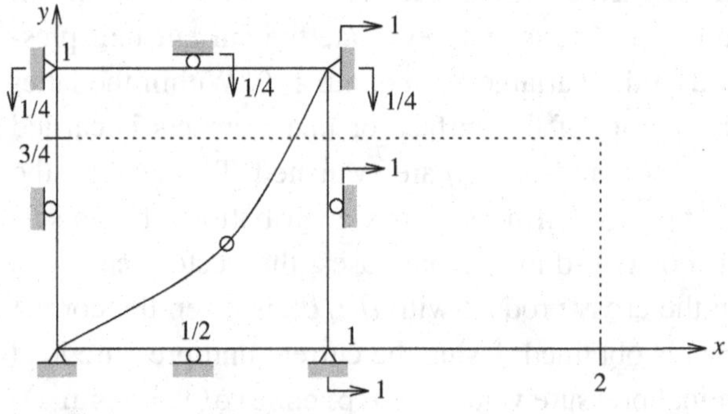


**FIGURE 10.5** Patch test.

### 10.12.2 Nonlinear Truss

**Plane stress truss:** The nonlinear truss example described in Section 1.3.2 can now be re-examined. For this purpose, a single four-noded element is used with the support conditions shown in Figure 10.6(a). In order to minimize the two-dimensional effects, a large length-to-thickness ratio of approximately 100 is used. The initial angle is 45° as in Figure 1.5 of Chapter 1 and the initial thickness is set to $1/\sqrt{2}$ so that the initial area of the truss is 1. Material 8 is used with $\mu = 1/3$, which for $\nu = 0.5$ implies $E = 1$. The displacement of the top node is prescribed either upward or downward in small increments of 0.5. The resulting vertical force is shown in Figure 10.6(b) and is practically identical to that obtained in Section 1.3.2 using the logarithmic strain equation and shown in Figure 1.5.

It is worth noting that when the truss is pushed downward and reaches the $-45°$ position, the program apparently fails to converge. The reason for this is simply that at this position the stresses in the truss, and hence all the reactions, are zero to within machine accuracy. Consequently, when the convergence norm is evaluated by dividing the norm of the residuals by the norm of the reactions, a practically random number is obtained. In order to escape from this impasse, an artificially large convergence norm must be chosen for the increment that corresponds to this position (0.98 has been used by the authors). The computation can then be continued using normal convergence norms for the rest of the increments. Note that the restarting facilities in FLagSHyP enable this artifice to be easily performed.

**Solid truss:** The elasto-plastic single truss element example given in Section 3.6.1 is now repeated using a single eight-node hexahedron element with the same dimensions and boundary conditions as those given in Figure 10.6(a). In order to maintain a value of Young's modulus of 210 000 kN/mm$^2$ and a Poisson's ratio of

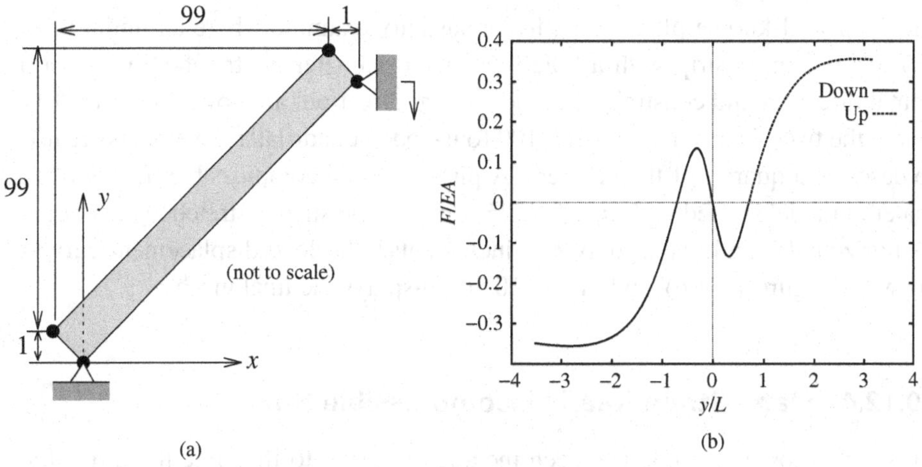


**FIGURE 10.6** Plane stress truss example: (a) Geometry; (b) Load–displacement curve.

0.3, the corresponding constants for material 17 are $\mu = 80\,769.23\,\text{kN/mm}^2$ and $\lambda = 121\,153.86\,\text{kN/mm}^2$, the yield stress and hardening parameters remaining the same at $\tau_y = 25\,000.0\,\text{kN/mm}^2$ and 1.0 respectively. The analysis was carried out using a fixed arc length of 5.0. Elastic and elasto-plastic load deflection curves are shown in Figure 10.7 (the Z axis now being the vertical direction), and coincide with the results using the single truss element, the elastic analysis being achieved with an artificially high yield stress.

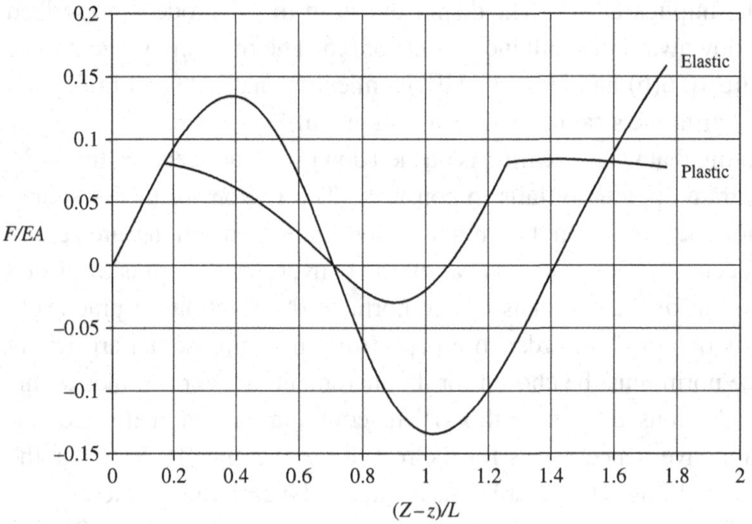


**FIGURE 10.7** Solid truss example: elastic and elasto-plastic load–displacement curves.

### 10.12.3 Strip with a Hole

This is a well-known plane stress hyperelasticity example where an initial $6.5 \times 6.5 \times 0.079\,\text{mm}^3$ strip with a hole 0.5 mm in diameter is stretched in the horizontal direction and constrained in the vertical direction, as shown in Figure 10.8. Given the two planes of symmetry, 100 four-noded quadrilateral elements are used to describe a quarter of the problem. A plane stress incompressible neo-Hookean material model is used with $\mu = 0.4225\,\text{N/mm}^2$. The strip is stretched to six times its horizontal length using only five increments. The load-displacement curve is shown in Figure 10.8(b), and Figure 10.8(c) displays the final mesh.

### 10.12.4 Plane Strain Nearly Incompressible Strip

This well-known example has been included in order to illustrate the difficulties that can be encountered using the penalty type of nearly incompressible plane

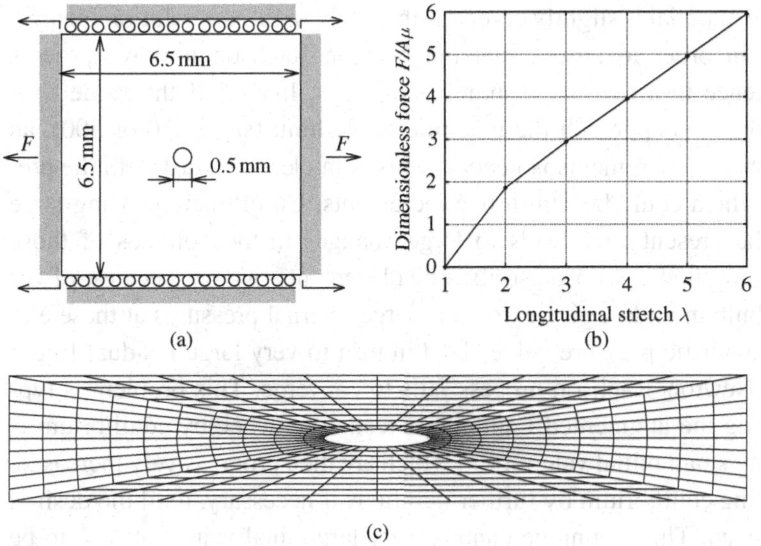


**FIGURE 10.8** Strip with hole: (a) Geometry; (b) Load–stretch curve; (c) Final mesh.

strain or three-dimensional materials implemented in FLagSHyP in conjunction with a displacement control process. In this case a $20 \times 20\,\text{mm}^2$ strip is clamped and stretched as shown in Figure 10.9. Because of the symmetry, only a quarter of the problem is actually modeled, using 256 four-noded mean dilatation quadrilateral elements. Material 5 is used with $\mu = 0.4225\,\text{N/mm}^2$ and $\kappa = 5\,\text{N/mm}^2$. The final mesh is shown in Figure 10.9(b), where for a total horizontal stretch of 3 a vertical stretch of 0.3711 is observed (smaller values are obtained as $\kappa$ is increased to reach the incompressible limit).

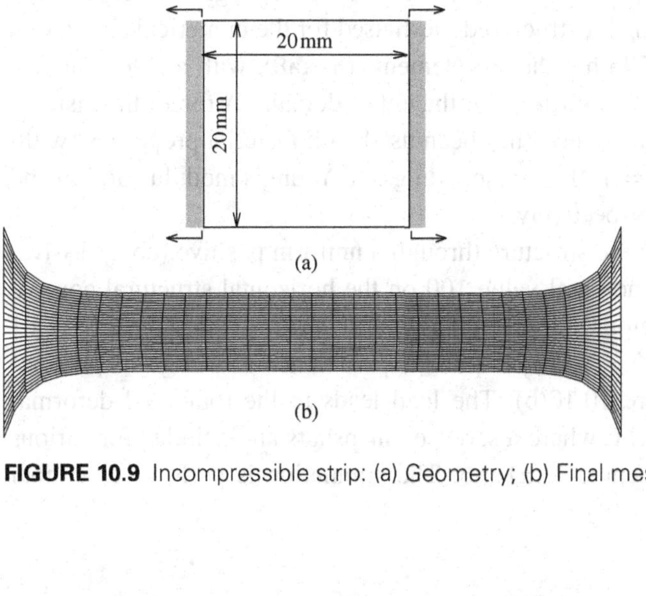


**FIGURE 10.9** Incompressible strip: (a) Geometry; (b) Final mesh.

Although the material is slightly compressible ($\nu \approx 0.46$), the solution requires 200 increments in order to ensure that the Newton–Raphson iterative process reaches convergence in a reasonable number of steps. In fact, if the value of $\kappa$ is increased in order to approach the incompressible limit (say to 50 or 500), an even larger number of increments is needed. This is in clear contrast with the previous example, which could be run in five increments. Unfortunately, using large increments in the present case leads to large changes in the volumes of those elements containing nodes with prescribed displacements, which, because of the relatively large bulk modulus, give extremely large internal pressures at these elements. These unrealistic pressure values lead in turn to very large residual forces from which the Newton–Raphson process fails to converge. This problem is typically solved using the augmented Lagrangian technique whereby equilibrium is first reached using small initial values of $\kappa$, which are then progressively increased, always maintaining equilibrium by further iterations if necessary, until the desired $\kappa/\mu$ ratio is reached. This technique enables very large final values of $\kappa/\mu$ to be used without increasing the number of increments needed. In order to keep the code as simple as possible, however, this technique has not been implemented in FLagSHyP, which implies that a very large number of increments is needed for plane strain or three-dimensional problems involving true incompressible materials and displacement control.

### 10.12.5 Twisting Column

This example shows the deformation pattern of a relatively complex structure when subjected to a torsion induced behavior via the application of surface loads. The structure is comprised of a horizontal short beam resting on top of a column. The geometry and dimensions (without units) are those shown in Figure 10.10(a). Also in that figure, the finite element structured mesh used for the numerical simulation is displayed, comprising 576 hexahedral elements (`hexa8`), with an element size chosen as $0.25 \times 0.25 \times 0.25$ throughout the entire domain. A three-dimensional compressible neo-Hookean material has been used with material properties (without units) $\mu = 100$ and $\lambda = 100$, corresponding to a Young's modulus of 250 and a Poisson's ratio of 0.25, respectively.

A torque is applied on the structure through a uniform positive (compressive) pressure follower load of nominal value 100 on the horizontal structural component, specifically in the regions defined by $\{0 \leq X \leq 1.5\} \cap \{Y = 0.5\} \cap \{Z \geq 6\}$ and $\{1.5 \leq X \leq 0\} \cap \{Y = -0.5\} \cap \{Z \geq 6\}$. A sketch showing this pressure load is displayed in Figure 10.10(b). The load leads to the rotational deformation depicted in Figure 10.11, where a series of snapshots are included for various values of the load increment `CON.xlamb`. The simulation is performed without

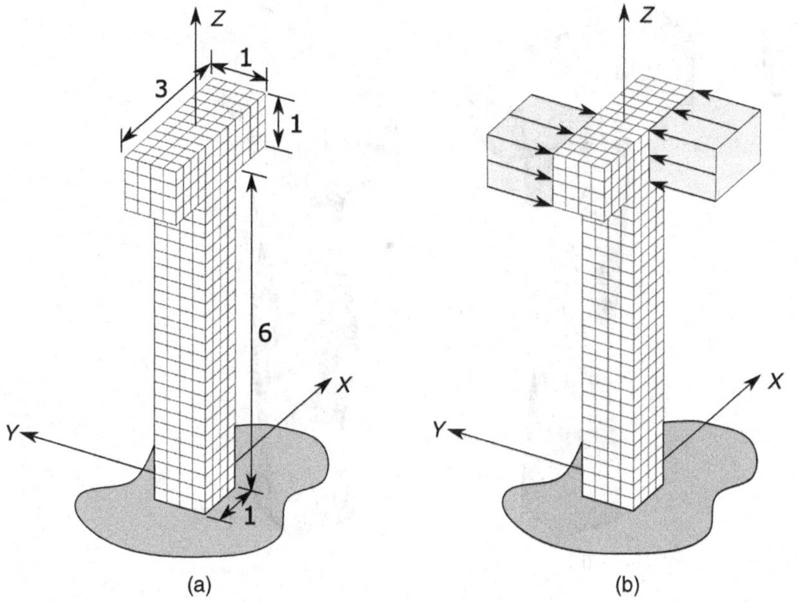


**FIGURE 10.10** Twisting column: (a) Geometry; (b) Pressure load distribution.

the need to resort to the line search or arc length algorithms. A variety of possible output plots are possible. However, to get a sense of the rotation of the column, Figure 10.12 shows the force–displacement diagram corresponding to the node defined by initial coordinates $[1, 0.5, 6.5]$. Specifically, graphical representation of the norm of the equivalent nodal force (due to the surface load) at the node is plotted against the component of the nodal displacement along the $OX$ direction.

### 10.12.6 Elasto-Plastic Cantilever

This example shows the elastic and elasto-plastic behavior of a cantilever undergoing finite deformations. The geometry is shown in Figure 10.13(a), where 20, 2, and 8 `hexa8` elements are used in the $X$, $Y$, and $Z$ directions respectively. The constants for material 17 are $\mu = 80\,769.23\,\text{kN/mm}^2$ and $\lambda = 121\,153.86\,\text{kN/mm}^2$, the yield stress and hardening parameters being $\tau_y = 2500.0\,\text{kN/mm}^2$ and 1.0 corresponding to a value of Young's modulus of $210\,000\,\text{kN/mm}^2$ and a Poisson's ratio of 0.3 respectively. The elastic analysis is carried out with an artificially high yield stress. In both elastic and elasto-plastic cases a variable arc length is used. Figures 10.13(b,d) show the load deflection behavior at the tip of the cantilever, while Figure 10.13(c) shows the development of plasticity at the support

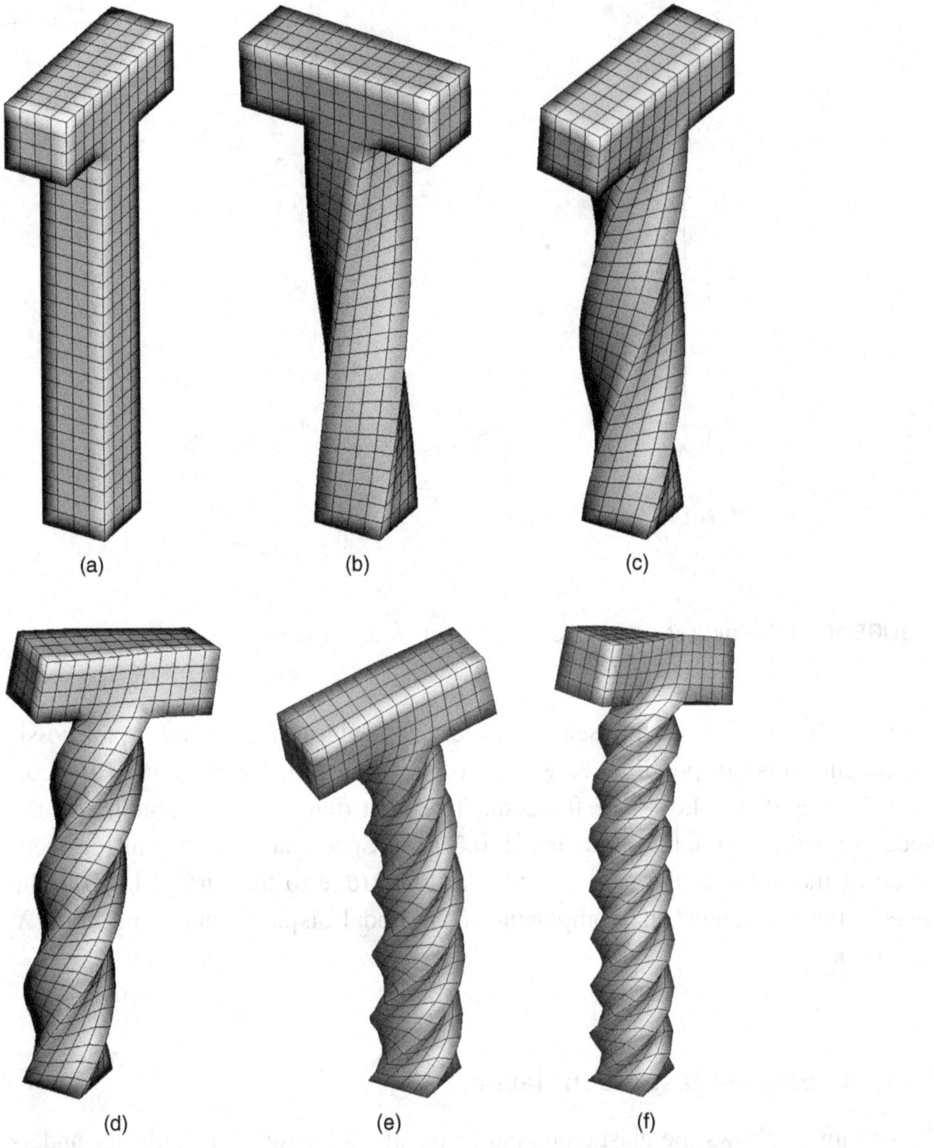


**FIGURE 10.11** Twisting column example: (a) `CON.xlamb`=0.00; (b) `CON.xlamb` = 0.16; (c) `CON.xlamb`=0.32; (d) `CON.xlamb`=0.60; (e) `CON.xlamb`=0.92; (f) `CON.xlamb`=1.24.

for a load of 788.18 N. Observe that Figure 10.13(c) shows coordinates and not the displacement at the loaded point. Also, note that an analysis will fail when the tangent matrix looses its positive definite nature for this material model. Essentially, the simple logarithmic stretch-based constitutive model is only valid for moderate deformations, even though displacements may be large.

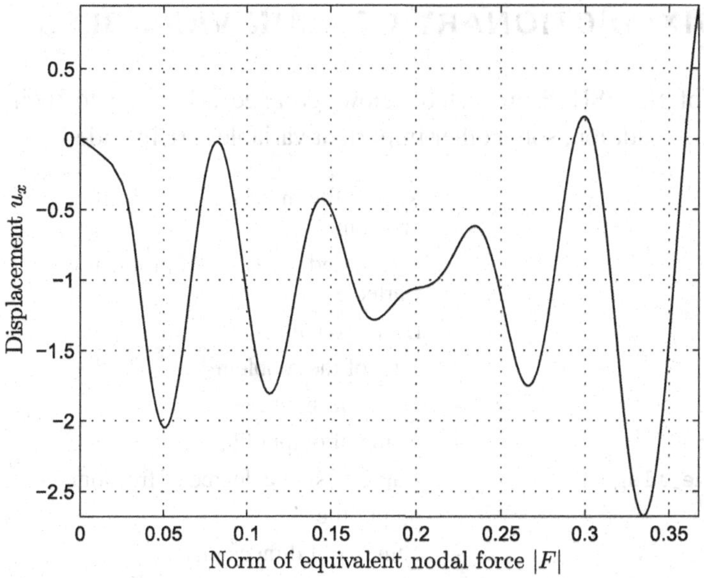


**FIGURE 10.12** Twisting column example: curve displaying the norm of the equivalent nodal force vs the $OX$ component of the displacement at node $[1, 0.5, 6.5]$.

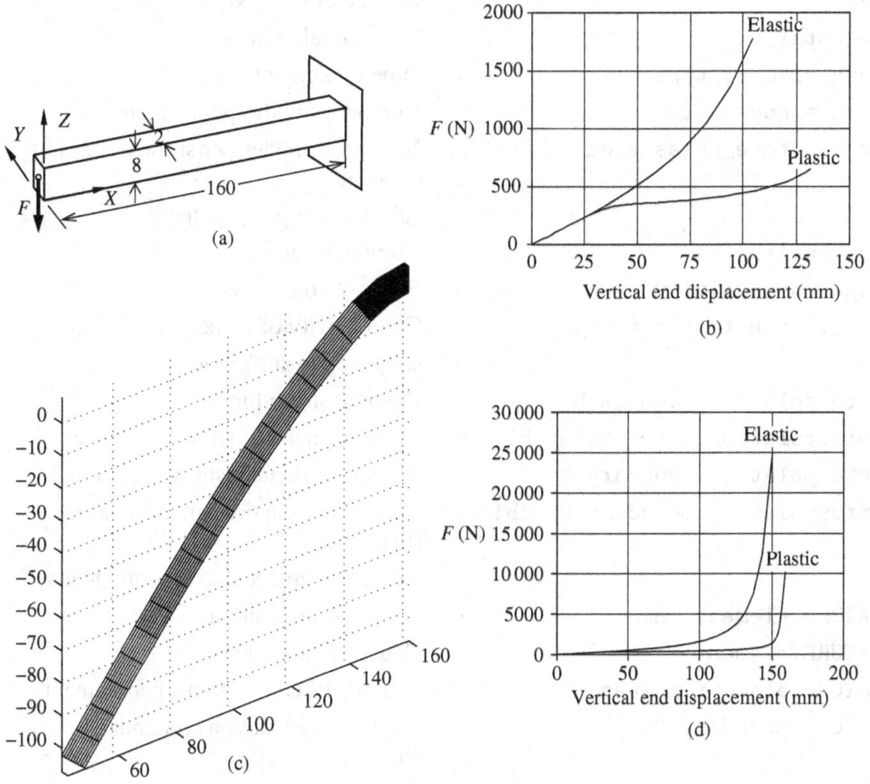


**FIGURE 10.13** Large deflection elasto-plastic behavior of a cantilever: (a) Geometry; (b) Detail of force deflection behavior; (c) Deformation showing yielding at the support at a load of 788.18 N; (d) Overall force deflection behavior.

## 10.13 APPENDIX: DICTIONARY OF MAIN VARIABLES

The main variables of FLagSHyP are listed below, grouped according to their parent data structure. In addition, some other important variables are included.

| | | |
|---|---|---|
| `PRO.rest` | - | Restart file indicator: `.true.` if problem is restarted, `.false.` if problem is started from scratch |
| `PRO.title` | - | Title of the problem |
| `PRO.inputfile_name` | - | Name of input file |
| `PRO.outputfile_name` | - | Name of output file |
| `PRO.outputfile_name_flagout` | - | Name of single degree of freedom output file |
| `GEOM.ndime` | - | Number of dimensions |
| `GEOM.npoin` | - | Number of nodes in the mesh |
| `GEOM.x` | - | Current coordinates |
| `GEOM.x0` | - | Initial coordinates |
| `GEOM.Ve` | - | Initial elemental volume |
| `GEOM.V_total` | - | Total initial volume |
| `FEM.mesh.element_type` | - | Type of element |
| `FEM.mesh.n_nodes_elem` | - | Number of nodes per element |
| `FEM.mesh.n_face_nodes_elem` | - | Number of nodes per surface (line) element subjected to pressure loads |
| `FEM.mesh.nelem` | - | Number of elements |
| `FEM.mesh.connectivity` | - | Element connectivity |
| `FEM.mesh.connectivity_faces` | - | Connectivity of surface (line) elements subjected to surface loads |
| `FEM.interpolation.element.N` | - | Element shape functions |
| `FEM.interpolation.element.DN_chi` | - | Element shape function derivatives |
| `FEM.interpolation.boundary.N` | - | Boundary element shape functions |
| `FEM.interpolation.boundary.DN_chi` | - | Shape function derivatives of surface (line) elements subjected to pressure loads |
| `QUADRATURE.element.Chi` | - | Gauss point locations |
| `QUADRATURE.element.W` | - | Gauss point weights |
| `QUADRATURE.element.ngauss` | - | Number of Gauss points per element |
| `QUADRATURE.boundary.Chi` | - | Gauss point locations (boundary element) |

| | | |
|---|---|---|
| QUADRATURE.boundary.W | - | Gauss point weights (boundary element) |
| QUADRATURE.boundary.ngauss | - | Number of Gauss points per boundary element |
| KINEMATICS.DN_x | - | Current Cartesian derivatives of shape functions |
| KINEMATICS.Jx_chi | - | Volume ratio (Jacobian) between current and nondimensional isoparametric domain |
| KINEMATICS.F | - | Deformation gradient tensor |
| KINEMATICS.J | - | Volume ratio (Jacobian) between current and initial domain ($J = \det\boldsymbol{F}$) |
| KINEMATICS.b | - | Left Cauchy–Green deformation tensor |
| KINEMATICS.Ib | - | First invariant of the left Cauchy–Green deformation tensor |
| KINEMATICS.lambda | - | Principal stretches |
| KINEMATICS.n | - | Principal directions |
| MAT.matno | - | Element material number identifier |
| MAT.nmats | - | Number of materials |
| MAT.props | - | Material properties: the first is always the initial density, the rest depend on the material type |
| MAT.matyp | - | Material number identifier |
| MAT.n_nearly_incompressible | - | Number of nearly incompressible materials |
| BC.icode | - | Nodal boundary condition identifier |
| BC.freedof | - | Nodal free degrees of freedom |
| BC.fixdof | - | Nodal fixed degrees of freedom |
| BC.n_prescribed_displacements | - | Number of prescribed displacements |
| BC.presc_displacement | - | Prescribed displacement vector |
| BC.dofprescribed | - | Prescribed degree of freedom |
| LOAD.n_pressure_loads | - | Number of pressure load surface elements |
| LOAD.gravt | - | Gravity vector |
| LOAD.pressure | - | Value of pressure in surface element |
| LOAD.pressure_element | - | Element subjected to pressure |
| GLOBAL.nominal_external_load | - | Nominal external load (without pressure) |
| GLOBAL.Residual | - | Global residual vector |
| GLOBAL.external_load | - | Current load vector |
| GLOBAL.nominal_pressure | - | Equivalent nodal force due to pressure |
| GLOBAL.T_int | - | Equivalent internal force vector |
| GLOBAL.K | - | Global stiffness matrix |
| GLOBAL.R_pressure | - | Equivalent internal force vector due to pressure |

| | | |
|---|---|---|
| `GLOBAL.K_pressure` | - | Global stiffness matrix due to pressure |
| `PLAST.yield.f` | - | Yield surface value |
| `PLAST.yield.Dgamma` | - | Incremental plastic multiplier |
| `PLAST.yield.nu_a` | - | Direction vector |
| `PLAST.trial.lambdae` | - | Trial elastic stretches |
| `PLAST.trial.tau` | - | Trial Kirchhoff stress tensor |
| `PLAST.trial.n` | - | Trial eigenvectors |
| `PLAST.UPDATED.invCp` | - | Updated plastic right Cauchy–Green tensor |
| `PLAST.UPDATED.epbar` | - | Updated equivalent plastic strain |
| `PLAST.stress.Cauchy` | - | Cauchy stress tensor |
| `PLAST.stress.Cauchyaa` | - | Cauchy principal stresses |
| `CON.nincr` | - | Number of load increments |
| `CON.incrm` | - | Current load increment |
| `CON.xlmax` | - | Maximum load parameter |
| `CON.dlamb` | - | Incremental load value |
| `CON.miter` | - | Maximum number of iterations per increment |
| `CON.niter` | - | Current iteration per increment |
| `CON.cnorm` | - | Convergence criterion for Newton-Raphson |
| `CON.searc` | - | Line search parameter |
| `CON.msearch` | - | Maximum number of line search iterations |
| `CON.incrm` | - | Current load increment |
| `CON.xlamb` | - | Current load parameter |
| `CON.ARCLEN.farcl` | - | Logical fixed arc length indicator |
| `CON.ARCLEN.arcln` | - | Arc length parameter (0.0 means no arc length) |
| `CON.ARCLEN.itarget` | - | Target iteration/increment for variable arc length option |
| `CON.ARCLEN.iterold` | - | Number of iterations in previous increment for arc length |
| `CON.ARCLEN.xincr` | - | Total displacement over the load increment |
| `CON.ARCLEN.afail` | - | Logical arc length failure indicator |
| `CON.OUTPUT.incout` | - | Output counter |
| `CON.OUTPUT.nwant` | - | Single output node |
| `CON.OUTPUT.iwant` | - | Output degree of freedom |
| `eta` | - | Scaling factor or displacement factor |
| `eta0` | - | Previous value of the parameter `eta` |
| `rtu` | - | Current dot product of $\mathbf{R}$ by $\mathbf{u}$ |
| `rtu0` | - | Initial dot product of $\mathbf{R}$ by $\mathbf{u}$ |
| `displ` | - | Newton–Raphson displacement vector $\mathbf{u}$ |
| `dispf` | - | load component of the displacement vector $\mathbf{u}_F$ |
| `rnorm` | - | Current residual norm |
| `c` | - | Fourth-order tensor |

| | | |
|---|---|---|
| `indexi` | - | Vector storing the row entry of the stiffness matrix coefficient |
| `indexj` | - | Vector storing the column entry of the stiffness matrix coefficient |
| `counter` | - | Counter to travel through the vectors storing the stiffness matrix coefficients |
| `global_stiffness` | - | Vector storing each coefficient of the stiffness matrix |
| `mu` | - | $\mu$ coefficient (material parameter) |
| `lambda` | - | $\lambda$ coefficient (material parameter) |
| `kappa` | - | $\kappa$ coefficient (material parameter) |
| `H` | - | strain hardening (material parameter) |

## 10.14 APPENDIX: CONSTITUTIVE EQUATION SUMMARY

To facilitate the understanding of the implementation of the various constitutive equations, the boxes below summarize the required constitutive and kinematic equations for each material type. These equations are presented here in an indicial form to concur with the code.

**Material 10.1: Three-dimensional or plane strain compressible neo-Hookean**

$$\sigma_{ij} = \frac{\mu}{J}(b_{ij} - \delta_{ij}) + \frac{\lambda}{J}(\ln J)\delta_{ij} \qquad (6.29)$$

$$\mathcal{C}_{ijkl} = \lambda'\delta_{ij}\delta_{kl} + \mu'\left(\delta_{ik}\delta_{jl} + \delta_{il}\delta_{jk}\right) \qquad (6.40)$$

$$\lambda' = \frac{\lambda}{J}; \qquad \mu' = \frac{\mu - \lambda \ln J}{J} \qquad (6.41)$$

**Material 10.2: One-dimensional stretch-based hyperelastic plastic**

$$\lambda_{\mathrm{e},n+1}^{\mathrm{trial}} = \frac{l_{n+1}}{l_{p,n}}; \quad J = \left(\lambda_{\mathrm{e},n+1}^{\mathrm{trial}}\right)^{(1-2\nu)} \qquad (3.67, 3.66)$$

$$\varepsilon_{\mathrm{e},n+1}^{\mathrm{trial}} = \ln \lambda_{\mathrm{e},n+1}^{\mathrm{trial}} \qquad (3.68\mathrm{a,b,c})$$

*(continued)*

**Material 10.2:** *(cont.)*

$$\tau_{n+1}^{\text{trial}} = E\,\varepsilon_{e,n+1}^{\text{trial}} \tag{3.77a,b}$$

$$f\left(\tau_{n+1}^{\text{trial}}, \bar{\varepsilon}_{p,n}\right) = \left|\tau_{n+1}^{\text{trial}}\right| - \left(\tau_y^0 + H\bar{\varepsilon}_{p,n}\right) \tag{3.78}$$

$$\Delta\gamma = \begin{cases} \dfrac{f\left(\tau_{n+1}^{\text{trial}}, \bar{\varepsilon}_{p,n}\right)}{E+H} & \text{if } f\left(\tau_{n+1}^{\text{trial}}, \bar{\varepsilon}_{p,n}\right) > 0 \\ 0 & \text{if } f\left(\tau_{n+1}^{\text{trial}}, \bar{\varepsilon}_{p,n}\right) \leq 0 \end{cases} \tag{3.85}$$

$$\tau_{n+1} = \tau_{n+1}^{\text{trial}} - E\Delta\gamma\ \text{sign}(\tau_{n+1}) \tag{3.81}$$

$$\textbf{If} \quad f > 0 \quad \frac{d\,\tau_{n+1}}{d\,\varepsilon_{n+1}} = \frac{EH}{E+H} \tag{3.90}$$

$$\textbf{Else} \quad \frac{d\,\tau_{n+1}}{d\,\varepsilon_{n+1}} = E \tag{3.15}$$

**Material 10.3: Three-dimensional or plane strain hyperelasticity in principal directions**

$$\sigma_{\alpha\alpha} = \frac{2\mu}{J}\ln\lambda_\alpha + \frac{\lambda}{J}\ln J \tag{6.94}$$

$$\sigma_{ij} = \sum_{\alpha=1}^{3} \sigma_{\alpha\alpha} T_{\alpha i} T_{\alpha j}; \qquad (T_{\alpha i} = \boldsymbol{n}_\alpha \cdot \boldsymbol{e}_i) \tag{6.81}$$

$$\mathcal{C}_{ijkl} = \sum_{\alpha,\beta=1}^{3} \frac{\lambda + 2(\mu - J\sigma_{\alpha\alpha})\delta_{\alpha\beta}}{J} T_{\alpha i} T_{\alpha j} T_{\beta k} T_{\beta l} + \sum_{\substack{\alpha,\beta=1 \\ \alpha\neq\beta}}^{3} \mu_{\alpha\beta}\left(T_{\alpha i} T_{\beta j} T_{\alpha k} T_{\beta l} + T_{\alpha i} T_{\beta j} T_{\beta k} T_{\alpha l}\right) \tag{6.90, 6.95}$$

$$\mu_{\alpha\beta} = \frac{\sigma_{\alpha\alpha}\lambda_\beta^2 - \sigma_{\beta\beta}\lambda_\alpha^2}{\lambda_\alpha^2 - \lambda_\beta^2}; \quad \text{if } \lambda_\alpha \neq \lambda_\beta \quad \text{or} \tag{6.90, 6.91, 6.95}$$

$$\mu_{\alpha\beta} = \frac{\mu}{J} - \sigma_{\alpha\alpha} \ \text{ if } \lambda_\alpha = \lambda_\beta$$

**Material 10.4: Plane stress hyperelasticity in principal directions**

$$\gamma = \frac{2\mu}{\lambda + 2\mu} \tag{6.116a,b$)_b$}$$

$$\bar{\lambda} = \gamma\lambda \tag{6.116a,b$)_a$}$$

$$J = j^{\gamma}\,; \qquad (J = dv/dV;\ j = da/dA) \tag{6.117}$$

$$\sigma_{\alpha\alpha} = \frac{2\mu}{J}\ln\lambda_{\alpha} + \frac{\bar{\lambda}}{J}\ln j \tag{6.118}$$

$$\sigma_{ij} = \sum_{\alpha=1}^{2}\sigma_{\alpha\alpha}T_{\alpha i}T_{\alpha j}; \qquad (T_{\alpha i} = \boldsymbol{n}_{\alpha}\cdot\boldsymbol{e}_i) \tag{6.81}$$

$$\begin{aligned}\mathcal{C}_{ijkl} = {} & \sum_{\alpha,\beta=1}^{2}\frac{\bar{\lambda} + 2(\mu - J\sigma_{\alpha\alpha})\delta_{\alpha\beta}}{J}T_{\alpha i}T_{\alpha j}T_{\beta k}T_{\beta l} \\ & + \sum_{\substack{\alpha,\beta=1\\ \alpha\neq\beta}}^{2}\mu_{\alpha\beta}\left(T_{\alpha i}T_{\beta j}T_{\alpha k}T_{\beta l} + T_{\alpha i}T_{\beta j}T_{\beta k}T_{\alpha l}\right)\end{aligned} \tag{6.90, 6.119}$$

$$\mu_{\alpha\beta} = \frac{\sigma_{\alpha\alpha}\lambda_{\beta}^{2} - \sigma_{\beta\beta}\lambda_{\alpha}^{2}}{\lambda_{\alpha}^{2} - \lambda_{\beta}^{2}}; \quad \text{if } \lambda_{\alpha}\neq\lambda_{\beta} \quad \text{or} \quad \mu_{\alpha\beta} = \frac{\mu}{J} - \sigma_{\alpha\alpha} \text{ if } \lambda_{\alpha} = \lambda_{\beta}$$

(6.90, 6.91, 6.119)

$$h = \frac{HJ}{j}$$

(Exercise 4 of Chapter 4)

**Material 10.5: Three-dimensional or plane strain nearly incompressible neo-Hookean**

$$\bar{J} = \frac{v^{(e)}}{V^{(e)}} \tag{8.53a}$$

$$p = \kappa(\bar{J} - 1) \tag{8.55}$$

$$\bar{\kappa} = \kappa\frac{v^{(e)}}{V^{(e)}} \tag{8.64}$$

*(continued)*

**Material 10.5:** *(cont.)*

$$\sigma'_{ij} = \mu J^{-5/3}\left(b_{ij} - \frac{1}{3}I_{\boldsymbol{b}}\delta_{ij}\right) \qquad (6.55)$$

$$\sigma_{ij} = \sigma'_{ij} + p\delta_{ij} \qquad (5.49a,b(a))$$

$$\hat{\mathcal{C}}_{ijkl} = 2\mu J^{-5/3}\left[\frac{1}{6}I_{\boldsymbol{b}}\left(\delta_{ik}\delta_{jl} + \delta_{il}\delta_{jk}\right) - \frac{1}{3}b_{ij}\delta_{kl} - \frac{1}{3}\delta_{ij}b_{kl} + \frac{1}{9}I_{\boldsymbol{b}}\delta_{ij}\delta_{kl}\right] \qquad (6.59a)$$

$$\mathcal{C}_{p,ijkl} = p\left(\delta_{ij}\delta_{kl} - \delta_{ik}\delta_{jl} - \delta_{il}\delta_{jk}\right) \qquad (6.59b)$$

**Material 10.6: Plane stress incompressible neo-Hookean (Exercise 1 of Chapter 6)**

$$J = 1; \quad (J = dv/dV,\ j = da/dA)$$

$$\sigma_{ij} = \mu(b_{ij} - j^{-2}\delta_{ij}); \quad \left(j^2 = \det_{2\times 2}\boldsymbol{b}\right)$$

$$\mathcal{C}_{ijkl} = \lambda'\delta_{ij}\delta_{kl} + \mu'\left(\delta_{ik}\delta_{jl} + \delta_{il}\delta_{jk}\right)$$

$$\lambda' = \frac{2\mu}{j^2}$$

$$\mu' = \frac{\mu}{j^2}$$

$$h = \frac{H}{j}$$

**Material 10.7: Nearly incompressible in principal directions**

$$\bar{J} = v^{(e)}/V^{(e)} \qquad (8.53a)$$

$$p = \frac{\kappa \ln \bar{J}}{\bar{J}} \qquad (6.103)$$

*(continued)*

**Material 10.7:** *(cont.)*

$$\bar{\kappa} = \bar{J}\frac{dp}{d\bar{J}} = \frac{\kappa}{\bar{J}} - p \qquad (8.62)$$

$$\sigma'_{\alpha\alpha} = -\frac{2\mu}{3J}\ln J + \frac{2\mu}{J}\ln\lambda_\alpha \qquad (6.107)$$

$$\sigma'_{ij} = \sum_{\alpha=1}^{3}\sigma'_{\alpha\alpha}T_{\alpha i}T_{\alpha j}; \qquad (T_{\alpha i} = \boldsymbol{n}_\alpha \cdot \boldsymbol{e}_i) \qquad (6.81)$$

$$\sigma_{ij} = \sigma'_{ij} + p\delta_{ij} \qquad (5.49\text{a,b})_a$$

$$\mathcal{C}_{p,ijkl} = p(\delta_{ij}\delta_{kl} - \delta_{ik}\delta_{jl} - \delta_{il}\delta_{jk}) \qquad (6.59\text{b})$$

$$\hat{\mathcal{C}}_{ijkl} = \sum_{\alpha,\beta=1}^{3}\frac{2}{J}\left[(\mu - J\sigma'_{\alpha\alpha})\delta_{\alpha\beta} - \frac{1}{3}\mu\right]T_{\alpha i}T_{\alpha j}T_{\beta k}T_{\beta l} + \sum_{\substack{\alpha,\beta=1\\ \alpha\neq\beta}}^{3}\mu_{\alpha\beta}\left(T_{\alpha i}T_{\beta j}T_{\alpha k}T_{\beta l} + T_{\alpha i}T_{\beta j}T_{\beta k}T_{\alpha l}\right) \qquad (6.110,\ 6.111)$$

$$\mu_{\alpha\beta} = \frac{\sigma'_{\alpha\alpha}\lambda_\beta^2 - \sigma'_{\beta\beta}\lambda_\alpha^2}{\lambda_\alpha^2 - \lambda_\beta^2}; \quad \text{if } \lambda_\alpha \neq \lambda_\beta \quad \text{or} \quad \mu_{\alpha\beta} = \frac{\mu}{J} - \sigma'_{\alpha\alpha} \ \text{if } \lambda_\alpha = \lambda_\beta \qquad (6.110,\ 6.91,\ 6.111)$$

**Material 10.8: Plane stress incompressible in principal directions**

$$\lambda \to \infty; \qquad \gamma = 0; \qquad \bar{\lambda} = 2\mu \qquad (6.116\text{a,b})$$

$$J = 1; \qquad (J = dv/dV; \qquad j = da/dA) \qquad (6.117)$$

$$\sigma_{\alpha\alpha} = 2\mu\ln\lambda_\alpha + \bar{\lambda}\ln j \qquad (6.118)$$

$$\sigma_{ij} = \sum_{\alpha=1}^{2}\sigma_{\alpha\alpha}T_{\alpha i}T_{\alpha j}; \quad (T_{\alpha i} = \boldsymbol{n}_\alpha \cdot \boldsymbol{e}_i) \qquad (6.81)$$

$$\mathcal{C}_{ijkl} = \sum_{\alpha,\beta=1}^{2}\left[\bar{\lambda} + 2(\mu - \sigma_{\alpha\alpha})\delta_{\alpha\beta}\right]T_{\alpha i}T_{\alpha j}T_{\beta k}T_{\beta l} + \sum_{\substack{\alpha,\beta=1\\ \alpha\neq\beta}}^{2}\mu_{\alpha\beta}\left(T_{\alpha i}T_{\beta j}T_{\alpha k}T_{\beta l} + T_{\alpha i}T_{\beta j}T_{\alpha l}T_{\beta k}\right) \qquad (6.90,\ 6.119)$$

*(continued)*

**Material 10.8:** *(cont.)*

$$\mu_{\alpha\beta} = \frac{\sigma_{\alpha\alpha}\lambda_\beta^2 - \sigma_{\beta\beta}\lambda_\alpha^2}{\lambda_\alpha^2 - \lambda_\beta^2}; \qquad \text{if } \lambda_\alpha \neq \lambda_\beta \quad \text{or} \quad \mu_{\alpha\beta} = \mu - \sigma_{\alpha\alpha} \text{ if } \lambda_\alpha = \lambda_\beta$$

(6.90, 6.91, 6.119)

$$h = \frac{H}{j}$$

(Exercise 4 of Chapter 4)

**Material 10.17: Three-dimensional or plane strain hyperelastic-plastic in principal directions**

$$\bar{J} = v^{(e)}/V^{(e)} \tag{8.53a}$$

$$p \approx \kappa \frac{\ln \bar{J}}{\bar{J}} \tag{6.103}$$

$$\nu_\alpha^{n+1} = \frac{\tau'^{\,\text{trial}}_{\alpha\alpha}}{\sqrt{\frac{2}{3}}\|\boldsymbol{\tau}'^{\,\text{trial}}\|} \tag{7.54a,b}$$

$$\Delta\gamma = \begin{cases} \dfrac{f(\boldsymbol{\tau}^{\text{trial}}, \bar{\varepsilon}_{p,n})}{3\mu + H} & \text{if } f(\boldsymbol{\tau}^{\text{trial}}, \bar{\varepsilon}_{p,n}) > 0 \\ 0 & \text{if } f(\boldsymbol{\tau}^{\text{trial}}, \bar{\varepsilon}_{p,n}) \leq 0 \end{cases} \tag{7.59}$$

$$\ln \lambda_{e,\alpha}^{n+1} = \ln \lambda_{e,\alpha}^{\text{trial}} - \Delta\gamma\nu_\alpha^{n+1} \tag{7.45a,b}$$

$$\tau'_{\alpha\alpha} = \left(1 - \frac{2\mu\Delta\gamma}{\sqrt{2/3}\|\boldsymbol{\tau}'^{\,\text{trial}}\|}\right)\tau'^{\,\text{trial}}_{\alpha\alpha} \tag{7.60}$$

$$\sigma'_{\alpha\alpha} = \frac{1}{J_{n+1}}\tau'_{\alpha\alpha} \tag*{(5.31a,b)$_b$}$$

$$\sigma_{\alpha\alpha} = \sigma'_{\alpha\alpha} + p \tag*{(5.49a,b)$_a$}$$

$$\sigma_{ij} = \sum_{\alpha=1}^{2} \sigma_{\alpha\alpha} T_{\alpha i} T_{\alpha j}; \quad (T_{\alpha i} = \boldsymbol{n}_\alpha \cdot \boldsymbol{e}_i) \tag{6.81}$$

$$\hat{\mathfrak{c}}_{ijkl} = \sum_{\alpha,\beta=1}^{3} \frac{1}{J}\,\mathfrak{c}_{\alpha\beta}\, T_{\alpha i}T_{\alpha j}T_{\beta k}T_{\beta l} - \sum_{\alpha=1}^{3} 2\sigma'_{\alpha\alpha}\, T_{\alpha i}T_{\alpha j}T_{\alpha k}T_{\alpha l}$$
$$+ \sum_{\substack{\alpha,\beta=1\\ \alpha\neq\beta}}^{3} \frac{\sigma'_{\alpha\alpha}\left(\lambda_{e,\beta}^{\text{trial}}\right)^2 - \sigma'_{\beta\beta}\left(\lambda_{e,\alpha}^{\text{trial}}\right)^2}{\left(\lambda_{e,\alpha}^{\text{trial}}\right)^2 - \left(\lambda_{e,\beta}^{\text{trial}}\right)^2} \left(T_{\alpha i}T_{\beta j}T_{\alpha k}T_{\beta l} + T_{\alpha i}T_{\beta j}T_{\beta k}T_{\alpha l}\right)$$

(7.62a)

*(continued)*

**Material 10.17:** *(cont.)*

$$f(\boldsymbol{\tau}^{\text{trial}}, \bar{\varepsilon}_{p,n}) \leq 0 \qquad (7.20\text{a,b})$$

$$\textbf{If} \quad f > 0 \;\; \mathbf{c}_{\alpha\beta} = \left(1 - \frac{2\mu\,\Delta\gamma}{\sqrt{2/3}\,\|\boldsymbol{\tau}'^{\,\text{trial}}\|}\right)\left(2\mu\delta_{\alpha\beta} - \frac{2}{3}\mu\right) - 2\mu\,\nu_\alpha\nu_\beta\left(\frac{2\mu}{3\mu + H} - \frac{2\mu\sqrt{2/3}\,\Delta\gamma}{\|\boldsymbol{\tau}'^{\,\text{trial}}\|}\right) \qquad (7.67)$$

$$\textbf{Else} \quad \mathbf{c}_{\alpha\beta} = 2\mu\delta_{\alpha\beta} - \frac{2}{3}\mu \qquad (7.64)$$

# BIBLIOGRAPHY

BATHE, K-J., *Finite Element Procedures in Engineering Analysis*, Prentice Hall, 1996.

BELYTSCHKO, T., LIU, W. K., and MORAN, B., *Nonlinear Finite Elements for Continua and Structures,* John Wiley & Sons, 2000.

BONET, J. and BHARGAVA, P., The incremental flow formulation for the analysis of 3-dimensional viscous deformation processes: Continuum formulation and computational aspects, *Int. J. Num. Meth. Engrg.*, **122**, 51–68, 1995.

BONET, J., GIL, A. J., and ORTIGOSA, R., A computational framework for polyconvex large strain elasticity, *Comput. Meths. Appl. Mech. Engrg.*, **283**, 1061–1094, 2015.

BONET, J., WOOD, R. D., MAHANEY, J., and HEYWOOD, P., Finite element analysis of air supported membrane structures, *Comput. Meths. Appl. Mech. Engrg.*, **190**, 579–595, 2000.

CRISFIELD, M. A., *Non-Linear Finite Element Analysis of Solids and Structures*, John Wiley & Sons, Volume 1, 1991.

ETEROVIĆ, A. L. and BATHE, K-L., A hyperelastic-based large strain elasto-plastic constitutive formulation with combined isotropic-kinematic hardening using logarithmic stress and strain measures, *Int. J. Num. Meth. Engrg.*, **30**, 1099–1114, 1990.

GONZALEZ, O. and STUART, A. M., *A First Course in Continuum Mechanics,* Cambridge University Press, 2008.

GURTIN, M., *An Introduction to Continuum Mechanics*, Academic Press, 1981.

HOLZAPFEL, G. A., *Nonlinear Solid Mechanics: A Continuum Approach for Engineering*, John Wiley & Sons, 2000.

HUGHES, T. J. R., *The Finite Element Method*, Prentice Hall, 1987.

HUGHES, T. J. R. and PISTER, K. S., Consistent linearization in mechanics of solids and structures, *Compt. & Struct.*, **8**, 391–397, 1978.

LUBLINER, J., *Plasticity Theory*, Macmillan, 1990.

MALVERN, L. E., *Introduction to the Mechanics of a Continuous Medium*, Prentice Hall, 1969.

MARSDEN, J. E. and HUGHES, T. J. R., *Mathematical Foundations of Elasticity*, Prentice Hall, 1983.

MIEHE, C., Aspects of the formulation and finite element implementation of large strain isotropic elasticity, *Int. J. Num. Meth. Engrg.*, **37**, 1981–2004, 1994.

ODEN, J. T., *Finite Elements of Nonlinear Continua*, McGraw-Hill, 1972. Also Dover Publications, 2006.

OGDEN, R. W., *Non-Linear Elastic Deformations*, Ellis Horwood, 1984.

PERIĆ, D., OWEN, D. R. J., and HONNOR, M. E., A model for finite strain elasto-plasticity based on logarithmic strains: Computational issues, *Comput. Meths. Appl. Mech. Engrg.*, **94**, 35–61, 1992.

REDDY, J. N., *An Introduction to Nonlinear Finite Element Analysis,* Oxford University Press, 2004.

SCHWEIZERHOF, K. and RAMM, E., Displacement dependent pressure loads in non-linear finite element analysis, *Compt. & Struct.*, **18**, 1099–1114, 1984.

SIMMONDS, J. G., *A Brief on Tensor Analysis*, Springer, 2nd edition, 1994.

SIMO, J. C., A framework for finite strain elasto-plasticity based on a maximum plastic dissipation and the multiplicative decomposition: Part 1. Continuum formulation, *Comput. Meths. Appl. Mech. Engrg.*, **66**, 199–219, 1988.

SIMO, J. C., Algorithms for static and dynamic multiplicative plasticity that preserve the classical return mapping schemes of the infinitesimal theory, *Comput. Meths. Appl. Mech. Engrg.*, **99**, 61–112, 1992.

SIMO, J. C. and HUGHES, T. J. R., *Computational Inelasticity*, Springer, 1997.

SIMO, J. C. and ORTIZ, M., A unified approach to finite deformation elastoplastic analysis based on the use of hyperelastic constitutive equations, *Comput. Meths. Appl. Mech. Engrg.*, **49**, 221–245, 1985.

SIMO, J. C. and TAYLOR, R. L., Quasi-incompressible finite elasticity in principal stretches. Continuum basis and numerical algorithms, *Comput. Meths. Appl. Mech. Engrg.*, **85**, 273–310, 1991.

SIMO, J. C., TAYLOR, R. L., and PISTER, K. S., Variational and projection methods for the volume constraint in finite deformation elasto-plasticity, *Comput. Meths. Appl. Mech. Engrg.*, **51**, 177–208, 1985.

SPENCER, A. J. M., *Continuum Mechanics*, Longman, 1980.

WEBER, G. and ANAND, L., Finite deformation constitutive equations and a time integration procedure for isotropic, hyperelastic-viscoplastic solids, *Comput. Meths. Appl. Mech. Engrg.*, **79**, 173–202, 1990.

ZIENKIEWICZ, O. C. and TAYLOR, R. L., *The Finite Element Method*, McGraw-Hill, 4th edition, Volumes 1 and 2, 1994.

# INDEX

Printed in the United States
by Baker & Taylor Publisher Services

Printed in the United States
by Baker & Taylor Publisher Services